MICROBIAL SIGNALLING AND COMMUNICATION

Microbial chemical signals have been found to mediate the regulation of diverse metabolic reactions and processes such as antibiotic production, pathogenesis, sexual conjugation, sporulation and differentiation. Their study has the potential to secure advances in our ability to control microbial processes to our benefit. This volume presents information at the forefront of knowledge in this exciting field and includes contributions on a range of organisms (both prokaryote and eukaryote, unicellular and multicellular) and signalling molecules. As such, it will provide an invaluable resource for professional microbiologists and an excellent reference text for advanced students.

REG ENGLAND is a Senior Lecturer in the Department of Biological Sciences at the University of Central Lancashire.

GLYN HOBBS is a Senior Lecturer in the School of Biomolecular Sciences at Liverpool John Moores University.

NIGEL BAINTON is a Senior Lecturer in the School of Biological Sciences at the University of Surrey.

DAVE ROBERTS is Head of the Zoology Department at the Natural History Museum, London.

SYMPOSIA OF THE
SOCIETY FOR GENERAL MICROBIOLOGY

Series editor (1996–2001): Dr D. Roberts, Zoology Department, The Natural History Museum, London
  Volumes currently available:

MICROBIAL SIGNALLING AND COMMUNICATION

EDITED BY

R. R. ENGLAND, G. HOBBS, N. J. BAINTON AND
D. McL. ROBERTS

FIFTY-SEVENTH SYMPOSIUM OF THE
SOCIETY FOR GENERAL MICROBIOLOGY
HELD AT THE UNIVERSITY OF EDINBURGH
APRIL 1999

Published for the Society for General Microbiology

CAMBRIDGE
UNIVERSITY PRESS

PUBLISHED BY THE PRESS SYNDICATE OF THE UNIVERSITY OF CAMBRIDGE
The Pitt Building, Trumpington Street, Cambridge, United Kingdom

CAMBRIDGE UNIVERSITY PRESS
The Edinburgh Building, Cambridge CB2 2RU, UK http://www.cup.cam.ac.uk
40 West 20th Street, New York, NY 10011-4211, USA http://www.cup.org
10 Stamford Road, Oakleigh, Melbourne 3166, Australia

First published 1999

Printed in the United Kingdom at the University Press, Cambridge

Typeset in Monotype Times 10/12pt, in 3B2 [PN]

A catalogue record for this book is available from the British Library

ISBN 0 521 65261 8 hardback

The Society would like to dedicate this volume to the memory of Professor Gordon S. A. B. Stewart who died on 27th February 1999, aged 47.

He worked in many areas of microbiology, including rapid methods and the use of bioluminescent reporter genes. Gordon was instrumental in the discovery that bacteria communicate by small signalling molecules, and his group and their many collaborators have made significant contributions to the field of bacterial quorum sening using *N*-acyl homoserine lactones. The main focus of h's later work was directed at breaking bacterial communication by blocking the small molecule language to attenuate bacterial pathogenicity, thereby providing a new concept for anti-infectives.

CONTENTS

CONTRIBUTORS

ATKINSON, S. Institute of Infections and Immunity, Queen's Medical Centre, University of Nottingham, Nottingham NG7 2UH, UK

BAINTON, N. J. Microbial Physiology Group, School of Biological Sciences, University of Surrey, Guildford, Surrey GU2 5XH, UK

BALDWIN, T. J. Institute of Infections and Immunity, and School of Clinical Laboratory Sciences, University of Nottingham, Nottingham, UK

BATCHELOR, M. Department of Biochemistry, Imperial College of Science, Technology and Medicine, London SW7 2AZ, UK

BURKHOLDER, J. M. Department of Botany, North Carolina State University, Box 7612, Raleigh, NC 27695, USA

CAMPBELL, T. A. Department of Molecular and Cell Biology, Institute of Medical Sciences, University of Aberdeen, Aberdeen AB25 2ZD, UK

CORNELIS, G. R. Microbial Pathogenesis Unit, Christian de Duve International Institute of Cellular Pathology, and Faculté de Médécine, Université Catholique de Louvain, 74 Avenue Hippocrate, PO Box 74-49, B-1200 Brussels, Belgium

DALLON, J. C. Department of Mathematics, Heriot-Watt University, Edinburgh EH14 4AS, UK

DAVEY, J. Department of Biological Sciences, University of Warwick, Coventry CV4 7AL, UK

DELAHAY, R. M. Department of Biochemistry, Imperial College of Science, Technology and Medicine, London SW7 2AZ, UK

DOUGAN, G. Department of Biochemistry, Imperial College of Science, Technology and Medicine, London SW7 2AZ, UK

DOWNIE, J. A. Department of Genetics, John Innes Centre, Norwich, UK

DUNNY, G. M. Department of Microbiology, University of Minnesota Medical School, 1460 Mayo Building, Box 196, UMHC, 420 Delaware Street, SE, Minneapolis, MN 55455, USA

ENGLAND, R. R. Department of Biological Sciences, Maudland Building, University of Central Lancashire, Corporation Street, Preston PR1 2HE, UK

ERLANDSEN, S. Department of Cell Biology and Neuroanatomy, University of Minnesota Medical School, Minneapolis, MN 55455, USA

FRANKEL, G. M. Department of Biochemistry, Imperial College of Science, Technology and Medicine, Exhibition Road, London SW7 2AZ, UK

GOODAY, G. Department of Molecular and Cell Biology, Institute of Medical Sciences, University of Aberdeen, Aberdeen AB25 2ZD, UK

Gow, N. A. R. Department of Molecular and Cell Biology, Institute of Medical Sciences, University of Aberdeen, Aberdeen AB25 2ZD, UK

Greenberg, E. P. Department of Microbiology, University of Iowa, Iowa City, IA 52242, USA

Hale, C. Department of Biochemistry, Imperial College of Science, Technology and Medicine, London SW7 2AZ, UK

Hartland, E. L. Department of Biochemistry, Imperial College of Science, Technology and Medicine, London SW7 2AZ, UK

Hirt, H. Department of Microbiology, University of Minnesota Medical School, Minneapolis, MN 55455, USA

Hobbs, G. School of Biomolecular Sciences, Liverpool John Moores University, Byrom Street, Liverpool L3 3AF, UK

Höfer, T. Institute of Biophysics, Humboldt University of Berlin, Invalidenstrasse 42, D-10115 Berlin, Germany

Iglewski, B. H. Department of Microbiology and Immunology, University of Rochester School of Medicine and Dentistry, Rochester, NY 14642, USA

Isherwood, K. E. DERA, CBD, Porton Down, Salisbury, Wiltshire SP4 0JQ, UK

Kaiser, D. Department of Biochemistry and Developmental Biology, Stanford University School of Medicine, Stanford, CA 94305, USA

Kaprelyants, A. S. Bakh Institute of Biochemistry, Russian Academy of Sciences, Leninsky pr.33, 11707 Moscow, Russia

Kell, D. B. Institute of Biological Sciences, University of Wales, Aberystwyth, Ceredigion SY23 3DD, UK

Knutton, S. Institute of Child Health, University of Birmingham, Birmingham B4 6NH, UK

Kormer, S. S. Bakh Institute of Biochemistry, Russian Academy of Sciences, Leninsky pr.33, 11707 Moscow, Russia

McGowan, S. J. Department of Biochemistry, University of Cambridge, Tennis Court Road, Cambridge CB2 1QW, UK

Maini, P. K. Centre for Mathematical Biology, Mathematical Institute, 24–29 St Giles', Oxford OX1 3LB, UK

Morris, B. M. Department of Molecular and Cell Biology, Institute of Medical Sciences, University of Aberdeen, Aberdeen AB25 2ZD, UK

Mukamolova, G. V. Bakh Institute of Biochemistry, Russian Academy of Sciences, Leninsky pr.33, 11707 Moscow, Russia

Neyt, C. Microbial Pathogenesis Unit, Christian de Duve Institute of Cellular Pathology and Faculté de Médécine, Université Catholique de Louvain, B-1200 Brussels, Belgium

Osborne, M. C. Department of Molecular and Cell Biology, Institute of Medical Sciences, University of Aberdeen, Aberdeen AB25 2ZD, UK

Oyston, P. C. F. DERA, CBD, Porton Down, Salisbury, Wiltshire SP4 0PQ, UK

PALLEN, M. J. Microbial Pathogenesis Research Group, Department of Medical Microbiology, St. Bartholomew's and the Royal London Schools of Medicine and Dentistry, London EC1A 7BE, UK

PESCI, E. C. Department of Microbiology and Immunology, East Carolina University School of Medicine, Greenville, NC 27858-4354, USA

REID, B. Department of Molecular and Cell Biology, Institute of Medical Sciences, University of Aberdeen, Aberdeen AB25 2ZD, UK

ROBERTS, D. McL. Department of Zoology, Natural History Museum, Cromwell Road, London SW7 5DB, UK

ROSENSHINE, I. Departments of Molecular Genetics and Biotechnology, and Clinical Microbiology, The Hebrew University, Faculty of Medicine, POB 12272, Jerusalem 9112, Israel

SALMOND, G. P. C. Department of Biochemistry, University of Cambridge, Tennis Court Road, Cambridge CB2 1QW, UK

SHEPHERD, S. J. Department of Molecular and Cell Biology, Institute of Medical Sciences, University of Aberdeen, Aberdeen AB25 2ZD, UK

SHERRATT, J. A. Department of Mathematics, Heriot-Watt University, Edinburgh EH14 4AS, UK

SPRINGER, J. J. Department of Marine, Earth and Atmosphere Sciences, College of Agriculture and Life Sciences, North Carolina State University, Raleigh, NC 27695, USA

STEWART, G. S. A. B. School of Pharmaceutical Sciences, University of Nottingham, University Park, Nottingham NG7 2RD, UK

SWIFT, S. Institute of Infections and Immunity, Queen's Medical Centre, University of Nottingham, Nottingham NG7 2UH, UK

VAN WEST, P. Department of Molecular and Cell Biology, Institute of Medical Sciences, University of Aberdeen, Aberdeen AB25 2ZD, UK

WEICHART, D. H. Institute of Biological Sciences, University of Wales, Aberystwyth, Ceredigion SY23 3DD, UK

WILLIAMS, P. Department of Pharmaceutical Sciences, University of Nottingham, University Park, Nottingham NG7 2RD, UK

WREN, B. Microbial Pathogenesis Research Group, Department of Medical Microbiology, St. Bartholomew's and the Royal London Schools of Medicine and Dentistry, London EC1A 7BE, UK

YAMADA, Y. Department of Biotechnology, Graduate School of Engineering, Osaka University, 2-1 Yamadaoka, Suita Osaka 565, Japan

YOUNG, M. Institute of Biological Sciences, University of Wales, Aberystwyth, Ceredigion SY23 3DD, UK

EDITORS' PREFACE

Inspection of the primary research journals reveals a healthy and increasing interest in microbial signalling and communication. It is therefore timely to bring together, in this volume, current understanding on this topic, in a range of microbial systems.

By way of introduction to the subject, it is appropriate to start with literal definitions. The word signal is defined as 'to send, notify, announce, communicate by means of signals', whereas communication is defined as 'that which is communicated, a letter, a message, information imparted by speech, writing, etc.' Taking these two interrelated dictionary definitions as they stand would suggest that, fundamental to each, there is an absolute requirement for a 'language', based upon a set of symbols, by which the signaller can communicate and be understood by the signalled. As human beings, we are constantly signalling and communicating in the form of words, gestures, symbols, etc., to ourselves and to each other. These communications allow us to carry out many diverse functions in a 'social' environment with relative speed and efficiency, hopefully enabling us to enjoy and survive another day. At a simpler level, it is known that, for successful cell division to occur within a culture of mammalian cells, there is a requirement for extracellular growth factors called cytokines, which act as chemical signals. It is becoming clear that similar chemicals also occur in higher plants, multicellular invertebrates and ciliates. Within the world of micro-organisms, signalling, communication, and hence information flow, also occur.

Language is the common factor among all methods of communication used by biological organisms. Certain chapters within this volume will attempt to decode and translate the different languages and, by definition, the vocabularies (chemical signal molecules) utilized by a wide range of different micro-organisms within various environmental situations. For some micro-organisms we know the chemical structure of the signal molecule(s) utilized; however, in others the structures are far less clear. As you read through the book you will learn how, and under what conditions, micro-organisms communicate with each other and also with other biological cells, and how, in some instances, we can exploit this knowledge.

The opening chapter by Williams demonstrates the diverse nature of signalling and communication throughout the microbial world and sets the agenda for the rest of the book. One feature that is highlighted in the first chapter, and also in the chapter by Kaprelyants *et al.*, is the recent rapid advances that have been made in the field of bacterial cell–cell communication. This has been facilitated by the discovery of the chemical nature of the signal molecules involved. In most cases they have been shown to be small peptides or a modified form of homoserine lactone. These types of signal molecules have often been referred to as 'pheromones' or 'autoinducers'. If we accept the definition of a pheromone, as 'substances which are secreted to the outside by an individual and received by a second individual of the same species, which elicit a behavioural or a developmental process', then their *raison d'être* becomes clearer. In most cases this can be viewed as a density-dependent or quorum sensing process, by which a signal molecule that cannot be detected by an individual bacterium, or even by low numbers of bacteria, is released into the local environment. Only when bacteria are at relatively high numbers, or within a confined environment, will a threshold level of signal molecule be reached that can initiate specific gene expression required for the 'survival' mechanisms peculiar to the genus of bacterium involved. Thus we have the captivating situation of intercellular communication, by signalling, from bacteria that may not be in close physical contact. The chapter by Greenberg provides an introduction to the role of acylhomoserine lactone (AHL) and quorum sensing in Gram-negative bacteria. This is followed by Swift *et al.* who describe the role of AHL and quorum sensing circuits in two genera of pathogenic Gram-negative bacteria: *Aeromonas*, which is an emerging pathogen of fish and man; and *Yersinia*, a genus that includes pathogens known to infect humans and also certain fish, such as trout. The chapter by Pesci and Iglewski continues the theme of signalling and communication in pathogenic Gram-negative bacteria by reviewing the two separate quorum sensing systems (*las* and *rhl*) in *P. aeruginosa*.

It is probably reasonable to say that less is known about signalling and communication involving peptide molecules. The chapter by Dunny *et al.* reviews the work being carried out on the many roles of sex pheromone peptides in *Enterococcus* conjugation. This is followed with a chapter by Kaiser, in which he describes his elegant work on peptide signalling molecules and on their role in multicellular differentiation of the gliding bacterium, *Myxococcus xanthus*.

Exploitation of the knowledge available to us is becoming distinctly possible, for example, understanding the role of acylhomoserine lactones, in terms of influencing production of particular natural products such as antibiotic production (carbapenems) in *Erwinia carotovora* (McGowan and Salmond) and butyrolactones controlling antibiotic production (viginiamycin) in *Streptomyces virginiae* (Yamada).

Pheromones are not only produced by bacteria. Events in pheromone pathways of yeasts are similar to those found in higher eukaryotes. The fission yeast, *Schizosaccharomyces pombe*, has proved to be an excellent organism for studying the communication processes. Davey describes the production and action of peptide hormones on target cells; also how the cell recovers from the effects of stimulation and returns to a resting state. Continuing the eukaryotic theme, chemical communication between fungal hyphae is discussed. Gooday explains how pheromones involved in the cross-talk between hyphae are structurally very diverse, ranging from oxygen to complex peptides. In all cases these molecules interact with specific chemo-receptors, coupled to signal transduction pathways within the hyphae.

The life cycle of the slime mould *Dictyostelium discoideum* incorporates key features of morphogenesis found in higher organisms, e.g. chemotaxis, cellular differentiation and multicellular organization. The chapter by Sherratt *et al.* shows how mathematical modelling of cell streams in *D. discoideum* can provide reassuring evidence that we do now appear to understand the fundamentals of signal mechanisms.

Microbial–plant cell communication is addressed in the book and is discussed both from pathogenic and symbiotic aspects. The signalling molecules involved in bacterial–plant cell communication can be broadly classified as: synthesized metabolites, e.g. syringolides produced by *Pseudomonas syringae* that infects soybean; secreted proteins, e.g. non-specific plant-degrading enzymes that in some bacteria are regulated via quorum sensing; proteins that are delivered into plant cells causing a hypersensitive response, which eventually kills the invasive bacteria; and nodulation signalling proteins produced by the symbiont *Rhizobium* spp.

Gow *et al.* discuss plant pathogenicity and describe the signalling interactions between the eukaryotes *Phytophthora* and *Pythium* and their host-plant cells. The hallmark of these organisms is their ability to form zoospores that are required for the dispersal of the organism through films of water within wet soils. The signalling systems involve chemical and electrical signals generated by the host plant to guide zoospores to the plant which eventually leads to invasion of the plant cells. Much of the work described deals with zoospore–root and zoospore–zoospore interactions.

Understanding the mechanisms by which plant-associated pathogens/symbionts produce/regulate synthesis of signalling molecules or respond to plant-induced signals will be of immense benefit to the agricultural industry. It could lead to the development of blocking or enhancing agents, either *ex planta* or *in planta*, depending on the particular requirement.

Moving on from plant-associated micro-organisms, another extremely important topic that is addressed is bacterial–animal cell communication. It is recognised that infective bacteria are able to alter eukaryotic signal transduction pathways and thus host-cell functions. As a consequence, invasive pathogenic bacteria are able to overcome the defence mechanisms of their animal host and to reproduce in the tissues. Within the last few years there have been considerable advances in the molecular detail of communication and signalling between pathogenic bacteria and animal host cells. In particular, the mammalian cell targets of some of the bacterial effector proteins have been investigated. To help illustrate the advances made in this important area, work is presented on the interaction of enteropathogenic *Escherichia coli* (EPEC) and enterohaemorrhagic *E. coli* (EHEC) with mammalian intestinal enterocytes (Frankel *et al.*) and on the Yop system of

Yersinia spp. that obstructs a cellular immune response (Neyt and Cornelis). Clearly, a better understanding of pathogenic bacteria–host cell communication would allow the rational design/development of drugs that could block bacterial effector protein action and/or synthesis.

Another group of organisms that is discussed, and for many people will be new to them, is the dinoflagellates. These organisms dominate the plankton of the subtropics in the world's oceans and consequently are important ecologically and economically. However, very little is known about signalling mechanisms that have been proposed to mediate cellular processes including encystment, cell division and bioluminescence. Cell-to-cell recognition of endosymbiotic relationships between the coral–dinoflagellate associations is only just beginning to be understood (Burkholder and Springer).

The authors have presented an excellent range of current reviews on microbial signalling and communication, which hopefully will encourage more scientists from widely diverse disciplines (academic, medical and industrial) to embrace the topic. We thank all contributors for the considerable amount of effort and time afforded to produce each chapter.

R. R. England, G. Hobbs, N. J. Bainton and D. McL Roberts
January 1998

BACTERIAL CROSSTALK – COMMUNICATION BETWEEN BACTERIA, PLANT AND ANIMAL CELLS

PAUL WILLIAMS[1,2,3], TOM J. BALDWIN[1,3] AND J. ALLAN DOWNIE[4]

[1] *Institute of Infections and Immunity,* [2] *School of Pharmaceutical Sciences and* [3] *School of Clinical Laboratory Sciences, University of Nottingham, Nottingham, UK and* [4]*Department of Genetics, John Innes Centre, Norwich, UK*

INTRODUCTION

Prokaryotes have evolved elaborate signal transduction mechanisms to perceive sensory information and so facilitate adaptation to the prevailing growth environment. Many of the intracellular events which take place in response to changes in osmolarity, pH, temperature, and nutrient availability have been the subject of much molecular scrutiny. In this context, a major recurring theme is the so-called 'two-component' system in which information is relayed via phosphoryl transfer from sensor to regulator proteins (Hoch & Silhavy, 1995; see Frankel, this volume). Such systems closely resemble the signal transduction mechanisms operating in higher organisms, implying that bacteria are capable of exhibiting more complex patterns of multicellular behaviour than would perhaps be predicted for unicellular microorganisms. Bacteria respond to, and process, external signals from neighbouring cells whether they are of bacterial, plant or animal origin. Such intercellular communication may involve small diffusible signal molecules and secreted proteins as well as surface-associated macromolecules. With respect to the former, the last decade has seen a tremendous growth in our understanding of the mechanisms bacteria use to coordinate their behaviour in concert with their own population size. 'Quorum sensing' as this cell density sensing mechanism has become known, is now accepted as a generic phenomenon within the bacterial kingdom (for reviews, see Fuqua *et al.*, 1994, 1996; Salmond *et al.*, 1995; Swift *et al.*, 1996; Williams *et al.*, 1992; Williams, 1994). Essentially, quorum sensing enables a bacterium to function in a multicellular manner by coupling gene expression to the attainment of a critical ('quorate') population size. Quorum sensing depends on the activation of a response regulator by a 'self-generated' diffusible signal molecule, and several chemically distinct families of such molecules have now been identified. Moreover, some quorum sensing systems are themselves controlled by external host-derived signals, as in the case of *Agrobacterium tumefaciens* where plant derived opines are required to trigger the quorum sensing-dependent conjugal transfer of Ti plasmids (Zhang *et al.*, 1993). Conversely, bacterial quorum sensing signal molecules *per se* can influence the host immune response as exemplified by the immunomodulatory properties of the *Pseudomonas aeruginosa* autoinducer *N*-(3-oxododecanoyl) homoserine lactone (Telford *et al.*, 1998). Furthermore, several pathogens and symbionts are capable of manipulating eukaryotic host cell signal transduction pathways either by targeting host-cell receptors, whose function is to receive hormonal, cytokine or other signals or via the direct delivery of bacterial proteins into eukaryotic cells by surface-attached bacteria.

In this chapter, an overview of this intercellular information flow and its potential exploitation will be presented which, in the case of bacterial cell–cell communication, has largely become apparent through identifying the signal molecules involved in systems where their effect can easily be detected. In the sections on bacterial–plant (see McGowan & Salmond, this volume) and bacterial–animal cell (see also Neyt & Cornelis, this volume) communication, examples have been chosen to illustrate different themes that are found within this cross-talk. Parallels can be drawn between the ways in which different classes of signalling molecules are recognized by plant and animal cells, although often such communication is a consequence of the eukaryotic cells identifying bacterial components in order to defend themselves against pathogenic attack. This is particularly true with regard to recognition of proteins secreted by bacteria and there are very close similarities between the mechanisms used by bacteria to deliver proteins to (or into) both plant and animal cells. It is beyond the scope of this chapter to review all of the literature related to prokaryotic–eukaryotic interactions, but specific examples of the types of signalling events that occur will be given, while a more extensive treatment of the subversion of host cell signal transduction pathways by pathogenic *Yersinia* spp. can be found in the chapter by Neyt and Cornelis.

THE 'LANGUAGES' OF BACTERIAL CELL–CELL COMMUNICATION

Bacteria, like higher differentiated organisms, are likely to derive considerable survival benefits by coordinating the behaviour of their populations, whether in pure or mixed cultures (Shapiro, 1988). Such benefits may include improved access to complex nutrients or environmental niches, collective defence against other competitive microorganisms or eukaryotic host defence mechanisms, and optimization of population survival by differentiation into morphological forms better adapted to combating an environmental threat. Examples of the latter include the coordinated differentiation and migration of groups of highly specialized swarmer cells in the rapid colonization of a surface, sporulation, the formation of multicellular fruiting bodies and adoption of the biofilm mode of growth.

Similarly, with regard to conjugal plasmid transfer, the ability of a potential donor to monitor the availability of recipients contributes to the decision of whether or not to make the energetically expensive investment required for cell–cell conjugation.

The coordination of this multicellular behaviour in prokaryotes frequently centres upon the generation of small diffusible signal molecules, which are sometimes referred to as microbial 'hormones' or 'pheromones' or 'auto-inducers' (Wirth *et al.*, 1996; Swift *et al.*, 1996), which constitute one of the few ways in which an individual bacterial cell can obtain information about the status of other members of the same species. By far the most intensively investigated family of bacterial quorum sensing systems utilizes *N*-acyl-homoserine lactones (AHLs) as signalling molecules, although it is now apparent that alternative signalling molecule 'languages' exist (Table 1).

Discovering AHL-mediated cell–cell communication in Gram-negative bacteria

In 1992 Bainton *et al.* (1992*a,b*) discovered that a diffusible signal molecule present in cell-free stationary phase culture supernatants regulated the production of the *β*-lactam antibiotic 1-carbapen-2-em-3-carboxylic acid in the plant pathogen *Erwinia carotovora*. This molecule was identified as *N*-(3-oxohexanoyl)-L-homoserine lactone (OHHL; Fig. 1) and its significance relates to the fact that OHHL had been identified many years earlier as the 'autoinducer' of bioluminescence in *Vibrio fischeri* (Eberhard *et al.*, 1981). In this Gram-negative marine bacterium, OHHL synthesis requires the *luxI* gene product and the *luxICDABE* genes are transcribed in a cell density-dependent manner by the transcriptional activator protein, LuxR (Sitnikov *et al.*, 1995; see Greenberg, this volume). LuxR is activated by binding OHHL, and homologues of LuxI and LuxR termed CarI (ExpI) and CarR were subsequently identified in *E. carotovora* (Swift *et al.*, 1993; Pirhonen *et al.*, 1993; McGowan *et al.*, 1995; McGowan & Salmond, this volume). In contrast to the *lux* operon, *carI* is not linked to the *car* structural genes but is located elsewhere on the chromosome (McGowan *et al.*, 1995; 1996). Furthermore, mutations in *carI(expI)* downregulate exoenzyme production thus rendering the organism avirulent *in planta* unless supplied with the exogenous OHHL (Jones *et al.*, 1993; Pirhonen *et al.*, 1993).

These findings stimulated the development and use of biosensors to screen cell-free culture supernatants from a wide range of bacterial species for the presence of OHHL or related AHLs (Bainton *et al.*, 1992*a*; Pearson *et al.*, 1994; Pearson *et al.*, 1995; Shaw *et al.*, 1997; Swift *et al.*, 1993; Throup *et al.*, 1995; Winson *et al.*, 1995; Winson *et al.*, 1998). Such biosensors usually contain an AHL-activated promoter fused to a reporter gene(s) such as *lacZ* or *luxCDABE*, together with a LuxR homologue but lack the AHL synthase. Consequently, activation of the reporter depends on the presence of an exogenous AHL. Using these approaches, Gram-negative bacterial species

Table 1. *Some examples of diffusible bacterial cell–cell signalling molecules and the phenotypes they control*

Bacterial species	Signal molecule	Phenotype
Aeromonas salmonicida	BHL/HHL	Exoproteases
Aeromonas hydrophila	BHL/HHL	Exoproteases, biofilms
Agrobacterium tumefaciens	OOHL	Ti plasmic conjugation
Bacillus subtilis	ComX peptide, CSF	Competence
Bacillus subtilis	Oligopeptide	Sporulation
Chromobacterium violaceum	HHL	Violacein pigment, HCN, exoenzymes
Erwinia carotovora	OHHL	Carbapenem antibiotic, exoenzymes
Erwinia stewartii	OHHL	Exopolysaccharide
Enterococcus faecalis	Oligopeptides	Conjugal plasmid transfer
Myxococcus xanthus	A-signal	Fruiting body formation
Pseudomonas aeruginosa	BHL/HHL/OdDHL	Exoenzymes, HCN, pyocyanin, secretion, RpoS, biofilms
Pseudomonas aureofaciens	HHL	Phenazine antibiotics
Ralstonia solanacearum	3-OH PAME	Exopolysaccharide, exoenzymes
Rhizobium leguminosarum	7,8-*cis*-HtDHL	Growth inhibition
Rhodobacter sphaeroides	7,8-*cis*-tDHL	Aggregation
Serratia liquefaciens	BHL/HHL	Swarming
Staphylcoccus aureus	Cyclic octapeptides	Exotoxins, exoenzymes, coagulase, protein A, cell wall proteins
Streptococcus pneumoniae	Heptadecapeptide (CSP)	Competence
Vibrio anguillarum	ODHL	Haemolysin
Vibrio fischeri	OHHL	Bioluminescence
Vibrio harveyi	HBHL	Bioluminescence, polyhydroxybutyrate
Xanthomonas campestris	DSF	Exoenzymes; exopolysaccharide
Yersinia pseudotuberculosis	OHHL/HHL/OHL	Motility

The abbreviations used are: OHHL, N-(e-oxohexanoyl)homoserine lactone; HBHL, N-(3-hydroxybutanoyl)homoserine lactone; HHL, N-hexanoyl homoserine lactone; BHL, N-butanoylhomoserine lactone; OOHL, N-(3-oxooctanoyl)homoserine lactone; OHL, N-octanoylhomoserine lactone; ODHL, N-(3-oxodecanoyl)homoserine lactone; OdDHL, N-(3-oxododecanoyl)homoserine lactone; 7,8-*cis*-HtDHL, 7,8-*cis*-N-(3-hydroxytetradecenoyl) homoserine lactone; 7,8-*cis*-tDHL, 7,8-*cis*-N-(tetradecenoyl)homoserine lactone. DSF, an uncharacterized diffusible extracellular factor; CSP, competence stimulating peptide; CSF, competence stimulating factor oligopeptide; 3-OH PAME, 3-hydroxypalmitic acid methyl ester.

belonging to the genera *Agrobacterium, Erwinia, Enterobacter, Pseudomonas, Vibrio, Yersinia, Aeromonas, Serratia, Chromobacterium, Hafnia, Rahnella,* and *Obesumbacterium* (Bainton *et al.*, 1992*a*; McClean *et al.*, 1997; Milton *et al.*, 1997; Pearson *et al.*, 1994, 1995; Shaw *et al.*, 1997; Swift *et al.*, 1993, 1997; Throup *et al.*, 1995; Winson *et al.*, 1995, 1998) have all given positive results with LuxR-based biosensors. However, although LuxR proteins respond most sensitively to their natural AHL, they are sufficiently flexible to detect a range of related AHL analogues such that the identity of the activating molecule(s) cannot be deduced without further chemical analysis. This usually requires the partitioning of the active compound(s) into an organic solvent prior to purification by preparative reverse phase HPLC followed by mass spectrometry and nuclear magnetic resonance spectroscopy. Elucida-

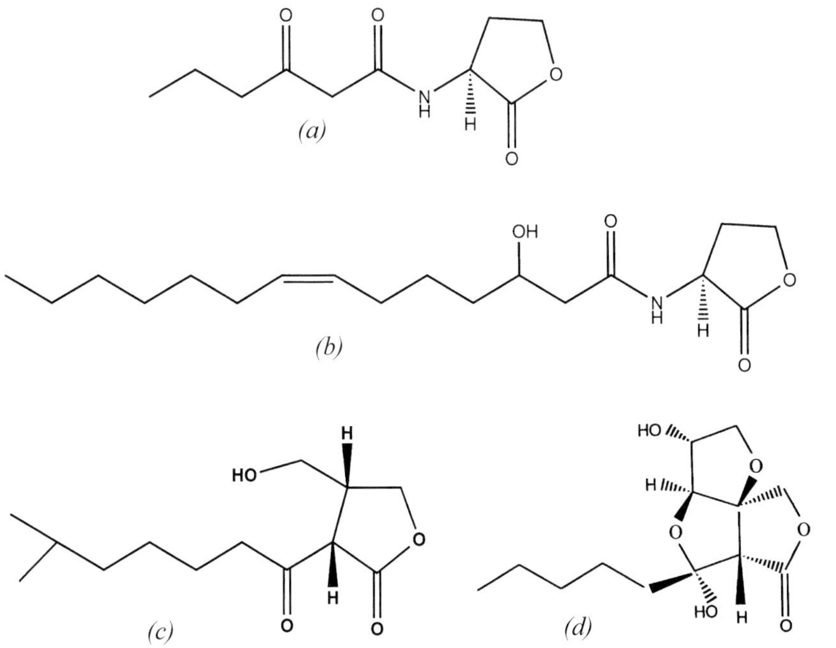

Fig. 1. Structures of (a) *N*-(3-oxohexanoyl)homoserine lactone; (b) 7,8-*cis*-*N*-(3-hydroxytetra-decenoyl)homoserine lactone; (c) A-factor and (d) Syringolide-1.

tion of the structure, stereochemistry and synthesis of the AHL responsible for regulating the production of the carbapenem antibiotic in *Erwinia carotovora* is comprehensively presented in Bainton *et al.* (1992*b*).

Using the approaches described above, AHLs have been identified with *N*-acyl side chains of 4, 6, 8, 10, 12 and 14 carbons with either an oxo- or hydroxy- or no substituent at the C3 position of the *N*-linked acyl chain. To date, only two compounds with acyl chains containing double bonds have been identified. These are 7,8-cis-*N*-(3-hydroxytetradecenoyl)homoserine lactone (Fig. 1) and 7,8-*cis*-*N*-(tetradecenoyl) homoserine lactone produced by *Rhizobium leguminosarum* (Schripsema *et al.*, 1996; Gray *et al.*, 1996) and *Rhodobacter sphaeroides* (Puskas *et al.*, 1997), respectively. The former compound has recently also been identified in *Pseudomonas fluorescens* (B. Laue, G. S. A. B. Stewart & P. Williams, unpublished data).

This AHL-based quorum sensing 'language' family is involved in the regulation of diverse cell-density associated phenotypes including antibiotic biosynthesis, plasmid conjugal transfer, swarming, cessation of cell growth, aggregation, nodulation, capsular polysaccharide production, protein secre-tion, biofilm maintenance and differentiation, in the production of exo-enzyme virulence determinants and cytotoxins in human, plant and animal pathogens (Fuqua *et al.*, 1994, 1996; Williams *et al.*, 1992, 1994; Salmond *et al.*, 1995; Swift *et al.*, 1996). More detailed descriptions of these AHL-dependent regulatory circuits in *V. fischeri, P. aeruginosa, E. carotovora, Aeromonas* and *Yersinia* spp. will be presented in the following chapters by Greenberg, Pesci and Iglewski, McGowan and Salmond and Swift *et al.*

Non-AHL mediated Gram-negative bacterial cell–cell signalling

Cell density-dependent gene regulation, mediated via small diffusible signalling molecules appears to be a common theme throughout the bacterial kingdom. Although AHLs constitute only one such 'quorum sensing language' and appear to be restricted to Gram-negative bacteria, lactones feature as components of other bacterial cell–cell signalling molecules. In certain *Streptomyces* species for example, the γ-butyrolactones (Fig. 1) have long been recognized as autoregulatory signal molecules involved in the control of antibiotic biosynthesis, resistance and differentiation (Horinouchi & Beppu, 1992; see Yamada, this volume). Butyrolactones with antifungal activity, have also been isolated from spent culture supernatants of *Pseudomonas aureofaciens* (Gamard *et al.*, 1997) a Gram-negative bacterium which employs AHL-mediated quorum sensing to control the synthesis of the antifungal phenazine antibiotics (Wood *et al.*, 1997). Furthermore, the plant pathogen *Ralstonia solanacearum* possesses a AHL-based quorum sensing circuit which is itself regulated by the LysR-type transcriptional activator, PhcA and a novel volatile signalling molecule, 3-hydroxypalmitic acid methyl ester (3-OH PAME; Flavier *et al.*, 1997*a,b*). 3-OH PAME is required for the expression of PhcA-regulated virulence factors and for virulence *in planta* The volatility of this signalling molecule is particularly interesting, especially since plants have been suggested to use volatile signalling molecules such as methyl-jasmonate, ethylene and methylsalicylate in the generation of intra- and inter-plant systemic responses following wounding or infection (Mur *et al.*, 1997).

In the cabbage pathogen *Xanthomonas campestris* pv. *campestris*, mutations in the *rpf* gene cluster result in down-regulation of exoenzyme and exopolysaccharide synthesis and reduced virulence (Barber *et al.*, 1997). The phenotype of mutants in one of these genes, *rpfF*, was restored by a diffusible extracellular factor (DSF) present in the culture supernatants of the parent strain. Although DSF appears not to be an AHL, and *rpfF* mutants do not respond to the presence of exogenous AHLs, it may be a fatty acid derivative, since decanoic and dodecanoic acids and their hydroxy derivatives, acid hydrolysates of lipopolysaccharide and chloroform–methanol extracted lipids all restore, to some extent, endoglucanase activity in the *X. campestris rpfF* mutant (Barber *et al.*, 1997). DSF is however distinct from the 3-OH PAME produced by *R. solanacearum*, since it is not volatile and the corresponding methyl esters of decanoic and dodecanoic acids do not restore exoenzyme production in *X. campestris*. Apart from other xanthomonads and a strain of *Erwinia herbicola*, culture supernatants or chloro-form-methanol cell extracts from other phytopathogenic bacteria including *Erwinia, Pseudomonas* and *Ralstonia* strains failed to induce endoglucanase synthesis. DSF may therefore represent a novel class of signalling molecule which awaits full chemical characterization. However, DSF is clearly not the only diffusible signal molecule produced by *X. campestris* since a second diffusible factor has been implicated in the regulation of both extracellular polysaccharide and the yellow pigment xanthomonadin indicating the existence of two separate intercellular signalling systems with overlapping roles (Poplawsky *et al.*, 1998).

Although biosensors containing a LuxR homologue, together with an AHL-activated promoter fused to a reporter gene, have proved extremely valuable in the search for AHLs, a second class of putative cell–cell signalling molecules capable of activating *luxRI'::luxCDABE* biosensors has been identified during the screening of Gram-negative bacterial cell-free super-natants for AHLs (S. R. Chhabra, M. T. G. Holden, P. Stead, N. J. Bainton, P. J. Hill, G. P. C. Salmond, G. S. A. B. Stewart, B. W. Bycroft & P. Williams, unpublished observations). Elucidation of the structures of these molecules revealed that they constitute a family of diketopiperazines (DKPs; Fig. 2) i.e. cyclic dipeptides including cyclo(ΔAla-L-Val), and cyclo(L-Pro-L-Tyr). While both of these DKPs are produced by *P. aeruginosa*, cyclo(ΔAla-L-Val) is also made by *Proteus mirabilis, Citrobacter freundii* and *Enterobacteragglomerans*. A third DKP, cyclo(L-Phe-L-Pro) has also been identified in *P. fluorescens* and *Pseudomonas alcaligenes* (S. R. Chhabra, M. T. G. Holden, P. Stead, N. J. Bainton, P. J. Hill, G. P. C. Salmond, G. S. A. B. Stewart, B. W. Bycroft & P. Williams, unpublished observations). Although they only weakly activate LuxR-based biosensors, these compounds are capable of antagonizing OHHL-mediated induction of bioluminescence, suggesting that they may compete for the same binding site on the LuxR protein target. The detection of these molecules via the LuxR-based AHL biosensor probably represents fortuitous cross-talk between distinct signal-ling systems rather than stimulation by an additional quorum sensing signal. However, it has been shown (R. de Nys, K. Yamamoto, M. Givskov, N. Kumar, R. Read, R. Utsumi & S. Kjelleberg; personal communication) that DKPs such as cyclo(L-Pro-L-Tyr) are produced by *E. coli* and can influence the transcription of specific stationary phase (RpoS) regulated genes such as *fic* and *bolA*, suggesting that this class of molecules has a signal transducing function. Cyclo(L-Phe-L-Pro) and cyclo(L-Pro-L-Tyr) possess significant phytotoxic activity on spotted knapweed (*Centaurea maculosa*) and are produced not only by the pseudomonads but also by the phytopathogenic fungus *Alternaria alternata* (Stierle *et al.*, 1988; Bobylev *et al.*, 1996). DKPs have also been shown to elicit pharmacological effects in humans and other mammals, acting on the central nervous system where they modulate a remarkable range of behaviours (Prasad, 1995). This is perhaps not that

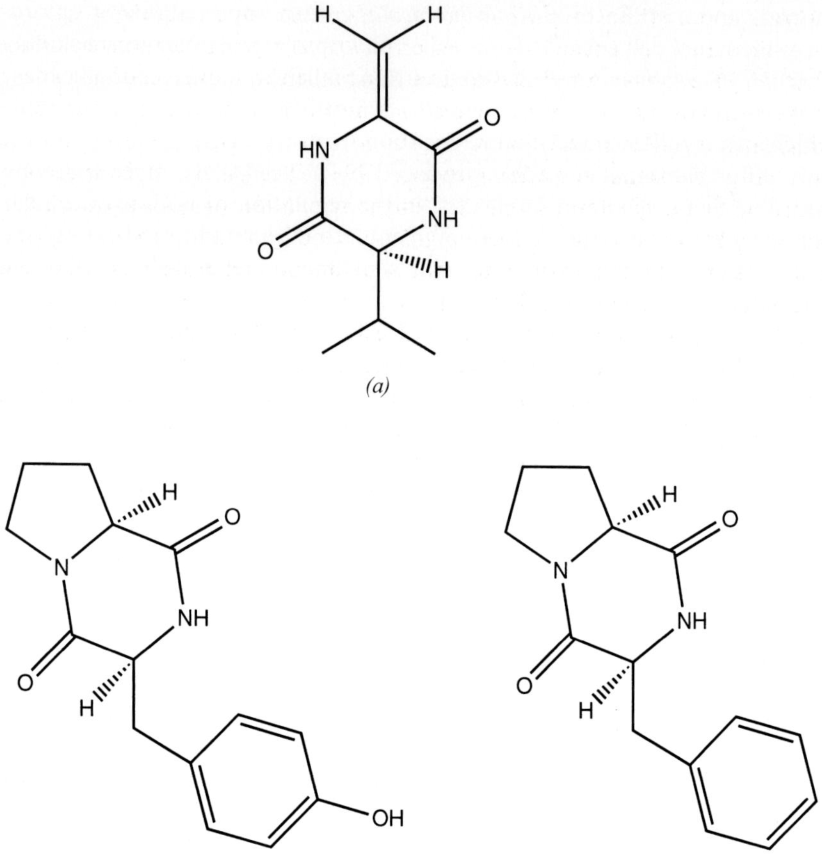

Fig. 2. Structures of (a) Cyclo(ΔAla-L-Val); (b) Cyclo(L-Pro-L-Tyr) and (c) Cyclo(L-Phe-L-Pro).

surprising considering the structural similarity shared between some cyclic dipeptides and endogenous signalling peptides such as thyrotropin-releasing hormone (Prasad, 1995). It is therefore conceivable that these bacterially generated cyclic dipeptides directly influence host–bacterial pathogen interactions.

Despite the significant advances made in understanding cell–cell communication in diverse Gram-negative bacterial genera, the existence of such signalling mechanisms in bacteria such as *E. coli* and *Salmonella typhimurium* have remained enigmatic. Although the *E. coli* genome sequence does not contain any known AHL synthases (i.e. members of either the LuxI or any LuxM/AinS families), both *E. coli* and *Salmonella* possess a LuxR homologue, SdiA, which is over 50% identical to RhlR (VsmR) from *P. aeruginosa*

(Ahmer *et al.*, 1988; Garcia-Lara *et al.*, 1996; Sitnikov *et al.*, 1996; Latifi *et al.*, 1995). In *E. coli*, SdiA has been proposed to regulate expression of the *ftsQAZ* gene cluster which is required for cell division (Garcia-Lara *et al.*, 1996; Sitnikov *et al.*, 1996), while the *Salmonella* homologue has been implicated in the expression of a number of genes including several putative virulence determinants (Ahmer *et al.*, 1998). The AHLs, OHHL, *N*-(3-hydroxybutanoyl) homoserine lactone and *N*-decanoylhomoserine-DL-lactone all weakly stimulate the SdiA-dependent P_2 promoter of the *ftsQAZ* gene cluster, although no obvious structure–function relationship is apparent (Sitnikov *et al.*, 1996). However, spent *E. coli* culture supernatant is more effective, implying the presence of a diffusible signalling molecule. Furthermore, *sdiA* itself has been reported to be regulated by an extracellular factor present in spent growth medium (Garcia-Lara *et al.*, 1996). More recently, Surette and Bassler (1998) have provided further evidence for their existence of a cell–cell signalling molecule(s) in both *E. coli* and *S. typhimurium*, based on the capacity of spent culture supernatants to stimulate bioluminescence in a *V. harveyi* mutant defective in the production of an as yet unidentified polar autoinducer molecule. The chemical identity and relationship between these putative *E. coli* and *Salmonella* signalling molecules and those responsible for activation of *sdiA* and the SdiA-driven *ftsQAZ* P_2 promoter remain to be established.

Cell–cell communication in Gram-positive bacteria

As yet, no AHLs have been identified in any Gram-positive bacterial genus, although several Gram-positives have been reported to weakly activate LuxR-based AHL biosensors (Williams, 1994). However, such activating molecules may well be DKPs. A number of Gram-positive organisms are, however, known to employ extracellular signalling molecules. These generally appear to be small, modified oligopeptides which interact with two-component histidine protein kinase signal transduction systems (Kleerebezem *et al.*, 1997; Wirth *et al.*, 1996; see Dunny, this volume). These systems regulate, for example, the development of genetic competence in *Bacillus subtilis* and *Streptococcus pneumoniae*, conjugation in *Enterococcus faecalis* and virulence in *Staphylococcus aureus*. In the latter, the expression of a number of cell-density dependent virulence factors is regulated by the global regulatory locus, *agr* (**a**ccessory **g**ene **r**egulator), which consists of two transcriptional units (Ji *et al.*, 1995, 1997*b*). The smaller unit encodes δ-haemolysin and RNAIII, while the larger encodes a sensor (*agrC*) and a response regulator (*agrA*) together with two genes (*agrD* and *agrB*) responsible for the generation of the quorum sensing signal molecule. This is an octapeptide cleaved from the middle of the *agrD* gene product and exported into the extracellular medium in a process which appears to involve AgrB (Ji *et al.*, 1997*b*). Furthermore, *S. aureus* strains can be divided into at least three groups on the basis of the ability of their peptide signalling molecules to cross-activate or -inhibit *agr* expression, i.e. the cognate peptides of one *S. aureus* group inhibit expression of *agr* in another group of strains (Ji *et al.*, 1997*b*). Although the peptide sequences for the three *S. aureus* groups are known, the linear molecules are inactive. By comparing the physical, spectroscopic and biological properties of the purified native *S. aureus* group I octapeptide and a series of synthetic analogues, the structure of the *S. aureus* pheromones have been unequivocally established as cyclic thiolactones in which a central cysteine residue is covalently linked to the carboxy terminus of the C-terminal amino acid (Z. Affas, P. W. McDowell, W. C. Chan, B. W. Bycroft, G. S. A. B. Stewart & P. Williams, unpublished observations; Fig. 3). In addition, synthetic group III cyclic thiolactone (Fig. 3) has been shown to effectively inhibit α-toxin and TSST-1 synthesis in group I *S. aureus* (Z. Affas, P. W. McDowell, W. C. Chan, B. W. Bycroft, G. S. A. B. Stewart & P. Williams, unpublished observations). While coagulase-negative staphylococci such as *Staphylococcus epidermidis* also make these cyclic thiolactone signalling molecules (Otto *et al.*, 1998), their contribution to virulence gene regulation has not been established, and it is conceivable that their function is to facilitate competition with *S. aureus* for particular ecological niches.

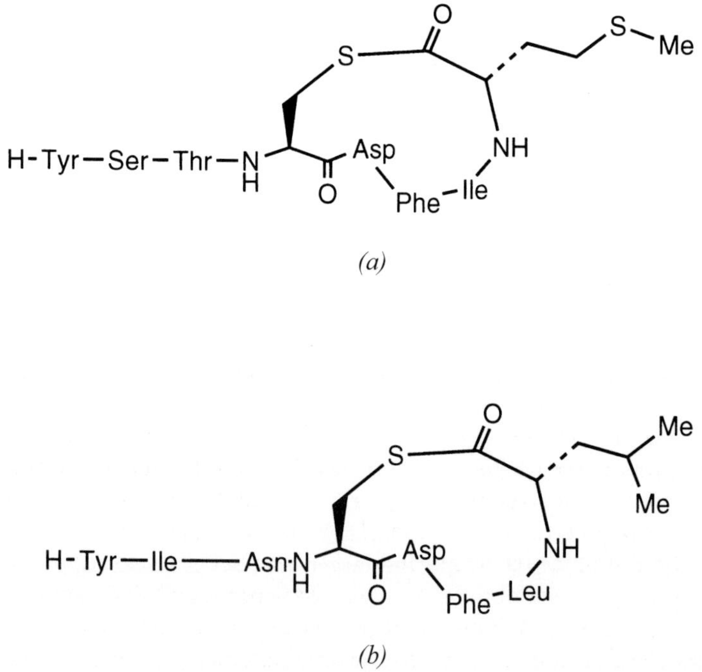

(a)

(b)

Fig. 3. Structures of (a) Group I and (b) Group III *S. aureus* cyclic thiolactone peptide pheromones.

In contrast to the AHLs which have to diffuse across both outer and inner membranes to reach their cytoplasmically located target LuxR homologue, the targets for these Gram-positive signalling molecules appear to be located on the outer surface of the cytoplasmic cell membrane and can easily be reached through the relatively porous structure of the cell wall. In the case *S. aureus*, the receptor is the transmembrane protein AgrC, which is auto-phosphorylated in response to the purified octapeptide signal molecule (Lina *et al.*, 1998). This, in turn, leads to phosphorylation of the response regulator, AgrA, resulting in the stimulation of RNAIII transcription. Since RNAIII is the effector molecule responsible for controlling both the expression of *S. aureus* virulence determinants and the *agrACDB* operon, this sets up an autoinduction cascade which leads to the production and export of more octapeptide (Ji *et al.*, 1995, 1997*b*).

The genetic organization of the *agrACDB* locus in *S. aureus* is very similar to the *comAPXQ* locus of *Bacillus subtilis* which controls the development of competence, a cell density-dependent process, which is regulated via an extracellular peptide (the competence stimulating factor peptide which is derived from the C-terminus of ComX) and a two-component system (Grossman, 1995). Similarly, accumulation of a phosphorylated response regulator (SpoA), which is required for the initiation of sporulation in *B. subtilis*, is mediated by another extracellular signal peptide. The development of genetic competence in *Streptococcus pneumoniae* is also coordinately regulated by a quorum sensing mechanism. Competence in this case is initiated by a processed heptadecapeptide (competence stimulating peptide) signal which is transduced via the ComDE two-component system (Pestova *et al.*, 1996).

BACTERIA–PLANT CELL–CELL COMMUNICATION

The signalling molecules that are made by bacteria and are recognized by plant cells can be classified into three broad categories: synthesized metabolites, secreted proteins and components of the bacterial cell surface. Aspects of this cell-to-cell communication are based on the ability of the plant to recognize bacterial components so that the plant can mount a defence response. However, there are several examples, particularly with plant–microbe symbioses, in which signalling molecules have been used to maintain and sustain the plant–bacterial interaction. Different examples from the three broad categories of signalling molecules have been chosen, but it must be recognized that, for any given plant–bacterial interaction, there are likely to be multiple signalling systems in operation and the expression of one can influence the effect of another.

Signalling metabolites

Pseudomonas–soybean pathogenesis

Pseudomonas syringae pv. *glycinea* is a soybean pathogen, which makes glycolipid elicitors that induce a defence response to resistant plants. These elicitors are called syringolides (Midland *et al.*, 1993) and their general structure is illustrated in Fig. 1. Some varieties of soybean are resistant to strains of *P. syringae* that make syringolide molecules and this resistance is determined by a single plant gene *Rpg4*. Addition of purified syringolide to resistant cells of soybean induces a Ca^{2+} influx, a K^+ efflux and a net extracellular alkalinisation (Atkinson *et al.*, 1996). The recognition is highly specific, since only those plant lines carrying the *Rpg4* resistance gene induce the response; it is thought that the 3-hydroxyl group of the syringolide may be crucial for recognition (Ji *et al.*, 1997*a*). Ca^{2+}-influx, K^+-efflux and extracellular alkalinization are proposed to be components of a syringolide-activated signalling pathway, although at this stage the signalling pathway has not been defined. The syringolides can be made in *E. coli* if the *P. syringae* pv. *glycinea* gene *avrD* (encoding syringolide synthesis) is present (Midland *et al.*, 1993) and it is noteworthy that the syringolide structure has some characteristics in common with AHLs and γ-butryolactones used in bacterial quorum sensing.

Rhizobium–legume symbiosis

Rhizobia such as *Rhizobium leguminosarum* bv. *viciae* and *Sinorhizobium meliloti* enter into a symbiotic interaction with legumes such as peas and alfalfa. The bacteria recognize compounds (flavonoids) exuded from legume roots and, in response, induce the formation of lipochitin–oligosaccharide nodulation (Nod) factors (Dénarié *et al.*, 1996). The formation of nitrogen-fixing nodules and the invasion of plants by rhizobia is totally dependent on these Nod factors. These factors can induce nodule morphogenesis on legume roots in the absence of bacteria and are highly specific signalling molecules. Subtle modifications such as changes to the fatty acyl group, acetylation and sulphation can influence the range of plants that are capable of recognizing the Nod factors. Thus, for example, alfalfa plants require a sulphate group on the reducing sugar, whereas in peas the presence of such a substituent prevents recognition (Roche *et al.*, 1991).

Nod factors induce plant responses at concentrations as low as 10^{-12} M (Dénarié *et al.*, 1996), implying the presence of very high-affinity plant receptors. Although the receptors have not yet been defined, there are strong indications that a Ca^{2+}-based signalling pathway is involved in the induction of plant genes in response to Nod factors (Franssen *et al.*, 1995). Electrophysiology studies have shown that, shortly after addition of Nod-factor to legume roots, there is a Ca^{2+}-influx accompanied by membrane depolarization and K^+-efflux (Ehrhardt *et al.*, 1992; Félle *et al.*, 1996, 1998), and increases in intracellular free Ca^{2+} (Gehring *et al.*, 1997). Subsequently the Nod factors can induce a periodic oscillation of intracellular Ca^{2+} which is thought to be associated with a plant commitment to developmental change (Ehrhardt *et al.*, 1996). The periodicity of the oscillation is around 1 minute and the type of oscillation is similar to that seen in animal cells in response to external stimuli (Fewtrell, 1993). Using plants carrying a transgenic reporter gene that reveals expression of plant early nodulation genes, Pingret *et al.* (1998) concluded that the Nod factors induce a G-protein-mediated signalling pathway. This was deduced on the basis of the effect of inhibitors and agonists. Thus, pertussis toxin, an inhibitor known to block G-protein-mediated signalling in animal cells inhibited the Nod factor-induced gene expression. Antagonists that interfere with phospholipase C activity or Ca^{2+} movement also blocked the Nod-factor induced signalling. Furthermore, compounds such as Mastoparan and Mas7, known G-Protein agonists, could induce expression of the plant early nodulation gene analysed.

Purified lipochitin oligosaccharide Nod-factors also induce cytoskeletal rearrangements, root cortical cell division, and have effects on the position of the nucleus (Timmers *et al.*, 1998). In root epidermal cells, these changes are associated with the initiation of a plant-made specialized structure, the infection thread (van Brussel *et al.*, 1992), that enables the bacteria to enter the legume root and gain access to the dividing cells that will eventually become a nodule. Such cytoskeletal changes are analogous to changes seen in animal cells in response to bacterial pathogens such as the enterovirulent *E. coli* (see below).

Secreted proteins

Many Gram-negative bacteria secrete exoproteins, a number of which are involved in signalling to eukaryotic cells of both plant and mammalian origin. Translocation of such secreted proteins across the outer membrane depends on one of three major secretion pathways referred to as the Type I, Type II and Type III (Wandersman, 1996; Filloux & Hardie, 1998). Type I secretion is a one-step pathway involving an ABC transporter in which the exported protein contains a C-terminal signal sequence which is not removed during translocation. In contrast, the general secretory pathway (Type II) is a two-step pathway in which the secreted protein bears an N-terminal signal peptide, which is removed during translocation across the inner membrane. Type III, in common with Type I, is again a single step pathway, but proteins translocated via this secretion machinery carry the N-terminal signal sequence which is not cleaved during export.

Type I secretion system

Both symbiotic and pathogenic plant-associated bacteria use proteins secreted via Type I systems to promote their interaction with plants.

R. leguminosarum bv. viciae secretes a nodulation signalling protein that, although not essential for the symbiosis, plays an important ancillary role in nodulation (Economou et al., 1994). This protein (NodO) is secreted via a system similar to that used for α-haemolysin and protease secretion (Finnie et al., 1997) and there is a degree of cross-complementation in that NodO can be secreted via the E. coli haemolysin secretion system (Scheu et al., 1992). NodO forms cation-selective pores in lipid bilayer membranes (Sutton et al., 1994) and it is thought to stimulate nodulation possibly by enhancing the Nod-factor-induced movement of ions across the plant plasma membrane. NodO may stimulate the infection process (Geurts et al., 1997) and its effects are likely to be localized to those cells that are in direct contact with the rhizobia (Sutton et al., 1994). There are limited similarities between the structure of NodO and that of α-haemolysins made by pathogens of animal cells. There may also be functional similarities since haemolysins can induce the formation of ion-selective channels and at low (sub-lytic) levels, haemo-lysins can induce the formation of leukotrienes by granulocytes (Konig et al., 1994).

Plant pathogens such as Erwinia chrysanthemi secrete multiple proteases via Type I secretion systems (Wandersman, 1996; Fath & Kolter, 1993). These proteases are thought to be involved in soft-rot diseases in degradation of plant cell wall proteins and of released cytoplasmic proteins. Analysis of the E. chrysanthemi secretion systems has given a good insight into the mechanism of action of the Type I secretion system, revealing it to be very similar to the system involved in secretion of haemolysin, protease and lipase by animal pathogens (Wandersmann, 1996).

Type II secretion system

Many plant pathogens secrete enzymes such as pectinases and cellulases that macerate plant cell walls. Such bacteria are often opportunistic pathogens that attack damaged plants, e.g. following attack by, e.g. nematodes. Typical examples of such pathogens include E. chrysanthemi, E. carotovora and X. campestris (Pugsley et al., 1997). The secreted proteins are relatively non-specific and in animal pathogens there are analogous pathogens such as P. aeruginosa which, e.g. secrete elastase and proteases that are also involved in tissue degradation. This type of attack is very non-specific and depends on relatively large numbers of bacteria to be successful. In this regard, it is relevant to note that, in both the plant and animal pathogens, the production of proteins secreted via the Type II systems is regulated via quorum sensing AHLs (Jones et al., 1993; Pirhonen et al., 1993; Cui et al., 1995; Harris et al., 1998; McGowan & Salmond, this volume).

Type III secretion system

Many bacterial plant pathogens are thought to have the ability to deliver proteins intracellularly into plant cells via a Type III secretion system. The Type III secretion systems appear to 'inject' proteins directly into the cytoplasm of eukaryotic cells, and the protein translocation machinery is highly conserved between plant and animal pathogens (Hueck, 1998; Vivian & Gibbon, 1997). Many of the proteins thought to be delivered into plant cells can induce a 'hypersensitive response'. This hypersensitive response is a localized plant cell death that causes death of the invading bacteria. Thus resistance to such bacterial pathogens is dependent on the ability of the plant to recognize bacterially made components. Several different bacterial genes encoding such recognized proteins (probably delivered into plant cells) have been identified, but their biochemical functions remain somewhat obscure. It is thought that the proteins may be virulence factors and thus be analogous to some of the proteins delivered by Type III secretion systems into animal cells (Vivian & Gibbon, 1997). The biochemical functions of several of the animal virulence factors delivered by Type III secretion systems are known (e.g. protein tyrosine phosphatase, protein serine/threonine kinase, ADP-ribosyl transferase, protein tyrosine phosphatase, see Hueck, 1998 for details). However, little is known about the function of the plant pathogen-related proteins. Nevertheless, it is clear from genetic work in plants and biochemical work in animals that Type III-mediated protein secretion is involved in a very specific recognition of bacteria by animal and plant cells.

Surface polysaccharides

The virulence of many bacterial plant pathogens is correlated with their ability to produce exopolysaccharide (EPS) (Denny, 1995), since mutants defective in EPS production can be significantly attenuated in pathogenicity. It is clear that subtle modifications to exopolysaccharide can decrease virulence of bacterial plant pathogens, but it is also evident that, for example, in *X. campestris* pv. *campestris* the xanthan exopolysaccharide is not essential for plant virulence, since a low level of virulence is retained in EPS-defective mutants (Katzen *et al.*, 1998). However, a precise role of the EPS on plant cells has not been determined.

Newman *et al.* (1995) have demonstrated that purified lipopolysaccharide from *X. campestris* pv. *campestris* can induce expression of plant defence–response genes when added to plant (turnip) leaves. It was also evident that the LPS from *Salmonella* and from *X. campestris* strains could be recognized by leaves of pepper. Unlike recognition by animal cells, the active component was not lipid A, but was lipid a attached to a disaccharide of 2 keto-3 deoxyoctulosonate (Newman *et al.*, 1997).

In the symbiotic interaction with legumes, analysis of LPS mutants of *Rhizobium* revealed that an intact LPS is needed for the symbiosis to proceed and the requirement was at a late stage in the infection process (Kannenberg & Brewin, 1994; Niehaus & Becker, 1998). However, the recognition mechanism of LPS fragments has not been investigated.

There appears to be a very specific interaction between *Rhizobium* EPS and legumes, and there is good evidence to suggest that low molecular weight EPS molecules act as signalling molecules required for the infection of legumes by rhizobia (Niehaus & Becker, 1998). Rhizobial mutants defective in EPS production failed to infect legumes (Finan *et al.*, 1985), and this was specific for the EPS structure since a mutant of *S. meliloti* defective in succinylation of EPS was also defective for infection, even though a normal amount of (non-succinylated) EPS was produced (Leigh *et al.*, 1987). Purified fragments of EPS can restore nodule invasion by EPS mutants and hetero-logous EPS (e.g. from a different species) did not restore infection (Djordje-vic *et al.*, 1987; Battisti *et al.*, 1992; Urzainqui & Walker, 1992). It appears that only relatively short oligomers of EPS fragments can function and these are effective at less than 10 pmol per plant (Gonzales *et al.*, 1996). The primary role of small amounts of these low molecular weight EPS fragments may be to suppress the defence system of plants. Niehaus *et al.* (1997) observed that the defence response (strong transient alkalinization of the culture media) of cultured alfalfa cells could be suppressed by addition of small amounts of low molecular weight EPS from *S. meliloti* (the symbiont of alfalfa). The suppression of defence was not observed with high molecular weight EPS or with EPS prepared from non-symbionts and the active EPS preparations had no suppression of defence responses on control plants such as tobacco or tomato that are not hosts for *S. meliloti*. Therefore, the EPS may be acting in some way to suppress defence responses thereby enabling the symbiotic bacteria to infect the plant (Niehaus *et al.*, 1997).

BACTERIA–ANIMAL CELL–CELL COMMUNICATION

Providing host defences can be circumvented, multicellular organisms offer ideal nutrient-rich protected habitats. Numerous bacterial species have evolved to exploit either plant or animal hosts for nutrition, proliferation and dissemination. Although there are examples such as the *Rhizobium*–legume symbiosis where the prokaryotic–eukaryotic interaction is mutually beneficial, the colonization of target organs and utilization of host resources by pathogenic bacteria are characteristically manifested as disease. It is now clear that signal transduction pathways which evolved to maintain coordination within the mammalian host, are targets for certain bacterial pathogens and their exoproducts. In this context, the enteropathogenic *Escherichia coli* (EPEC) have become the paradigm for studies of the subversion of

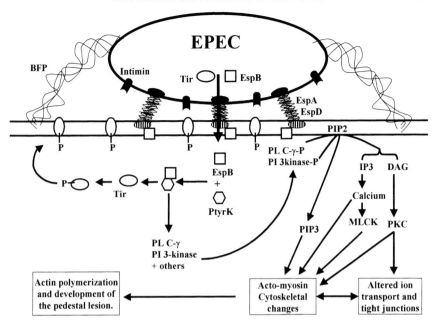

Fig. 4. Model of EPEC activation of host cell signal transduction. (**1**) EPEC initially attach to the host cell via fimbrial adhesins such as bundle forming pilli (BFP). (**2**) EPEC proteins EspA and EspD are secreted out of the bacteria cell and form the translocation apparatus, associated with the type III secretion machinery. (**3**) The translocation apparatus binds to receptors on the host cell surface, facilitating translocation of EPEC proteins Tir and EspB into the host cell. (**4**) EspB mediates the activation of a host tyrosine kinase (PtyrK) which in turn phosphorylates Tir causing its movement to the plasma membrane where it functions as the intimin receptor. (**5**) Phospholipase C-γ (PL C-γ) and phosphatidylinositol 3-kinase (PI 3-kinase) become activated through tyrosine phosphorylation and become associated with the plasma membrane where they generate the second messengers inositol tri-phosphate (IP3) and diacylglycerol (DAG), and phosphatidlyinositol tri-phosphate (PIP3), from phosphatidylinositol di-phosphate (PIP2), respectively. (**6**) IP3 mediates calcium release from intracellular stores which in turn activates myosin light chain kinase (MLCK), while DAG activates protein kinase C (PKC). (**7**) PKC alters ion secretion and the structure of tight junctions, which probably contributes to the development of diarrhoea. (**8**) PKC and MLCK differentially phosphorylate myosin light chains that are associated with acto-myosin structures in the cell. (**9**) PI3 kinase activity and the formation of PI3 contributes to the assembly of actin structures beneath attached bacteria and the development of the pedestal. ▨ : translocation syringe; ◧▥◨ : putative receptor for translocation syringe; ● : type III secretion system; ◯ : protein tyrosine kinase; ⬭ : unphosphorylated Tir; **P** ⬭ : tyrosine phosphorylated Tir; ☐ : EspB; ▮ : intimin or EaeA.

mammalian host cell signal transduction pathways, and a model of EPEC activation of host cell signal transduction is presented in Fig. 4.

Subversion of host cell signal transduction by EPEC

EPEC have attracted considerable interest since 1945 when they were first implicated in severe persistent diarrhoeal disease in infants and young children (Bray, 1945). Originally, they were identified simply on the basis of incriminating serotypes (Ewing *et al.*, 1957; Neter *et al.*, 1953) since no recognized toxin or pathology could be identified (Levine *et al.*, 1978). Later, it was observed that EPEC adhere to the human laryngeal carcinoma cell line, HEp-2, and the cervical carcinoma cell line, HeLa in a distinctive localized pattern (Scaletsky *et al.*, 1984, 1985). Ultrastructural examination of infected gut tissue and cultured cells revealed an elaborate lesion at the sites of EPEC attachment (Staley *et al.*, 1969). Microvilli are lost or effaced from infected cells and the denuded plasma membrane swells and cups around the attached bacteria to form a pedestal (Polotsky *et al.*, 1977; Knutton *et al.*, 1987). This attaching and effacing (A/E) lesion resembles the action of certain Ca^{2+} mobilizing hormones on microvilliated cells and has prompted intense investigation of the molecular basis of lesion formation.

Although cell-free culture supernatants do not appear to possess enterotoxic or cytotoxic activities (Levine *et al.*, 1978), EPEC strains do secrete a number of proteins in response to certain environmental signals (Haigh *et al.*, 1995; Kenny & Finlay, 1995; Kenny *et al.*, 1997*a*). These have been given the synonym Esp for **EPEC s**ecreted **p**roteins and comprise at least four proteins termed EspA, EspB, EspC and EspD (Jarvis *et al.*, 1995). These are mostly encoded on a region of the chromosome termed the locus of enterocyte effacement (LEE). This is a 35 kb segment of DNA (referred to as a 'pathogenicity island'), which also contains a type III secretion system and is inserted into the *selC* (selenocysteine tRNA) site located at 82 min on the *E. coli* chromosome (Jarvis *et al.*, 1995; McDaniel *et al.*, 1995).

Initially, EPEC attach to the tips of enterocyte microvilli via fimbria-like structures. A number of candidate filamentous adhesins have been suggested to mediate this event, including the plasmid-encoded, bundle forming pili known as BFP (Giron *et al.*, 1991). The loss of microvilli from infected cells through vesiculation occurs through an unknown process, but may involve influx of Ca^{2+} from outside the cell and breakdown of the actin core (Baldwin *et al.*, 1991). Effacement of microvilli is not mediated through a soluble toxin, since only cells with attached bacteria display the classical A/E-lesion. Once microvilli have been lost, EPEC attach to the body of the cell via the intimate adhesin, intimin, encoded by the *eaeA* gene (also found within LEE). EaeA is an integral membrane protein which facilitates intimate adhesion to host cells by binding to a specific plasma membrane receptor (Jerse *et al.*, 1991). Attachment of EPEC to the host cell surface triggers major perturbations in membrane architecture, which lead to pedestal formation through extensive actin polymerization. The pedestal structure is maintained by a complex of polymerized actin, myosin and other actin binding proteins (Knutton *et al.*, 1989).

Electron microscopy indicates that EspA is associated with the outer surface of the EPEC cell, possibly in conjunction with EspD, to form a filamentous organelle that is present on the bacterial surface during the early stage of A/E lesion formation (Knutton *et al.*, 1998). This organelle forms a physical bridge between bacterium and eukaryotic target cells, and is thought to mediate the translocation of other proteins from bacteria to host cell, rather like a syringe. Once inside the eukaryotic cell, the secreted proteins are thought to subvert signalling processes through interaction with host signal transduction machinery (Kenny & Finlay, 1995; Nataro & Kaper, 1998; Kaper 1998). Remarkably, one of these translocated proteins is the EPEC receptor for intimin, termed Tir (for translocated intimin receptor). Originally, Tir was thought to be a eukaryotic protein termed Hp90 (Rosenshine *et al.*, 1996) but is, in fact, a bacterial protein which is translocated into the host cell (Kenny *et al.*, 1997*b*). Here it is tyrosine-phosphorylated and then inserted into the host cell plasma membrane, where it forms the receptor for intimin-mediated intimate attachment.The coincident translocation of EspB is thought to mediate the tyrosine phosphorylation of Tir probably by direct or indirect activation of a specific host tyrosine kinase (Nataro & Kaper, 1998; Kaper, 1998). EspB also appears to be associated with the plasma membrane of the cell and may have other functions (Nataro & Kaper, 1998; Kaper, 1998).

In addition to Tir, there are a number of host proteins that are also tyrosine phosphorylated; among these are phospholipase C-γ (PLC-γ; Nataro & Kaper, 1998; Kaper, 1998; Kenny & Finlay, 1997) and phosphatidylinositol-3 kinase (PI$_3$-K; Baldwin *et al.*, 1996; Baldwin, 1998), both of which are involved in phosphatidylinositol-mediated host cell signalling processes. Phospholipid signalling is a component of most receptor-mediated signal transduction pathways. When a signalling molecule such as a hormone or neurotransmitter engages its receptor, this primary signal is amplified by effector enzymes. These effector enzymes convert phospholipids present in membranes into second messengers. Activation of PLC-γ by tyrosine phosphorylation, for example, causes cleavage of the membrane phospholipid phosphatidylinositol 4,5-bisphosphate (PIP$_2$) to yield two second messenger molecules (Baldwin, 1998; Baldwin *et al.*, 1996; Foubister *et al.*, 1994). The first, inositol-1,4,5 inositol tri-phosphate (IP$_3$) elevates intracellular Ca^{2+} by binding to receptors on the endoplasmic reticulum and mobilising Ca^{2+} stores (Nataro & Kaper, 1998; Baldwin *et al.*, 1996; Dytoc *et al.*, 1994; Foubister *et al.*, 1994). The second, diacylglycerol, is responsible for the activation of protein kinase C (PKC) which has major control functions in gene expression, cytoskeletal organisation, tight junction integrity and ion transport (Nishizuka, 1986; Baldwin *et al.*, 1990; Manjarrez-Hernandez *et al.*, 1996). PI$_3$-K phosphorylates membrane PIP$_2$ at the 3-position generating 1,3,4,5-phosphatidylinositol tri-phosphate (PIP$_3$), which itself acts as a potent second messenger involved in many aspects of cellular function and regulation including organisation of cellular actin structures (Baldwin, 1998; Baldwin *et al.*, 1996). As a consequence of IP$_3$ production and Ca^2 elevation, the protein kinase, myosin light chain kinase (MLCK), is activated and phosphorylates several host proteins. In conjunction with PKC, MLCK differentially phosphorylates myosin light chain (MLC) within the developing pedestal (Baldwin, 1998; Manjarrez-Hernandez et al., 1991). By studying phospho-peptide maps of MLC extracted from infected cells at various times after infection, it is evident that protein kinases are differentially active. Early in infection MLC is phosphorylated by PKC and is entirely soluble. As the infection progresses, the PKC phosphorylation events are dephosphorylated and MLC becomes phosphorylated by MLCK, causing MLC to become associated with the actin of the pedestal (Baldwin, 1998; Manjarrez-Hernandez et al., 1996). PKC-mediated phosphorylation also affects the structure of the tight junctions between cells, possibly contributing to the ensuring diarrhoea by causing leakage of fluids between cells (Manjarrez-Hernandez et al., 1991, 1996). In addition, PKC is known to regulate ion secretion through phosphorylation of ion transport proteins, which is possibly responsible for the early onset of a secretory type diarrhoea (Baldwin et al., 1990; Manjarrez-Hernandez, 1996; Crane & Ho, 1997).

Some controversy surrounds the nature and consequence of elevated Ca^{2+} in EPEC-infected cells (Baldwin, 1998). Since IP_3 is generated, it seems reasonable to propose that there is an accompanying early Ca^{2+} spike, which acts as the signal for downstream Ca^{2+} responsive events. However, reports of a prolonged Ca^{2+} elevation are inconsistent with the present state of knowledge of Ca^{2+} signalling and may represent influx of Ca^{2+} from extracellular sources in response to the generation of IP_4 from IP_3. Prolonged Ca^{2+} elevation is consistent with the observed loss of viability in heavily infected cells (Baldwin, 1998; Baldwin et al., 1993). However, an involvement of transient Ca^{2+} signalling is also possible, since there clearly is phophatidylinositol lipid metabolism and the generation of IP_3 (Foubister et al., 1994).

Other examples of bacteria capable of subverting or utilizing animal host signalling pathways include Listeria monocytogenes, S. typhimurium and Campylobacter jejuni. Adherence and invasion of host target cells by L. monocytogenes, triggers tyrosine phosphorylation of two mitogen activated protein (MAP) kinases, MEK-1 and Erk-2 (Tang et al., 1994; Kuhn & Goebel, 1998). Specific inhibition of tyrosine kinases using genistein or MEK-1 completely abolishes the invasion of cultured cells by Listeria (Ireton et al., 1996). Invasion is also blocked by wortmannin, a potent inhibitor of PI_3 kinase, which is involved in the regulation of actin architecture and cytochalasin D, which inhibits actin polymerization (Ireton et al., 1996). Salmonella typhimurium also activates MAP kinase pathways via ERK, JNK and p38 (Hobbie et al., 1997). Activation of these pathways is associated with synthesis of interleukin-8 and inflammation of infected tissues. Campylobacter jejuni invasion of cultured gut cells is also inhibited by wortmannin indicating the probable involvement of PI_3 kinase and possibly actin polymerization in the invasion of gut tissue (Wooldridge *et al.*, 1996).

Bacterial responses to mammalian signalling molecules

A number of studies have identified bacterial receptors for the hormones of higher organisms and there is some evidence to suggest that the growth of pathogenic bacteria in the gut depends on host-derived hormonal signals (Kaprelyants & Kell, 1996). For example, Lyte and co-workers (1996*a,b*) have reported that, in enterovirulent *E. coli*, catecholamines such as nor-adrenaline stimulate growth and enterotoxin production whilst others have shown that interleukin-2 (IL-2) and granulocyte-macrophage colony stimu-lating factor (GM-CSF) promote the growth of virulent, but not avirulent, *E. coli* strains (Denis *et al.*, 1991). The molecular mechanism(s) by which *E. coli* responds to host neurohormonal signals is not yet known. However, treatment of low density *E. coli* cultures with noradrenaline induces the production of a diffusible antoinducer which mediates the accelerated growth and altered gene expression observed (Lyte *et al.*, 1996*b*). The chemical identity of this putative inducer and its relationship, if any, to the *E. coli* autoinducer molecules described by Sitnikov *et al.* (1996), Garcia-Lara *et al.* (1996) and Surette and Bassler (1998) remains to be established.

EXPLOITATION OF BACTERIAL SIGNALLING STRATEGIES

Understanding how bacterial cells communicate with each other and with eukaryotic cells has a number of important practical implications for the control of disease-causing organisms, for the screening and exploitation of bacteria as biological factories for the production of antibiotics and other high value products, and perhaps for aiding the recovery of bacteria which cannot be cultured in the laboratory on conventional growth media.

In the context of anti-infective agents, the realization that many bacterial pathogens employ diffusible signal molecules to coordinate the control of multiple virulence and survival genes offers a new chemotherapeutic target. Interference with transmission of the molecular message by a small molecule antagonist, which competes for the pheromone binding site of the transcrip-tional activator protein (e.g. a LuxR homologue) thereby switching off virulence gene expression, so attenuating the pathogen, is an attractive strategy. In this context, the ability of various AHL analogues to inhibit the action of the cognate AHL *in vivo* has been demonstrated (Eberhard *et al.*, 1986; Schaefer *et al.*, 1996*a*; Milton *et al.*, 1997; Swift *et al.*, 1997). Furthermore, Givskov *et al.* (1996) provided evidence that furanone com-pounds produced by the Australian macroalgae, *Delisea pulchra* inhibit AHL-regulated processes including swarming in *S. liquefaciens* and biolumi-nescence in *V. fischeri*. As the inhibitory effect could be overcome by the addition of an excess of the cognate AHL, the furanones appear to be acting as competitive antagonists. Alternatively, the significant recent advances in defining the enzymatic activity and substrate requirements of LuxI homologues (Moré et al., 1996; Schaefer et al., 1996b; Jiang et al., 1998) also emphasizes the potential of the AHL synthase as an antimicrobial target.

AHL-mediated cell–cell communication is not the sole target for anti-infective agents which block quorum sensing. The ubiquity of pheromone-based gene regulation indicates that the concept of blocking quorum sensing may offer a generic target for controlling bacterial infection. An important threat in this context are methicillin resistant *Staphylococcus aureus* (MRSA) which are resistant to virtually all clinically available antibiotics including vancomycin (Hiramatsu et al., 1997; Speller et al., 1997). However, the work of Ji et al. (1997b) and our own work (Z. Affas, P. W. McDowell, W. C. Chan, B. W. Bycroft, G. S. A. B. Stewart & P. Williams, unpublished observations) showing that exotoxin virulence factor production can be blocked by synthetic peptide pheromone antagonists, offers the real possibility of a novel approach to anti-staphylococcal chemotherapy which will not be susceptible to conventional antibiotic resistance mechanisms.

Such quorum sensing blocking agents are, however, likely to have a much narrower spectrum of activity than conventional broad spectrum antibiotics and will require new diagnostic assays. Clearly the physician will need rapid assays to establish the signalling 'language' used by the infecting pathogen in order to choose the correct blocking agent. In addition, conventional minimum inhibitory concentration (MIC) assays for susceptibility will not be possible for agents targetting pathogenicity *per se*. While such agents would have obvious value in prophylaxis, their therapeutic potential for the treatment of established infections, especially in immunocompromised patients, is less apparent, since an intact host defence is likely to constitute a necessary pre-requisite for clearing the infection.

The discovery that many plant-associated bacteria employ quorum sensing to control phenotypic traits associated with either pathogenicity or symbiosis suggests that enhancing or blocking quorum sensing may be beneficial in agriculture. Quorum sensing blocking agents could be used for preventing, for example, *Erwinia*-mediated damage to root crops such as carrots and potatoes. Alternatively, the production of transgenic plants capable of generating bacterial quorum sensing signalling molecules *in planta* would offer novel opportunities for disease control and for manipulating plant–microbe interactions. In this case, transgenic plants producing the appropriate signalling molecules might enhance the establishment of an antifungal environment, allow the use of 'disarmed' strains for crop protection or improve the nitrogen-fixing 'biofertilizer' qualities of bacteria such as the rhizobia. The production of AHLs *in planta*, for example, would require the introduction and expression of a *luxI* homologue and the availability of the correct substrates (*S*-adenosylmethionine and the appropriately charged acyl acylcarrier protein or acyl coenzyme A; (Moré *et al.*, 1996; Schaefer *et al.*, 1996*b*; Jiang *et al.*, 1998). This possibility is beginning to be realized; introduction of the *Y. enterocolitica luxI* homologue, *yenI* into tobacco plants resulted in the production, *in planta*, of the same AHLs as in the parent bacterium at levels sufficient to restore biocontrol activity to an AHL-deficient *P. aureofaciens* strain (R. G. Fray, J. P. Throup, A. Wallace, P. Williams, G. S. A. B Stewart & D. Grierson, unpublished observations). Whether this exciting new technology will prove applicable to more vulnerable food crops remains to be established.

ACKNOWLEDGEMENTS

Work from the authors' laboratories is supported by the BBSRC, MRC, Wellcome Trust and the European Union which are gratefully acknowledged. We would also like to thank Gordon Stewart, Barrie Bycroft, Rupert Fray, Allan Collmer, Max Dow and Nick Brewin for helpful discussions.

REFERENCES

Ahmer, B. M., van Reeuwijk, J., Timmers, C. D., Valentine, P. J. & Heffron, F. (1998). *Salmonella typhimurium* encodes an SdiA homolog, a putative quorum sensor of the LuxR family that regulates genes on the virulence plasmid. *Journal of Bacteriology*, **180**, 1185–93.

Atkinson, M. M., Midland, S. L., Sims, J. J. & Keen, N. T. (1996). Syringolide 1 triggers Ca^{2+} influx, K^+ efflux, and extracellular alkalization in soybean cells carrying the disease-resistance gene *Rpg4*. *Plant Physiology*, **112**, 297–302.

Bainton, N. J., Bycroft, B. W., Chhabra, S. R., Stead, P., Gledhill, L., Hill, P. J., Rees, C. E. D., Winson, M. K., Salmond, G. P. C., Stewart, G. S. A. B. & Williams, P. (1992*a*). A general role for the *lux* autoinducer in bacterial cell signalling: control of antibiotic synthesis in *Erwinia*. *Gene*, **116**, 87–91.

Bainton, N. J., Stead, P., Chhabra, S. R., Bycroft, B. W., Salmond, G. P. C., Stewart, G. S. A. B. & Williams, P. (1992*b*). *N*-(3-oxohexanoyl)-L-homoserine lactone regulated carbapenem antibiotic production in *Erwinia carotovora*. *Biochemical Journal*, **288**, 997–1004.

Baldwin, T. J. (1998). Pathogenicity of enteropathogenic *Escherichia coli*. *Journal of Medical Microbiology*, **47**, 283–93.

Baldwin, T. J., Brooks, S. F., Knutton, S., Hernandez, H. A. M., Aitken, A. & Williams, P. H. (1990). Protein phosphorylation by protein kinase C in HEp-2 cells infected with enteropathogenic *Escherichia coli*. *Infection and Immunity*, **58**, 761–5.

Baldwin, T. J., Knutton, S., Haigh, R., Williams, P. H., Palmer, H. M., Aitken, A. & Borriello, S. P. (1996). Hijacking host-cell signal transduction mechanisms during infection with enteropathogenic *Escherichia coli*. *Biochemical Society Transactions*, **24**, 552–8.

Baldwin, T. J., Ward, W., Aitken, A., Knutton, S. & Williams, P. H. (1991). Elevation of intracellular free calcium in HEp-2 cells infected with enteropathogenic *Escherichia coli*. *Infection and Immunity*, **59**, 1599–604.

Barber, C. E., Tang, J. L., Feng, J. X., Pan, M. Q., Wison, T. J. G., Slater, H., Dow, J. M., Williams, P. & Daniels, M. J. (1997). A novel regulatory system required for pathogenicity *Xanthomonas campestris* is mediated by a small diffusible signal molecule. *Molecular Microbiology*, **24**, 555–66.

Battisti, L., Lara, J. C. & Leigh, J. A. (1992). Specific oligosaccharide form of the *Rhizobium meliloti* exopolysaccharide promotes nodule invasion in alfalfa. *Proceedings of the National Academy of Sciences, USA*, **89**, 5625–9.

Bobylev, M. M., Bobyleva, L. I. & Strobel, G. A. (1996). Synthesis and bioactivity of analogs of maculosin, a host-specific phytotoxin produced by *Alternaria alternata* on spotted knapweed (*Centaurea maculosa*). *Journal of Agricultural and Food Chemistry*, **44**, 3960–4.

Bray, J. (1945). Isolation of antigenically homogeneous strains *Bact. coli neapolitanum* from summer diarrhoea of infants. *Journal of Pathology and Bacteriology*, **57**, 239–47.

Crane, J. K. & Ho, J. S. (1997). Activation of host cell protein kinase C by enteropathogenic *Escherichia coli*. *Infection and Immunity*, **65**, 3277–85.

Cui, Y., Chatterjee, A., Liu, Y., Korsi Dumenyo, C. & Chatterjee, A. K. (1995). Identification of a global repressor gene, *rsmA* of *Erwinia carotovora* ssp. *carotovora* that controls extracellular enzymes, *N*-(3-oxohexanoyl)-L-homoserine lactone and pathogenicity in soft-rot *Erwinia*. *Journal of Bacteriology*, **177**, 5108–15.

Dénarié, J., Debellé, F. & Promé, J. C. (1996). Rhizobium lipo-chitooligosaccharide nodulation factors: signalling molecules mediating recognition and morphogenesis. *Annual Reviews of Biochemistry*, **65**, 503–35.

Denis, M., Campbell, D. & Gregg, E. O. (1991). Interleukin-2 and granulocyte-macrophage colony-stimulating factor stimulate the growth of a virulent strain of *Escherichia coli*. *Infection and Immunity*, **59**, 1853–6.

Denny, T. P. (1996) Involvement of bacterial polysaccharides in plant pathogenesis. *Annual Reviews of Phytopathology*, **33**, 173–97.

Djordjevic, S. P., Chen, H., Batley, M., Redmond, J. W. & Rolfe, B. G. (1987). Nitrogen fixation ability of exopolysaccharide synthesis mutants of *Rhizobium* sp. NGR234 and *Rhizobium trifolii* is restored by the addition of homologous exopolysaccharides. *Journal of Bacteriology*, **169**, 114–26.

Dytoc, M., Fedorko, L. & Sherman, P. M. (1994). Signal transduction in human epithelial cells infected with attaching and effacing *Escherichia coli in vitro*. *Gastroenterology*, **106**, 1150–61.

Eberhard, A., Burlingame, A. L., Kenyon, G. L., Nealson, K. H. & Oppenheimer, N. J. (1981). Structural identification of autoinducer of *Photobacterium fischeri* luciferase. *Biochemistry*, **20**, 2444–9.

Eberhard, A., Widrig, C. A., McBath, P. & Schineller, J. B. (1986). Analogs of the autoinducer of bioluminesence in *Vibrio fischeri*. *Archives of Microbiology*, **146**, 35–40.

Economou, A., Davies, A. E., Johnson, A. W. B. & Downie, J. A. (1994). The *Rhizobium leguminosarum* biovar *viciae nodO* gene can enable a *nodE* mutant of *Rhizobium leguminosarum* biovar *trifolii* to nodulate vetch. *Microbiology*, **140**, 2341–7.

Ehrhardt, D. W., Atkinson, E. M. & Long, S. R. (1992). Depolarization of alfalfa root hair membrane potential by *Rhizobium meliloti* Nod factors. *Science*, **256**, 998–1000.

Ehrhardt, D. W., Wais, R. & Long, S. R. (1996). Calcium spiking in plant root hairs responding to *Rhizobium* nodulation signals. *Cell*, **85**, 673–81.

Ewing, W. H., Tatum, H. W. & Davis, B. R. (1957). The occurrence of *Escherichia coli* serotypes associated with diarrhoeal disease in the United States. *Public Health Laboratory*, **15**, 118–38.

Fath, M. J. & Kolter, R. (1993). ABC transporters: bacterial exporters. *Microbiological Reviews*, **57**, 995–1017.

Féllé, H. H., Kondorosi, E., Kondorosi, A. & Schultze, M. (1996). Rapid alkalinization in alfalfa root hairs in response to rhizobial lipochitooligosaccharide signals. *Plant Journal*, **10**, 295–301.

Féllé, H. H., Kondorosi, E., Kondorosi, A. & Schultze, M. (1998). The role of ion fluxes in Nod factor signalling in *Medicago sativa*. *Plant Journal*, **13**, 455–63.

Fewtrell, C. (1993). Calcium oscillations in nonexcitable cells. *Annual Reviews of Physiology*, **55**, 427–54.

Filloux, A. & Hardie, K. R. (1998). A systematic approach to the study of protein secretion in Gram-negative bacteria. In *Bacterial Pathogenesis – Methods in Microbiology*, ed. P. Williams, J. Ketley & G. Salmond, vol. 27, pp. 301–18. Academic Press.

Finan, T. M., Hirsch, A. M., Leigh, J. A., Johansen, E., Kuldav, G. A., Deegan, S., Walker, G. C. & Signer, E. R. (1985). Symbiotic mutants of *Rhizobium meliloti* that uncouple plant from bacterial differentiation. *Cell*, **40**, 869–77.

Finnie, C., Hartley, N. M., Findlay, K. C. & Downie, J. A. (1997). The *Rhizobium leguminosarum prsDE* genes are required for secretion of several proteins, some of which influence nodulation, symbiotic nitrogen fixing and exopolysaccharide modification. *Molecular Microbiology*, **25**, 135–46.

Flavier, A. B., Clough, S. J., Schell, M. A. & Denny, T. P. (1997a). Identification of 3-hydroxypalmitic acid methyl ester as a novel autoregulator controlling virulence in *Ralstonia solanacearum*. *Molecular Microbiology*, **26**, 251–9.

Flavier, A. B., Ganova-Raeva, L. M., Schell, M. A. & Denny, T. P. (1997b). Hierarchical autoinduction in *Ralstonia solanacearum*: control of acyl-homoserine lactone production by a novel autoregulatory system responsive to 3-hydroxypalmitic acid methyl ester. *Journal of Bacteriology*, **179**, 7089–97.

Foubister, B., Rosenshine, I. & Finlay, B. B. (1994). A diarrhoeal pathogen, enteropathogenic *Escherichia coli* (EPEC), triggers a flux in inositol phosphates in infected cells. *Journal of Experimental Medicine*, **179**, 993–8.

Franssen, H., Mylona, P., Pawlowski, K., van de Sande, K., Heidstra, R., Geurts, R., Kozik, A., Matvienko, M., Yang, W-C., Hadri, A-E., Martinez-Abarca, F. & Bisseling, T. (1995). Plant genes involved in root-nodule development on legumes. *Philosophical Transactions of the Royal Society of London Series B*, **350**, 101–7.

Fuqua, W. C., Winans, S. C. & Greenberg, E. P. (1994). Quorum sensing in bacteria: the LuxR-LuxI family of cell density responsive transcriptional regulators. *Journal of Bacteriology*, **176**, 269–75.

Fuqua, W. C., Winans, S. C. & Greenberg, E. P. (1996). Census and consensus in bacterial ecosystems: the LuxR–LuxI family of quorum sensing transcriptional regulators. *Annual Reviews in Microbiology*, **50**, 727–51.

Gamard, P., Sauriol, F., Benhamou, N., Belanger, R. R. & Paulitz, T. C. (1997). Novel butyrolactones with antifungal activity produced by *Pseudomonas aureofaciens* strain 63-28. *Journal of Antibiotics*, **50**, 742–9.

Garcia-Lara, J., Shang, L. H. & Rothfield, L. I. (1996). An extracellular factor regulates expression of sdiA, a transcriptional activator of cell division genes in *Escherichia coli*. *Journal of Bacteriology*, **178**, 2742–8.

Gehring, C. A., Irving, H. R., Kabbara, A. A., Parish, R. W., Boukli, N. M. & Broughton, W. J. (1997). Rapid, plateau-like increases in intracellular free calcium are associated with nod-factor-induced root-hair deformation. *Molecular Plant–Microbe Interactions*, **10**, 791–802.

Geurts, R., Heidstra, R., Hadri, A-Z., Downie, J. A., Franssen, H., Van Kammen, A. & Bisseling, T. (1997). Sym2 of pea is involved in a nodulation factor-perception mechanism that controls the infection process in the epidermis. *Plant Physiology*, **115**, 351–9.

Giron, J. A., Ho, A. S. Y. & Schoolnik, G. K. (1991). An inducible bundle-forming pilus of enteropathogenic *Escherichia coli* (EPEC). *Science*, **254**, 710–13.

Givskov, M., de Nys, R., Manefield, M., Gram, L., Maximilien, R., Eberl, L., Soren, M., Steinberg, P. D. & Kjelleberg, S. (1996). Eukaryotic interference with homoserine lactone mediated prokaryotic signalling. *Journal of Bacteriology*, **178**, 6618–22.

Gonzales, J. E., York, G. M. & Walker, G. C. (1996). Low molecular weight EPS II of *Rhizobium meliloti* allows nodule invasion in *Medicago sativa*. *Proceedings of the National Academy of Sciences, USA*, **93**, 8636–41.

Gray, K. M., Pearson, J. P., Downie, J. A., Boboye, B. B. & Greenberg, E. P. (1996). Cell-to-cell signaling in the symbiotic nitrogen-fixing bacterium *Rhizobium leguminosarum*: autoinduction of stationary phase and rhizosphere-expressed genes. *Journal of Bacteriology*, **178**, 372–6.

Grossman, A. D. (1995). Genetic networks controlling the initiation of sporulation and the development of competence in *Bacillus subtilis*. *Annual Reviews of Genetics*, **29**, 477–508.

Haigh, R., Baldwin, T. J., Knutton, S. & Williams, P. H. (1995). Carbon dioxide regulated secretion of EaeB protein of enteropathogenic *Escherichia coli*. *FEMS Microbiology Letters*, **129**, 63–8.

Harris, S. J., Shih, Y.-L., Bentley, S. D. & Salmond, G. P. C. (1998). The *hexA* gene of *Erwinia carotovora* encodes a LysR homologue and regulates motility and the expression of multiple virulence determinants. *Molecular Microbiology*, **28**, 705–17.

Hiramatsu, K., Aritaka, N., Hanaki, H., Kawasaki, S., Hosoda, Y., Hopri, S., Fukuchi, Y. & Kobayashi, I. (1997). Dissemination in Japanese hospitals of strains of *Staphylococcus aureus* heterogeneously resistant to vancomycin. *Lancet*, **350**, 1670–3.

Hobbie, S., Chen, L. M., Davis, R. J. & Gallan, J. E. (1997). Involvement of mitogen activated protein kinase in the nuclear responses to cytokine production induced by *Salmonella typhimurium* in cultured intestinal epithelial cells. *Journal of Immunology*, **159**, 5550–9.

Hoch, J. A. & Silhavy, T. J. (1995). *Two-component Signal Transduction*. Washington: ASM Press.

Horinouchi, S. & Beppu, T. (1992). Autoregulatory factors and communication in *Actinomycetes Annual Reviews in Microbiology*, **46**, 377–98.

Hueck, C. J. (1998). Type III protein secretion systems in bacterial animal and plant pathogens. *Microbiology and Molecular Biology Reviews*, **62**, 379–435.

Ireton, K., Payrastre, B., Chap, H., Ogawa, W., Sakaue, H., Kasuga, M. & Cossart, P. (1996). A role for phosphoinositide 3-kinase in bacterial invasion. *Science*, **274**, 780–2.

Jarvis, K. G., Giron, J. A., Jerse, A. E., McDaniel, T. K., Donnenberg, M. S. & Kaper, J. B. (1995). Enteropathogenic *Escherichia coli* contains a specialised secretion system necessary for the export of proteins involved in attaching and effacing lesion formation. *Proceedings of the National Academy of Sciences, USA*, **92**, 7996–8000.

Jerse, A. E. & Kaper, J. B. (1991). The *eae* gene of enteropathogenic *Escherichia coli* encodes a 94-kilodalton membrane protein, the expression of which is influenced by the EAF plasmid. *Infection and Immunity*, **59**, 4302–9.

Ji, G., Beavis, R. C. & Novick, R. P. (1995). Cell density control of staphylococcal virulence mediated by an octapeptide pheromone. *Proceedings of the National Academy of Sciences, USA*, **92**, 12055–9.

Ji, C., Okinaka, Y., Takeuchi, Y., Tsurushima, T., Buzzell, R. I., Sims, J. J., Midland, S. L., Slaymaker, D., Yoshikawa, M., Yamaoka, N. & Keen, N. T. (1997*a*). Specific binding of the syringolide elicitors to a soluble protein fraction from soybean leaves. *Plant Cell*, **9**, 1425–33.

Ji, G., Beavis, R. & Novick, R. P. (1997*b*). Bacterial interference caused by autoinducing peptide variants. *Science*, **276**, 2027–30.

Jiang, Y., Camara, M., Chhabra, S. R., Hardie, K. R., Bycroft, B. W., Lazdunski, A., Salmond, G. P. C., Stewart, G. S. A. B. & Williams, P. (1998). *In vitro* biosynthesis of the *Pseudomonas aeruginosa* quorum sensing signal molecule *N*-butanoyl-L-homoserine lactone. *Molecular Microbiology*, **28** 192–203.

Jones, S., Yu, B., Bainton, N. J., Birdsall, M., Bycroft, B. W., Chhabra, S. R., Cox, A. J. R., Golby, P., Reeves, P. J., Stephens, S., Winson, M. K., Salmond, G. P. C., Stewart, G. S. A. B. & Williams, P. (1993). The *lux* autoinducer regulates the production of exoenzyme virulence determinants in *Erwinia carotovora* and *Pseudomonas aeruginosa*. *EMBO Journal*, **12**, 2477–82.

Kannenberg, E. L. & Brewin, N. J. (1994). Host-plant invasion by *Rhizobium*: the role of cell-surface components. *Trends in Microbiology*, **2**, 277–83.

Kaper, J. B. (1998). EPEC delivers the goods. *Trends in Microbiology*, **6**, 169–72.

Kaprelyants, A. S. & Kell, D. B. (1996). Do bacteria need to communicate with each other for growth? *Trends in Microbiology*, **4**, 237–42.

Katzen, F., Ferreiro, D. U., Oddo, C. G., Ielmini, M. V., Becker, A., Puhler, A. & Ielpi, L. (1998). *Xanthomonas campestris* pv. campestris *gum* mutants: effects on Xanthan biosynthesis and plant virulence. *Journal of Bacteriology*, **180**, 1607–17.

Kenny, B. & Finlay, B. B. (1995). Protein secretion by enteropathogenic *Escherichia coli* is essential for transducing signals to epithelial cells. *Proceedings of the National Academy of Sciences, USA*, **92**, 7991–5.

Kenny, B. & Finlay, B. B. (1997). Intimin-dependant binding of enteropathogenic *Escherichia coli* to host cells triggers novel signalling events including tyrosine phosphorylation of phospholipase C-γ 1. *Infection and Immunity*, **65**, 2528–36.

Kenny, B., Abe, A., Stein, M. & Finlay, B. B. (1997*a*). Enteropathogenic *Escherichia coli* protein secretion is induced in response to conditions similar to those in the gastrointestinal tract. *Infection and Immunity*, **65**, 2606–12.

Kenny, B., DeVinney, R., Stein, M., Reinscheid, D. J., Frey, E. A. & Finlay, B. B. (1997*b*). Enteropathogenic *E. coli* (EPEC) transfers its receptor for intimate adherence into mammalian cells. *Cell*, **91**, 511–20.

Kleerebezem, M., Quadri, L. E. N., Kuipers, O. P. & de Vos, W. M. (1997). Quorum sensing by peptide pheromones and two component signal transduction systems in Gram-positive bacteria. *Molecular Microbiology*, **24**, 895–904.

Knutton, S., Baldwin, T. J., Williams, P. H. & McNeish, A. S. (1989). Actin accumulation at the site of bacterial adhesion for enteropathogenic and enterohaemorrhagic *Escherichia coli*. *Infection and Immunity*, **57**, 1290–8.

Knutton, S., Lloyd, D. R. & McNeish, A. S. (1987). Adhesion of enteropathogenic *Escherichia coli* to human intestinal enterocytes and cultured human intestinal mucosa. *Infection and Immunity*, **55**, 69–77.

Knutton, S., Rosenshine, I., Pallen, M. J., Nisan, I., Neves, B. C., Bain, C., Wolff, C., Dougan, G. & Frankel, G. A. (1998). Novel EspA-associated surface organelle of enteropathogenic *Escherichia coli* involved in protein translocation into epithelial cells. *EMBO Journal*, **17**, 2166–76.

Konig, B., Ludwig, A., Goebel, W. & Konig, W. (1994). Pore formation by the *Escherichia coli* alpha-hemolysin: role for mediator release from human inflammatory cells. *Infection and Immunity*, **62**, 4611–17.

Kuhn, M. & Goebel, W. (1998). Host cell signalling during *Listeria monocytogenes* infection. *Trends in Microbiology*, **6**, 11–15.

Latifi, A., Winson, M. K., Foglino, M., Bycroft, B. W., Stewart, G. S. A. B., Lazdunski, A. & Williams, P. (1995). Multiple homologues of LuxR and LuxI control expression of virulence determinants and secondary metabolites through quorum sensing in *Pseudomonas aeruginosa* PAO1. *Molecular Microbiology*, **17**, 333–43.

Leigh, J. A., Reed, J. W., Hanks, J. F., Hirsch, A. M. & Walker, G. C. (1987). *Rhizobium meliloti* mutants that fail to succinylate their calcofluor-binding exopolysaccharide are defective in nodule invasion. *Cell*, **51**, 579–87.

Levine, M. M., Bergquist, E. J., Nalin, D. R., Waterman, D. H., Hornick, R. B., Young, C. R., Sotman, S. & Rowe, B. (1978). *Escherichia coli* strains that cause diarrhoea but do not produce heat-labile or heat-stable enterotoxins and are non-invasive. *Lancet*, **i**, 1119–22.

Lina, G., Jarraud, S., Ji, G., Greenland, T., Pedraza, A., Etienne, J., Novick, R. P. & Vandenesch, F. (1998). Transmembrane topology and histidine protein kinase activity of AgrC, the *agr* signal receptor in *Staphylococcus aureus*. *Molecular Microbiology*, **28**, 655–62.

Lyte, M., Arunlanandam, B. P. & Frank, C. D. (1996*a*). Production of Shiga-like toxins by *Escherichia coli* O157:H7 can be influenced by the neuroendocrine hormone, norepinephrine. *Journal of Laboratory and Clinical Medicine*, **128**, 392–8.

Lyte, M., Frank, C. D. & Green, B. T. (1996*b*). Production of an autoinducer of growth by norepinephrine cultured *Escherichia coli* O157-H7. *FEMS Microbiology Letters*, **39**, 155–9.

McClean, K. H., Winson, M. K., Fish, L., Taylor, A., Chhabra, S. R., Cámara, M., Daykin, M., Swift, S., Lamb, J., Bycroft, B. W., Stewart, G. S. A. B. & Williams, P. (1997). Quorum sensing and *Chromobacterium violaceum*: exploitation of violacein production and inhibition for the detection of *N*-acylhomoserine lactones. *Microbiology*, **143**, 3703–11.

McDaniel, T. K., Jarvis, K. G., Donnenberg, M. S. & Kaper, J. B. (1995). A genetic locus of enterocyte effacement conserved among diverse enterobacterial pathogens. *Proceedings of the National Academy of Sciences, USA*, **92**, 1664–8.

McGowan, S., Sebaihia, M., Jones, S., Yu, B., Bainton, N., chan, P. F., Bycroft, B., Stewart, G. S. A. B., Williams, P. & Salmond, G. P. C. (1995). Carbapenem antibiotic production in *Erwinia carotovora* is regulated by CarR, a homologue of the LuxR transcriptional activator. *Microbiology*, **141**, 541–50.

McGowan, S., Sebaihia, M., Porter, L. E., Stewart, G. S. A. B., Williams, P., Bycroft, B. W. & Salmond, G. P. C. (1996). Analysis of bacterial carbapenem antibiotic production genes reveals a novel *β*-lactam biosynthesis pathway. *Molecular Microbiology*, **22**, 415–26.

Manjarrez-Hernandez, H. A., Baldwin, T. J., Aitken, A., Knutton, S. & Williams, P. H. (1991). Intestinal epithelial cell protein phosphorylation in enteropathogenic *Escherichia coli* diarrhoea. *Lancet*, **339**, 521–3.

Manjarrez-Hernandez, H. A., Baldwin, T. J., Williams, P. H., Haigh, R., Knutton, S. & Aitken, A. (1996). Phosphorylation of myosin light chain at distinct sites and its association with the cytoskeleton during enteropathogenic *Escherichia coli* infection. *Infection and Immunity*, **64**, 2368–70.

Midland, S. L., Keen, N. T., Sims, J. J., Midland, M. M., Stayton, M. M., Burton, V., Smith, M. J., Mazzola, E. P., Graham, K. J. & Clardy, J. (1993). The structures of syringolides 1 and 2, novel C-glycosidic elicitors from *Pseudomonas syringae* pv. *tomato*. *Journal of Organic Chemistry*, **58**, 2940–5.

Milton, D. L., Hardman, A., Camara, M., Chhabra, S. R., Bycroft, B. W., Stewart, G. S. A. B. & Williams, P. (1997). *Vibrio anguillarum* produces multiple *N*-acylhomoserine lactone signal molecules. *Journal of Bacteriology*, **179**, 3004–12.

More, M. I., Finger, L. D., Stryker, J. L., Fuqua, C., Eberhard, A. & Winans, S. C. (1996). Enzymatic synthesis of a quorum sensing autoinducer through use of defined substrates. *Science*, **272**, 1655–8.

Mur, L. A. J., Kenton, P. & Draper, J. (1997). Something in the air: volatile signals in plant defence. *Trends in Microbiology*, **5**, 297–300.

Nataro, J. P. & Kaper, J. B. (1998). Diarrheagenic *Escherichia coli*. *Clinical Microbiology Reviews*, **11**, 143–201.

Neter, E., Korns, R. F. & Trussell, R. E. (1953). Association of *Escherichia coli* serogroup O111 with two hospital outbreaks of epidemic diarrhoea of new-born infants in New York State during 1947. *Pediatrics*, **12**, 377–83.

Newman, M-A., Daniels, M. J. & Dow, J. M. (1995). Lipopolysaccharide from *Xanthomonas campestris* induces defense-related gene expression in *Brassica campestris*. *Molecular Plant–Microbe Interactions*, **8**, 778–80.

Newman, M-A., Daniels, M. J. & Dow, J. M. (1997). The activity of lipid A and core components of bacterial lipopolysaccharides in the prevention of the hypersensitive response in pepper. *Molecular Plant–Microbe Interactions*, **10**, 926–8.

Niehaus, K. & Becker, A. (1998). The role of microbial surface polysaccharides in the *Rhizobium*–legume interaction. In *Subcellular Biochemistry, Vol. 29: Plant–Microbe Interactions*, ed. B. B. Biswas & H. K. Das, pp. 73–116. New York: Plenum Press.

Niehaus, K., Baier, R., Kohring, B., Flaschel, E. & Puhler, A. (1997). Symbiotic suppression of the *Medicago sativa* plant defence system by *Rhizobium meliloti* oligosaccharide. In *Biological Fixation of Nitrogen for Ecology and Sustainable Agriculture*, ed. A. Legocki, H. Bothe & A. Puhler, pp. 111–14. Heidelberg: Springer.

Nishizuka, Y. (1986). Studies and perspectives of protein kinase C. *Science*, **233**, 305–12.

Otto, M., Süßmuth, R., Jung, J. & Gotz, F. (1998). Structure of the pheromone peptide of the *Staphylococcus epidermidis agr* system. *FEBS Letters*, **424**, 89–94.

Pearson, J. P., Gray, K. M., Passador, L., Tucker, K. D., Eberhard, A., Iglewski, B. H. & Greenberg, E. P. (1994). Structure of the autoinducer required for expression of *Pseudomonas aeruginosa* virulence genes. *Proceedings of the National Academy of Sciences, USA*, **91**, 197–201.

Pearson, J. P., Passador, L., Iglewski, B. H. & Greenberg, E. P. (1995). A second *N*-acylhomoserine lactone signal produced by *Pseudomonas aeruginosa*. *Proceedings of the National Academy of Sciences, USA*, **92**, 1490–4.

Pestova, E. V., Håvarstein, L. S. & Morrison, D. A. (1996). Regulation of competence for genetic transformation in *Streptococcus pneumoniae* by an auto-induced peptide pheromone and a two-component regulatory system. *Molecular Microbiology*, **21**, 853–62.

Pingret, J-L., Journet, E-P. & Barker, D. G. (1998). *Rhizobium* nod factor signaling: evidence for a G protein-mediated transduction mechanism. *Plant Cell*, **10**, 659–71.

Pirhonen, M., Flego, D., Heikinheimo, R. & Palva, E. T. (1993). A small diffusible signal molecule is responsible for the global control of virulence and exoenzyme production in *Erwinia carotovora*. *EMBO Journal*, **12**, 2467–76.

Polotsky, Y. E., Dragunskaya, E. M., Seliverstova, V. G., Avdeeva, T. A., Chakhutinskaya, M. G., Ketyi, I., Vertenyi, A., Ralovich, B., Emody, L., Malovics, I., Safonova, N. V., Snigirevskaya, E. S. & Karyagina, E. I. (1977). Pathogenic effect of enterotoxigenic *Escherichia coli* and *Escherichia coli* causing infantile diarrhoea. *Acta Microbiologica Academica Scientifica Hungarica*, **24**, 221–36.

Poplawsky, A. R., Chun, W., Slater, H., Daniels, M. J. & Dow, J. M. (1998). Synthesis of extracellular polysaccharide, extracellular enzymes, and xanthomonadin in *Xanthomonas campestris*: evidence for the involvement of two intercellular regulatory signals. *Molecular Plant–Microbe Interactions*, **11**, 68–70.

Prasad, C. (1995). Bioactive cyclic dipeptides. *Peptides*, **16**, 151–64.

Pugsley, A. P., Francetic, O., Possot, O. M., Sauvonnet, N. & Hardie, K. R. (1997). Recent progress and future directions in studies of the main terminal branch of the general secretory pathway in Gram-negative bacteria – a review. *Gene*, **192**, 13–19.

Puskas, A., Greenberg, E. P., Kaplan, S. & Schaefer, A. L. (1997). A quorum-sensing system in the free-living photosynthetic bacterium *Rhodobacter sphaeroides*. *Journal of Bacteriology*, **179**, 7530–7.

Roche, P., Debellé, F., Maillet, F., Lerouge, P., Faucher, C., Truchet, G., Dénarié, J. & Prome, J-C. (1991). Molecular basis of symbiotic host specificity in *Rhizobium meliloti*: *nodH* and *nodPQ* genes encode the sulfation of lipo-oligosaccharide signals. *Cell*, **67**, 1131–43.

Rosenshine, I., Ruschkowski, S., Stein, D., Reinscheid, D., Mills, S. D. & Finlay, B. B. (1996). A pathogenic bacterium triggers epithelial cell signals to form a functional bacterial receptor that mediates actin pseudopod formation. *EMBO Journal*, **15**, 2613–24.

Salmond, G. P. C., Bycroft, B. W., Stewart, G. S. A. B. & Williams, P. (1995). The bacterial enigma: cracking the code of cell–cell communication. *Molecular Microbiology*, **16**, 615–24.

Scaletsky, I. C. A., Silva, M. L. M., Toledo, M. R. F., Davis, B. R., Blake, P. A. & Trabulsi, L. R. (1985). Correlation between adherence to HeLa cells and serogroups, serotypes, and biotypes of *Escherichia coli*. *Infection and Immunity*, **49**, 528–32.

Scaletsky, I. C. A., Silva, M. L. M. & Trabulsi, L. R. (1984). Distinctive patterns of adherence of enteropathogenic *Escherichia coli* to HeLa cells. *Infection and Immunity*, **45**, 534–6.

Schaefer, A. L., Hanzelka, B. L., Eberhard, A. & Greenberg, E. P. (1996a). Quorum sensing in *Vibrio fischeri*: probing autoinducer LuxR interactions with autoinducer analogs. *Journal of Bacteriology*, **178**, 2897–901.

Schaefer, A. L., Val, D. L., Hanzelka, B. L., Cronan, J. E. & Greenberg, E. P. (1996b). Generation of cell-to-cell signals in quorum sensing: acylhomoserine lactone synthase activity of a purified *Vibrio fischeri* LuxI protein. *Proceedings of the National Academy of Sciences, USA*, **93**, 9505–9.

Schripsema, J., de Rudder, K. E. E., van Vliet, T. B., Lankhorst, P. P., de Vroom, E., Kijne, J. W. & van Brussel, A. A. (1996). Bacteriocin *small* of *Rhizobium leguminosarum* belongs to the class of *N*-acyl-L-homoserine lactone molecules known as autoinducers and as quorum sensing co-transcriptional factors. *Journal of Bacteriology*, **178**, 366–71.

Shapiro, J. A. (1988). Bacteria as multicellular organisms. *Scientific American*, **256**, 82–9.

Shaw, P. D., Ping, G., Daly, S., Cronan, J. E., Rhinehart, K. & Farrand, S. K. (1997). Detecting and characterizing acyl-homoserine lactone signal molecules by thin layer chromatography. *Proceedings of the National Academy of Sciences, USA*, **94**, 6036–41.

Scheu, A. K., Economou, A., Hong, G-F., Ghelani, S., Johnston, A. W. B. & Downie, J. A. (1992). Secretion of the *Rhizobium leguminosarum* nodulation protein NodO by haemolysin-type systems. *Molecular Microbiology*, **6**, 231–8.

Sitnikov, D. M., Schineller, J. B. & Baldwin, T. O. (1995). Transcriptional regulation of bioluminescence genes from *Vibrio fischeri*. *Molecular Microbiology*, **17**, 801–12.

Sitnikov, D. M., Schineller, J. B. & Baldwin, T. O. (1996). Control of cell division in *Escherichia coli*: regulation of transcription of ftsQA involves both *rpoS* and SdiA-mediated autoinduction. *Proceedings of the National Academy of Sciences, USA*, **93**, 336–41.

Speller, D. C. E., Johnson, A. P., James, D., Marples, R. R., Charlett, A. & George, R. C. (1997). Resistance to methicillin and other antibiotics in isolates of

Staphylococcus aureus from blood and cerebrospinal fluid in England and Wales, 1989–1995. *Lancet*, **350**, 323–5.

Staley, J. D., Jones, E. W. & Gorley, L. D. (1969). Attachment and penetration of *E. coli* into the intestinal epithelium of the ileum in new-born pigs. *American Journal of Pathology*, **56**, 371–92.

Stierle, A. C., Cardellina, J. H. & Strobel, G. A. (1988). Maculosin, a host-specific phytotoxin for spotted knapweed from *Alternaria alternata*. *Proceedings of the National Academy of Sciences, USA*, **85**, 8008–11.

Surette, M. G. & Bassler, B. L. (1998). Quorum sensing in *Escherichia coli* and *Salmonella typhimurium*. *Proceedings of the National Academy of Sciences, USA*, **95**, 7046–50.

Sutton, J. M., Lea, E. J. A. & Downie, J. A. (1994). The nodulation-signaling protein NodO from *Rhizobium leguminosarum* biovar *viciae* forms ion channels in membranes. *Proceedings of the National Academy of Sciences, USA*, **91**, 9990–4.

Swift, S., Karlyshev, A. V., Durant, E. L., Winson, M. K., Williams, P., Macintyre, S. & Stewart, G. S. A. B. (1997). Quorum sensing in *Aeromonas hydrophila* and *Aeromonas salmonicida*: Identification of the LuxRI homologues AhyRI and AsaRI and their cognate signal molecules. *Journal of Bacteriology*, **179**, 5271–81.

Swift, S., Throup, J. P., Williams, P., Salmond, G. P. C. & Stewart, G. S. A. B. (1996). Quorum sensing: a population density component in the determination of bacterial phenotype. *Trends in Biochemical Sciences*, **21**, 214–19.

Swift, S., Winson, M. K., Chan, P. F., Bainton, N. J., Birdsall, M., Reeves, P. J., Rees, C. E. D., Chhabra, S. R., Hill, P. J., Throup, J. P., Bycroft, B. W., Salmond, G. P. C., Williams, P. & Stewart, G. S. A. B. (1993). A novel strategy for the isolation of *luxI* homologues: evidence for the widespread distribution of a LuxR:LuxI superfamily in enteric bacteria. *Molecular Microbiology*, **10**, 511–20.

Tang, P., Rosenshine, I. & Finlay, B. B. (1994). *Listeria monocytogenes* an invasive bacterium stimulates MAP kinase upon attachment to epithelial cells. *Molecular Biology of the Cell*, **5**, 455–64.

Telford, G., Wheeler, D., Williams, P., Tomkins, P. T., Appleby, P., Sewell, H., Stewart, G. S. A. B., Bycroft, B. W. & Pritchard, D. I. (1998). The *Pseudomonas aeruginosa* quorum sensing signal molecule, *N*-(3-oxododecanoyl)-L-homoserine lactone has immunomodulatory activity. *Infection and Immunity*, **66**, 36–42.

Throup, J., Camara, M., Briggs, G., Winson, M. K., Chhabra, S. R., Bycroft, B. W., Williams, P. & Stewart, G. S. A. B. (1995). Characterisation of the *yenI/yenR* locus from *Yersinia enterocolitica* mediating the synthesis of two *N*-acylhomoserine lactone signal molecules. *Molecular Microbiology*, **17**, 345–56.

Timmers, A. C. J., Auriac, M. C., Debilly, F. & Truchet, G. (1998). Nod factor internalization and microtubular cytoskeleton changes occur concomitantly during nodule differentiation in alfalfa. *Genes and Development*, **125**, 339–49.

Urzainqui, A. & Walker, G. C. (1992). Exogenous suppression of the symbiotic deficiencies of *Rhizobium meliloti exo* mutants. *Journal of Bacteriology*, **174**, 3403–6.

Van Brussel, A. A. N., Bakhuizen, R., van Spronsen, P. C., Spaink, H. P., Tak, T., Lugtenberg, B. J. J. & Kijne, J. W. (1992). Induction of pre-infection thread structures in the leguminous host plant by mitogenic lipo-oligosaccharides in *Rhizobium*. *Science*, **257**, 70–2.

Vivian, A. & Gibbon, M. J. (1997). Avirulence genes in plant-pathogenic bacteria signals or weapons? *Microbiology*, **143**, 693–704.

Wandersman, C. (1996). Secretion across the bacterial membrane. In *Escherichia coli and Salmonella: Cellular and Molecular Biology*, ed. F. C. Neidhart, pp. 955–966. ASM Press.

Williams, P. (1994). Compromising bacterial communication skills. *Journal of Pharmacy and Pharmacology*, **46**, 252–60.

Williams, P., Bainton, N. J., Swift, S., Chhabra, S. R., Winson, M. K., Stewart, G. S. A. B., Salmond, G. P. C. & Bycroft, B. W. (1992). Small molecule-mediated density-dependent control of gene expression in prokaryotes: bioluminescence and the biosynthesis of carbapenem antibiotics. *FEMS Microbiology Letters*, **100**, 161–7.

Winson, M. K., Camara, M., Latifi, A., Foglino, M., Chhabra, S. R., Daykin, M., Bally, M., Chapon, V., Salmond, G. P. C., Bycroft, B. W., Lazdunski, A., Stewart, G. S. A. B. & Williams, P. (1995). Multiple *N*-acyl-L-homoserine signal molecules regulate production of virulence determinants and secondary metabolites in *Pseudomonas aeruginosa*. *Proceedings of the National Academy of Sciences, USA*, **92**, 9427–31.

Winson, M. K., Swift, S., Fish, L., Throup, J. P., Jorgensen, F., Chhabra, S. R., Bycroft, B. W. & Stewart, G. S. A. B. (1998). Construction and analysis of *luxCDABE*-based plasmid sensors for investigating *N*-acyl homoserine lactone-mediated quorum sensing. *FEMS Microbiology Letters*, **163**, 185–92.

Wirth, R., Muscholl, A. & Wanner, G. (1996). The role of pheromones in bacterial interactions. *Trends in Microbiology*, **4**, 96–103.

Wood, D. W., Gong, F., Daykin, M., Williams, P. & Pierson, L. S. (1997). *N*-acylhomoserine lactone-mediated regulation of phenazine gene expression by *Pseudomonas aureofaciens* 30-84 in the wheat rhizosphere. *Journal of Bacteriology*, **179**, 7663–70.

Wooldridge, K. G., Williams, P. H. & Ketley, J. M. (1996). Host signal-transduction and endocytosis of *Campylobacter jejuni*. *Microbial Pathogenesis*, **21**, 299–305.

Zhang, L. H., Murphy, P. J., Kerr, A. & Tate, M. E. (1993). *Agrobacterium* conjugation and gene regulation by *N*-acyl-L-homoserine lactones. *Nature*, **362**, 446–8.

INTERCELLULAR SIGNALLING AND THE MULTIPLICATION OF PROKARYOTES: BACTERIAL CYTOKINES

ARSENY S. KAPRELYANTS[1],
GALINA V. MUKAMOLOVA[1], SVETLANA S.
KORMER[1], DIETER H. WEICHART[2], MICHAEL
YOUNG[2] AND DOUGLAS B. KELL[2]

[1] *Bakh Institute of Biochemistry, Russian Academy of Sciences,
Leninsky pr.33, 117071 Moscow, Russia*
[2] *Institute of Biological Sciences, University of Wales, Aberystwyth,
Ceredigion SY23 3DD, UK*

INTRODUCTION AND BACKGROUND

Tissue cultures of cells taken from higher, differentiated organisms normally need complex (and mainly polypeptidic or proteinaceous) extracellular growth factors for successful cell division (and even survival (Raff, 1992)). These factors are nowadays usually referred to as cytokines (Callard & Gearing, 1994; Hardie, 1991), and their role is generally understood (cf. Levine & Prystowsky, 1995) to involve binding at the cell membrane and the production of second messengers such as cGMP which serve to activate various segments of primary metabolism, which may include those responsible for their own synthesis (Alberts *et al.*, 1989). Although best known in mammalian systems, such polypeptide growth factors, which can be transported between cells and thus also have the properties of chemical signals, are currently being discovered and identified in higher plants (Matsubayashi & Sakagami, 1996; Matsubayashi *et al.*, 1997; van de Sande *et al.*, 1996), in multicellular invertebrates (Ottaviani *et al.*, 1996), and even in unicellular eukaryotes such as ciliates (Christensen *et al.*, 1998; Luporini *et al.*, 1995).

Where, in evolution, such signalling systems may have appeared in a recognizable form is uncertain (Beck & Habicht, 1994; Brown, 1998; Cooper *et al.*, 1994; Csaba, 1994; Janssens, 1988; Lenard, 1992; LeRoith *et al.*, 1986; Pertseva, 1991; Roth *et al.*, 1986), and the apparently conflicting molecular phylogenies (and thus the likely extensive horizontal gene transfer; Koonin & Galperin, 1997; Koonin *et al.*, 1997; Doolittle & Logsdon, 1998) recently revealed via comparative genomics suggests that no individual phylogenetic tree is likely to give an unambiguous answer in the short term (Forterre, 1997*a*).

Nevertheless, it is usually assumed in prokaryotic microbiology that each bacterial cell in an axenic culture can multiply independently of other bacteria, provided that appropriate concentrations of substrates, vitamins and trace elements are present in the culture medium, and that implicitly there is no such requirement for autocrine or paracrine, polypeptide/ proteinaceous growth factors. Current laboratory experience seems to be consistent with this, in that the development of bacterial colonies from single cells on agar plates is commonplace, and the Most Probable Number method is based on the apparently correct assumption that a test tube containing but one viable cell will, in due time, display visible growth or turbidity. While it is already clear that axenic bacterial cultures do not remotely represent a statistically homogeneous population (Davey & Kell, 1996; Kell *et al.*, 1991; Koch, 1987), these observations tend to be, and are most simply, interpreted as being in favour of 'autonomous' growth.

However, such growth is almost always analysed in the presence of culture supernatants (or cell-adherent molecules) that were introduced with the inoculum, and it is at least possible that these inocula may contain potent autocrine or paracrine growth factors, produced by the cells during their previous growth phase, which are necessary for the initiation of regrowth and whose presence would tend not to be recognized in conventional physiological experiments, partly due to the extremely low active concentrations. The recently discovered oligopeptide plant hormone ENOD40 modulates cell division in tobaco cell cultures at concentrations as low as 10^{-16} M (van de Sande *et al.*, 1996), a potency matched only by that of the glycoprotein sex pheromone of the green alga *Volvox carteri* (Hallmann *et al.*, 1998)!

An increasing body of evidence has pointed up the widespread importance of chemically mediated intercellular communications in bacterial cultures for such specific events as sporulation, conjugation, virulence, bioluminescence and so on. Thus, it is now clear that a variety of different autocrine chemical signals (pheromones; Stephens, 1986), which are produced as secondary metabolites (Bu'lock, 1961; Kell *et al.*, 1995), are responsible for these types of prokaryotic social behaviour, as are exhibited under conditions of obvious cellular differentiation (for review, see Fuqua *et al.*, 1994; Greenberg *et al.*, 1996; Kaiser & Losick, 1993; Kell *et al.*, 1995; Kleerebezem *et al.*, 1997; Salmond *et al.*, 1995; Swift *et al.*, 1994, and many other authors in this volume). Interestingly, while the Gram-negative organisms often use *N*-acyl homoserine lactone derivatives, the Gram-positives – with the exception of the butanolides of streptomycetes – tend to use proteins and polypetides as their signals (Greenberg *et al.*, 1996; Kell *et al.*, 1995; Kleerebezem *et al.*, 1997). The important properties of such molecules in this context, which discriminates them from nutrients, are that (i) they are produced by the organisms themselves, (ii) they are active at very low concentrations, and (iii) leaving aside the cleavage of prohormones their metabolism is not necessary for activity (although they may ultimately be degraded).

The question then arises as to whether similar types of signalling may be of significance not only for differentiation but for cell multiplication in growing bacterial cultures generally. Our purposes are thus to: (i) develop the idea that this is most probably so, (ii) summarize and bring together the relevant experimental evidence for the involvement of hormones and pheromones in prokaryotic growth and division, and (iii) emphasize the important conse-quences of this view for a number of apparent (and rather fundamental) microbiological puzzles connected with viability, culturability, dormancy and growth.

THE DEVELOPMENT OF BACTERIAL CULTURES DURING THE LAG PHASE

It is a matter of everyday experience that the duration of the lag phase in batch cultures often depends more or less inversely on the size of the inoculum, even when bacterial growth is monitored by counting viable cells (Penfold, 1914), and see below. (Note that in some studies this 'true' inoculum-dependent lag phase is confused with the 'apparent' lag when the lag phase is indirectly estimated from uncorrected optical density traces, which must necessarily be 'inoculum size dependent' since there is a minimally detectable or threshold change in OD of, say, 0.01 which equates to some 5×10^6 or 10^7 bacteria ml^{-1}.) The idea that such a dependence could reflect the accumulation of some growth inducer(s) secreted by cells during the lag phase has long been espoused (see, e.g. Hinshelwood, 1946), and the addition of supernatant from log phase bacterial cultures signifi-cantly shortened the inoculum-dependent lag phase in a number of cases (Dagley et al., 1950; Halman & Mager, 1967; Hinshelwood, 1946; Lankford et al., 1966) . However, an inoculum-dependent lag phenomenon may be observable only under a restricted range of conditions: in the case of Achromobacter delmarvae an inoculum-dependent lag was detected in a poor medium but not in a rich one (Shida et al., 1977). Similarly, the study of inoculum-dependent lags for various Bacillus spp. has been performed by using poor medium (Lankford et al., 1966). Dagley and colleagues (1950) found that the effect of supernatant on the inoculum-dependent lag itself depends on the size of the inoculum: the largest supernatant effect was observed with the smallest inoculum.

Any substance produced by cells in the culture supernatants may have growth-affecting properties, those of CO_2 being well known (Dixon & Kell, 1989). Notwithstanding the long history of the inoculum-dependent lag phase, however, little is known about the nature of the secreted substances involved in the phenomenon (Kaprelyants & Kell, 1996). Siderophores (formerly 'schizokinen') – iron transport compounds – were shown to act as growth factors influencing the inoculum-dependent lag in Bacillus cultures (Lankford et al., 1966; Mullis et al., 1971), and the activity of such compounds could be mimicked by autoclaved solutions of glucose or other carbohydrates and phosphate (Lankford et al., 1957) (although autoclaving glucose separately can have quite different physiological effects; Kell & Sonnleitner, 1995). Batchelor and colleagues (1997) reported on a 'dramatic reduction' of the lag phase during the regrowth of *Nitrosomonas europea* starved for up to 6 weeks consequent upon the addition of *N*-(3-oxo-hexanoyl)homoserine lactone (OHHL) to the growth medium, a compound known *inter alia* as an autoinducer of luminescence in *Vibrio fisheri* and of antibiotic production in *Erwinia carotovora* (Greenberg *et al.*, 1996; Kell *et al.*, 1995; Swift *et al.*, 1994). It is interesting that there was no lag phase for starved biofilms composed from the same *N. europea* cells, which strongly supports the density-dependent nature of the recovery process in this case (Batchelor *et al.*, 1997), although there are complex interactions between a variety of only partly characterized agonists and antagonists in these types of system (Givskov *et al.*, 1996; Kjelleberg *et al.*, 1997; Srinivasan *et al.*, 1998). Davies and colleagues (1998) found that *N*-(3-oxodecanoyl)-L-homoserine lactone, a signalling pheromone produced by *Pseudomonas aeruginosa*, was required for the proper development (but not the initiation) of biofilms of this organism, while Bloomquist and colleagues (1996) found that there is a significant stimulation of the initiation of streptococcal growth on the tooth surface when the cell concentration reaches a high enough density, and explained this behaviour as the autoinduction of cell growth via cell-to-cell signalling. However, the putative signal(s) has not been isolated.

It is relatively straightforward to demonstrate the excretion into bacterial supernatants of compounds which reduce the lag phase in cultures to which the supernatants are added. Figure 1 shows such activity in cultures of starved *E. coli* (D. H. Weichart & D. B. Kell, unpublished observations), and the authors consider it likely that many other such substances remain to be identified.

THE REGROWTH OF BACTERIA FOLLOWING STARVATION, SOME TERMINOLOGICAL QUESTIONS, AND THE PHENOMENA OF LIMITED DIVISIONS

'When I use a word,' Humpty Dumpty said, in a rather scornful tone, 'it means just what I choose it to mean, neither more nor less.'

'The question is', said Alice, 'whether you can make words mean so many different things.'

'The question is,' said Humpty Dumpty, 'which is to be master – that's all.'
(Carroll, 1974, orig. 1871)

'At present one must accept that the death of microbe can only be discovered retrospectively: a population is exposed to a recovery medium, incubated, and those individuals which do not divide to form progeny are taken to be dead ... there exist at present no short cuts which would permit assessment of the moment of

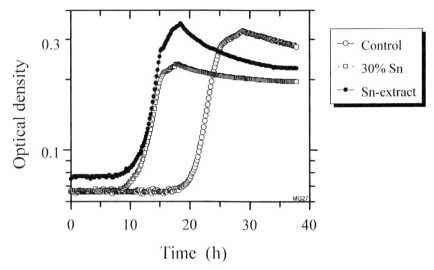

Fig. 1. Growth of a diluted 50-day-old stationary-phase culture of *Escherichia coli* ZK126 (W3110) in a MOPS-buffered minimal medium with the addition of 30% (v/v) supernatant (Sn) of a 2-day stationary culture of the same organism, or an addition of a crude extract of the same supernatant (Sn-extract), added in a corresponding concentration; control: untreated suspension. Growth was measured in a Bioscreen Microbiological Growth Analyser as in (Mukamolova *et al.*, 1998*a*). (DHW and DBK, unpublished observations).

death: vital staining, optical effects, leakage of indicator substances and so on are not of general validity' (p.5)

(Postgate, 1976)

'An organism is considered living or viable if it is capable of continued multiplication; if it is not so capable it is called dead or non-viable.' (p.31)

(Greenwood Peutherer, 1992)

Bacteria commonly face starvation in natural environments, and much information has recently become available concerning the physiological and biochemical changes accompanying bacterial starvation (for review, see Hengge-Aronis, 1996; Kjelleberg, 1993; Kolter *et al.*, 1993; Matin, 1994) which may result in (i) starvation survival (maintenance of cell viability for a prolonged time), (ii) cell death, (iii) sporulation (for some bacteria) and (iv) the formation of resting (dormant) forms of non-sporulating bacteria. Since it is desirable to avoid the Humpty Dumpty problem (above), it can be stated that, by 'dormancy', we mean *a reversible state of low metabolic activity, in which cells can persist for extended periods without division* (Kaprelyants *et al.*, 1993). Until recently, dormancy has mainly been connected with bacterial forms, which are obviously morphologically specialized, e.g. spores and cysts ('constitutive dormancy'; Sussman & Halvorson, 1966), structures which can be formed by only a limited number of bacterial species. However, there has recently been much discussion with regard to the possible existence of dormant states of vegetative, non-sporulating bacteria (Kaprelyants *et al.*, 1993).

In particular, Roszak and Colwell (1985, 1987) have proposed (Roszak *et al.*, 1984; Xu *et al.*, 1982) that, under some circumstances (mainly starvation), readily culturable bacteria may become non-culturable but retain 'viability'. Such putative 'viable- but non-culturable' (VBNC or VNC) bacteria (Oliver, 1993) have been proposed to represent some kind of resting (or in some usages active but non-dividing) bacterial state. However, the term VBNC is an oxymoron because the well-established convention (Postgate, 1967, 1969, 1976) is that a bacterial cell should only be considered as viable if it is capable of multiplying (Barer, 1997; Barer *et al.*, 1993; Kell *et al.*, 1998). In particular, it is necessary to consider that terms such as 'viability' can have both a conceptual meaning, in which a state of say 'viability' is ascribed to the organism itself, and an operational meaning, in which the state an organism is declared to have is a result only of the outcome of experimental analyses to which it is subjected (such that the 'state' could vary depending on the experimental outcomes). In the operational realm, such states as 'viability' are thus not an intrinsic property of a microbial cell. This crucial distinction, though perhaps unfortunately unfamiliar to many microbiologists, is well known both in the microscopic world of quantum physics (Primas, 1981), especially as the Schrödinger's Cat paradox (Kell *et al.*, 1998), and in the macroworld of human existence in which 'death' can be ascribed only *a posteriori* (Watson, 1987). *Only the operational definitions are free of paradoxes and difficulties.*

Notwithstanding that many of the problems accompanying such studies are semantic, it can at least be said that, if this putative 'VBNC' state of cells exists, it should be reversible if it is to be accepted as a specialized form of the bacterial life cycle (and these ostensibly non-culturable cells are not therefore simply dead). While many attemps to find conditions for the recovery ('resuscitation') of cells from so-called 'VBNC' states have been made, almost all published studies up to now unfortunately fail to discriminate adequately between resuscitation/recovery and the regrowth of any viable (culturable) cells initially present in the 'VBNC' population; only a few examples can be considered to have shown true resuscitation (as defined by a return to culturability) of cells under conditions in which the contribution of 'initially viable' cells is excluded (Bogosian *et al.*, 1998; Kell *et al.*, 1998).

Little is known about the processes that occur during the outgrowth of starved, 'VBNC' or ultramicrobacterial (Morita, 1988) cells, and where information is to hand the evidence suggests a profound differentiation in terms of the culture's ability to emerge smoothly from the lag phase. In addition, the problem of culture heterogeneity (the coexistence of viable, injured, dormant and dead cells) greatly complicates studies of this process. Nevertheless some peculiarities of cell behaviour during the 'outgrowth' phase could be of interest in the light of the problems discussed in this review.

Thus, it was found that a fraction of the cells in a population of starved marine bacteria ceased growth after one or a few cell divisions when placed on an agar surface; cells within such 'microcolonies' were morphologically different from cells which produced visible macrocolonies (Torrella & Morita, 1981). *Pseudomonas fluorescens* cells, starved in soil for 40 days, were tested for their ability to grow in different fresh media; an epifluorescence technique showed that, within the first 40 hours of incubation, cells divided only two to three times, after which cell division stopped (Binnerup *et al.*, 1993). A similar result was obtained when soil bacteria were subjected to cultivation on nutrient-poor media (Winding *et al.*, 1994).

It was found that *Micrococcus luteus* cells starved for 3–6 months in a prolonged stationary phase consist of 10–90% of dormant forms which could multiply several times (up to 10–17) when resuscitated in fresh liquid medium (which allows normally viable bacteria to grow rapidly to a high optical density). After this the cell growth stopped, with final cell concentrations of no more than 10^6 cells ml^{-1} (Mukamolova *et al.*, 1995), although the cells remained metabolically active as judged by flow cytometry (Votyakova *et al.*, 1994). A variety of cognate studies are reviewed elsewhere (Kaprelyants & Kell, 1996).

It is reasonable that the phenomenon of 'limited divisions' is actually rather common in microbiological practice, but because the threshold for optical detection of growth typically lies between 10^6 and 10^7 ml^{-1} it has not been registered in many cases (and we know of many unpublished or anecdotal cases in which it is known that such culture growth is detectable only with 'large' inocula). What kind of mechanism(s) may be responsible for this phenomenon?

Recently, the view has been expressed that the inability of some starved bacteria to grow under conditions which normally support their growth is due to imbalanced metabolism during the start of cell regrowth, as a result of which the accumulation of free radicals leads to 'self-destruction' by the cells (a suicide response; Bloomfield *et al.*, 1998), and it is true that some recovery experiments show that concentrations of nutrients which normally supported bacterial growth are too high for the recovery of starved cells, e.g. with *Vibrio* (MacDonell & Hood, 1982) or *M. luteus* (Mukamolova *et al.*, 1998*b*). The significance of aeration for the further resuscitation of dormant *M. luteus* has also been appreciated (Mukamolova *et al.*, 1998*b*), while it is interesting that embryos of *Artemia franciscana* can persist in a dormant state for several years when fully hydrated but under strong anoxia (Clegg, 1997). But, by definition, any such 'suicide response' cannot, for instance, properly explain the phenomenon of 'limited divisions' in which starved or dormant cells during resuscitation can make several divisions before growth ceases, nor resuscitation generally since (*pace*; Bloomfield *et al.*, 1998) dormancy is reversible but suicide is not (Barer *et al.*, 1998; Kell *et al.*, 1998).

Other mechanisms that may be responsible for causing the cessation of cell growth in the above circumstances include the accumulation of inhibitory substances during the growth of previously starved cells. In this context, the secretion of a 'killer factor' was demonstrated during the resuscitation of starved *M. luteus* cultures, which inhibited the growth of viable bacteria (Mukamolova *et al.*, 1995), while MacDonell and Hood (1982) described a recovery method for bacteria from estuarine waters, which included several passages of cells from plate to plate until they became able to produce visible colonies, consistent with the removal of an inhibitory substance. Possibly, these two mechanisms are linked, and accumulated inhibitors are products of cell destruction (by whatever mechanism).

It was shown in experiments with dormant *M. luteus* that the presence of a small fraction of viable cells at the onset of resuscitation facilitated the recovery of the majority of the remaining (dormant) cells. The cell density-dependence of the kinetics, or population effect, would suggest that this is due to the excretion of some factor(s) which promoted the transition of cells from a state incapable of growth and division to a colony-forming state (Votyakova *et al.*, 1994). In subsequent experiments, the addition of supernatants from growing *M. luteus* cultures to the starved culture relieved the inability of the cells to divide for more than a limited number of times, and allowed the resuscitation of cells to normal, colony-forming cells (Kaprelyants *et al.*, 1994; see Fig. 2). Possibly in this (and probably in other similar) cases, starved cells have a lowered ability, at the beginning of cell growth, to excrete an appropriate growth factor(s) needed for the stimulation of cell division. As a result, only some of the cells in a population can multiply, while the gradual accumulation of poisoning substances formed by non-dividing cells leads to a decreased excretion of the factor and eventually to the cessation of cell multiplication. More generally, there may be a limited quota of growth factor whose steady-state concentration is low in poor media and which is degrading over time, and being shared out between cells over generations until its concentration per cell is inadequate for division.

But overall, the main conclusion which can be made from this section is that, whilst in many cases starved cells become more resistant to environmental insults (Hengge-Aronis, 1996; Kjelleberg, 1993; Matin, 1991), starvation may be accompanied by an increased sensitivity of cells to specific chemical inducers or inhibitors of cell division which is not observable under normal growth conditions.

MICROENDOCRINOLOGY

The communication between bacterial cells in an axenic population ('quorum sensing') resembles the well-known chemically mediated interactions between cells in higher organisms, as mentioned above, and it seems logical to ask whether the 'bacterial' chemical language of communication might contain some homologous 'words' (molecules) or 'grammar' (modes of

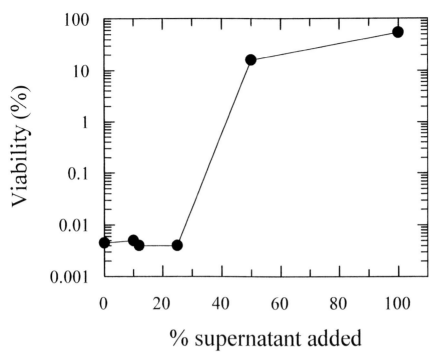

Fig. 2. Effect of culture supernatant on the resuscitation of dormant cells of *Micrococcus luteus*. *M. luteus* was starved for 3.5 months and its viability assessed using an MPN assay as described (Kaprelyants *et al.*, 1994). Supernatants were taken from a batch culture of the organism grown to an OD of 2, slightly before the beginning of stationary phase and mixed in the stated proportion with a lactate minimal medium containing 0.05% (w/v) yeast extract (Kaprelyants *et al.*, 1994).

action and regulation) as that of higher organisms? A number of studies have revealed the presence of vertebrate hormone-like substances in bacteria, which were specifically active on mammalian cells. The list includes steroid and polypeptide hormones (including insulin); moreover, specific, high-affinity binding proteins for many mammalian hormones were found in bacterial cells (for review, see Lenard, 1992; LeRoith *et al.*, 1986). In a most interesting study, an autocrine growth stimulation function of chorionic gonadotropin-like protein from *Xanthomonas maltophilia* has been associated with a fully sequenced 48 kDa membrane-bound protein (Grover *et al.*, 1995).

Lyte (1992) has developed the idea that the growth and virulence of pathogens *in vivo* depends strongly on host-derived hormonal signals, and it is certainly known that some hormones can stimulate the growth of micro-organisms; examples include catecholamines (in Gram-negative bacteria; Lyte *et al.*, 1996, 1997; Lyte & Bailey, 1997; Lyte & Ernst, 1992), serotonin (*S. faecalis*; Strakhovskaya *et al.*, 1993) and insulin (*Neurospora crassa*;

McKenzie *et al.*, 1988), while inhibition of growth has been observed for insulin in *Pseudomonas pseudomallei* cultures (Woods *et al.*, 1993). The stimulation of the growth of virulent strains of *E. coli* by interleukin-1 (Denis *et al.*, 1991*a*; Porat *et al.*, 1991) and by granulocyte-macrophage colony-stimulating factor (Denis *et al.*, 1991*a*) was reported in 1991. Later, the same effect was described for interleukin-6 on virulent *Mycobacterium avium* (Denis, 1992) and for transforming growth factor-β-1 which enhanced the intracellular growth of *M. tuberculosis* in monocytes (Hirsch *et al.*, 1994). A significant acceleration of growth of *Mycobacterium avium, M. tuberculosis* but not *M. smegmatis* was observed when they were cultured in the presence of 5–500 ng ml^{-1} epidermal growth factor (EGF), and a specific receptor for EGF was identified as a glyceraldehyde-3-phosphate dehydrogenase (Bermudez *et al.*, 1996), which has been shown to bind plasmin and some other proteins on the surface of streptococci (Pancholi & Fischetti, 1992). It is worth noting that, when the activity of the appropriate hormone or cytokine was tested in all these studies, the effect on bacterial growth rate (usually in an exponential phase) was not at all pronounced (excluding the last case with EGF). This is not very surprising, as competition experiments are far more sensitive in revealing small changes in growth rate or fitness (Baganz *et al.*, 1997; Dykhuizen, 1993; Dykhuizen & Hartl, 1983; Thatcher *et al.*, 1998), yet this problem was responsible for the apparently difficuties in reproducing some of these results (see Kim & Le, 1992; Porat *et al.*, 1992).

It important to note that, in most of these experiments, the conditions were actually the least suitable for detecting a stimulatory effect, e.g. the use of rich medium, a relatively large inoculum and the assessment of the effect during the log phase. Lag-phase studies (see above) indicate that the use of large inocula (with their attendant carry-over of stationary-phase supernatant) and rich medium can mask the stimulatory effect of growth factors. Lyte and Ernst (1992) found that catecholamines had a detectable effect on bacterial growth only if the starting concentration of cells was low. Table 1 summarizes cases in which either true pheromonal (autocrine and/or paracrine) effects on microbial growth have been observed or in which hormones usually considered characteristic of higher eukaryotes have been shown directly to stimulate microbial growth.

CYTOKINES OF UNICELLULAR EUKARYOTIC ORGANISMS AND CULTIVATED NUCLEATED CELLS

From the physiological point of view, the growth and development of cultures of bacteria and of nucleated unicellular organisms *in vitro* exhibit many similarities, and it could be expected that the mechanisms of control of cell multiplication – which might include the participation of cytokines – are also similar or even identical for the two kinds of cells. The need for factors synthesized by cells for their further multiplication has been noted for

Table 1. *Some autocrine/paracrine (pheromone) substances and some animal hormones which have been shown to stimulate the growth of bacteria and other unicellular organisms*

Organism	Role	Chemical nature	Reference
Autocrine factors			
Bacillus spp.	Reduction of lag phase, permit growth from small inoculum	Siderophores	(Lankford *et al.*, 1957, 1966)
Euplotes raikovi (ciliate)	Growth promotion, mating factor	Secreted polypeptides	(Vallesi *et al.*, 1995)
Micrococcus luteus	Resuscitation and stimulation of growth after dormancy	Secreted protein, 19 kDa	(Mukamolova *et al.*, 1998a)
Nitrosomonas europea	Reduction of lag phase	N-acyl homoserine lactone	(Batchelor *et al.*, 1997)
Paramecium tetraurelia (ciliate)	Stimulation of growth	Secreted protein 17 kDa	(Tanabe *et al.*, 1990)
Pasteurella (Francisella) tularensis	Permits growth from small inoculum	Not known, low MW	(Halman *et al.*, 1967; Halman & Mager, 1967)
Xanthomonas maltophila	Stimulation of growth	Chorionic gonadotropin-like ligand, membrane bound protein 48 kDa	(Carrell *et al.*, 1993; Grover *et al.*, 1995)
Mammalian hormones and cytokines			
Escherichia coli	Stimulation of growth	Interleukins	(Denis *et al.*, 1991a,b; Porat *et al.*, 1991)
Escherichia coli	Stimulation of growth	Granulocyte-macrophage colony-stimulating factor	(Denis *et al.*, 1991)
Escherichia coli and other Gram-negative bacteria	Stimulation of growth	Catecholamines	(Lyte, 1992; Lyte & Ernst, 1992)
Giardia lamblia	Stimulation of growth	Human insulin-like growth factor (IGF-II)	(Lujan *et al.*, 1994)
Mycobacterium avium	Stimulation of growth	Interleukin-6 epidermal growth factor	(Bermudez *et al.*, 1996; Denis, 1992)
Mycobacterium tuberculosis	Stimulation of growth	Transforming growth factor-beta-1 epidermal growth factor	(Bermudez *et al.*, 1996; Hirsch *et al.*, 1994)
Neurospora crassa	Stimulation of growth	Insulin	(McKenzie *et al.*, 1988)
Saccharomyces cerevisiae	Stimulation of growth	Insulin	(Berdicevsky & Mirsky, 1994)
Streptococcus (Enterococcus) faecalis	Stimulation of growth	Serotonin	(Strakhovskaya *et al.*, 1993)
Tetrahymena thermophila	Stimulation of growth	Lipids, alcohols, porphyrins, insulin, bovin serum albumin	(Christensen *et al.*, 1995, 1998; Wheatley *et al.*, 1993)
Tetrahymena pyriformis	Stimulation of growth	α2-macroglobulin	(Hosoya *et al.*, 1995)

unicellular organisms such as ciliates. If cells of *Tetrahymena thermophila* were inoculated into a poor medium at a concentration of < 750 cells per ml, they could not grow at all until either supernatant from growing culture or various 'inducers' including lipids, alcohols, insulin, haemin and porphyrins were added (Christensen *et al.*, 1995, 1998; Wheatley *et al.*, 1993). These authors pointed out the ability of *Tetrahymena* cells to produce 'insulin-related' material and speculated that insulin-like autocrine factors might be involved in the phenomenon (in the *Tetrahymena* model insulin is active at concentrations $< 10^{-15}$ M; Christensen *et al.*, 1995). At the same time it was found that the proliferation of *Tetrahymena periformis* was stimulated by $\alpha 2$-macroglobulin, a 120 kDa protein found in foetal calf serum, and specific antibodies to this protein detected a 180 kDa component in *Terahymena* extract (Hosoya *et al.*, 1995). The 'growth factor' for another ciliate, *Paramecium tetraurelia*, has been isolated from culture medium as a 17 kDa protein (Tanabe *et al.*, 1990), but remains regrettably unsequenced. It is interesting that crude samples of spent medium of *Terahymena* were active as growth factors for *Paramecium* cells (Tokusumi *et al.*, 1996), and the possible relationship between *Paramecium* and *Tetrahymena* cytokines remains to be established.

Cultivation of *S. cerevisiae* in a very poor medium from a small inoculum was not possible until glucose (0.001%) or the non-metabolizable 6-deoxy-glucose were added to the culture medium, and it was suggested that glucose here served as a signalling molecule for proliferation; some tetrapyrroles (which have a role in the regulation of cAMP levels) had a similar effect (Overgaard *et al.*, 1995).

The cytokine pheromones of the ciliate *Euplotes raikovi* seem to be the presently best characterized among unicellular species. These secreted pheromones are represented by several peptides (M_r about 4–5 kDa) which had traditionally been associated only with the organism's mating activity. However, it was found that they also possess growth-stimulating activity (Vallesi *et al.*, 1995). These cytokines are secreted in precursor forms, which undergo two proteolytic cleavages during exocytosis (Luporini *et al.*, 1994). Moreover, by an alternative splicing mechanism the cell secretes each pheromone in two forms: a membrane-bound isoform (M_r 14 kDa) and a soluble peptide. The first form has an extracellular (C-terminal) domain with an amino acid sequence identical to that of the soluble form (Luporini *et al.*, 1994). Unusually, this membrane-bound form serves as a receptor for soluble pheromone, and analysis of the crystal structure of the soluble forms revealed extensive helix–helix interactions between adjacent molecules which may mimic cytokine–receptor interaction (Weiss *et al.*, 1995), since the binding of effector to the receptor results in clustering of proteins (with subsequent activation of signal transduction pathways) and formation of tetrameric complexes on the cell surface (Vallesi *et al.*, 1995). In addition to a receptor function in the membrane, the membrane-bound form may play a role as a so-called juxtacrine growth factor by interacting with a receptor on the surface of a neighbouring cell. This type of ligand–receptor interaction has been proposed for vertebrate cytokines belonging to the epidermal growth factor (EGF) family and some other cytokines (colony-stimulating factor, tumour necrosis factor and others). It was suggested that 'juxtacrine' stimulation may play a role in the communication between closely adjacent cells, for example, for more precise targeting of growth factor in developing tissue when freely diffusing factors are less appropriate (Massague & Pandiella, 1993).

These similarities between the growth factors from ciliates and vertebrates allow interesting speculations regarding the evolutionary development of mechanisms of cell proliferation control (Luporini *et al.*, 1994), based on the fact that human interleukin IL-2 is a very active competitor for the binding of the Er-1 ciliate cytokine to its receptor. While these two signalling molecules are quite different in their overall composition, nevertheless one (short) conserved segment does reveal significant sequence similarity (Luporini *et al.*, 1994).

Similarities also occur between bacterial cells and cultivated cells of higher organisms. Thus, cultivation of pre-starved fibroblasts resulted in a low cell concentration at the end of cultivation in normal medium, irrespective of seeding densities (Pignolo *et al.*, 1994). Lens cells held in cultures with a low initial concentration of the cells die off rapidly, while conditioned medium from high-density cultures promoted their survival (Ishizaki *et al.*, 1993). It has been proposed that there is a universal mechanism for cell death, as in these cases, based on the idea that there may have been an absence or insufficient secretion of one or more factors which prevent normally growing cells from initiating apoptotic death (Raff, 1992). A similar mechanism probably works in *Tetrahymena* (Christensen *et al.*, 1995).

Thus autocrine polypeptide substances can help effect cellular growth by at least two general mechanisms: by stimulating growth *per se* and/or by inhibiting processes leading to stasis or to death.

ISOLATION AND PURIFICATION OF RESUSCITATION PROMOTING FACTOR (RPF) FROM *M. luteus*; THE FIRST BACTERIAL CYTOKINE?

Growing *M. luteus* cells secrete a resuscitation-promoting factor (Rpf) which (i) promotes the resuscitation of dormant cells and (ii) reduces the apparent lag phase of cultures of the same organism when inoculated at low density (see above and Kaprelyants *et al.*, 1993, 1994, 1995; Kaprelyants & Kell, 1992, 1993, 1996; Mukamolova *et al.*, 1995, 1998a,b; Votyakova *et al.*, 1994). Rpf was heat-labile, non-dialysable and trypsin-sensitive and it was purified to homogeneity from culture supernatants (for details see Mukamolova *et al.*, 1998a). The resuscitation and apparent lag phase-reducing activity corresponded to a protein with an apparent MW of 16–19 kDa as estimated by electrophoresis and gel filtration. Picomolar concentrations of Rpf increased the number of culturable *M. luteus* cells from dormant populations by several orders of magnitude and they also stimulated the growth of *M. luteus* in batch culture in a lactate minimal medium. In view of the low concentrations necessary for activity, a trivial nutritional role for (proteolytic degradation products of) Rpf was discounted. Moreover, Rpf was also active in the presence of yeast extract (0.05%) and in rich medium (Broth E) (unpublished data). Rpf therefore has the properties of a bacterial cytokine (more accurately Rpf is a signal which activates processes which eventually result in, and are necessary for, cell multiplication. The final proof that Rpf is really a cytokine awaits the elucidation of the relevant receptor and the pathways of signal transduction).

The gene encoding Rpf was isolated from *M. luteus* and sequenced (Mukamolova *et al.*, 1998*a*). It encodes a 220 amino acid product with a 38-residue signal sequence (Fig. 3). The predicted size of the secreted form of the gene product is 19 148 Da, and its predicted amino acid sequence agrees with protein microsequence data obtained from RPF purified from culture supernatants. To confirm cytokine activity, a histidine-tagged version of the secreted form of Rpf (i.e. lacking the signal sequence) was expressed in *E. coli* and purified to homogeneity by Ni^{2+} chelation chromatography. The recombinant Rpf reduced the apparent lag phase of viable cells of *M. luteus* at picomolar concentrations (Fig. 4) just as did Rpf isolated from *M. luteus* culture supernatants (Mukamolova *et al.*, 1998*a*).

Genes similar to *rpf* appear to be widely distributed among the high genomic G + C cohort of Gram-positive bacteria. Database searching has revealed that similar genes are present in *Mycobacterium tuberculosis* and *Mycobacterium leprae* (Fig. 3). Southern hybridization and/or PCR experiments with *rpf*-specific primers have revealed that similar genes are also detectable in several other organisms including *Mycobacterium smegmatis*, *Mycobacterium bovis* (BCG), *Corynebacterium glutamicum, Streptomyces coelicolor* and *Streptomyces rimosus*. On the other hand, *rpf*-like genes are not present in any of the other organisms whose genomes have been sequenced to date (*Aquifex aeolicus, Archaeoglobus fulvidus, Bacillus subtilis, Borrelia burgdorferi, Escherichia coli, Haemophilus influenzae, Helicobacter pylori, Methanococcus jannaschii, Methanococcus thermoautotrophicum, Mycoplasma genitalium, Mycoplasma pneumoniae, Saccharomyces cerevisiae, Staphylococcus aureus, Streptococcus pneumoniae, Synechocystis* PCC6803).

The N-terminal region of the secreted Rpf of *M. luteus* is substantially similar (42% of residues are identical or are conservatively substituted over a 69-residue segment) to the predicted products of several genes in *Mycobacterium tuberculosis* and *Mycobacterium leprae* (Fig. 3). The former organism encodes five Rpf-like gene products, according to the completed genome sequence (http://www.sanger.ac.uk/Projects/M_tuberculosis/). If it is assumed that these structurally similar mycobacterial proteins perform a

```
g2052146    mlrlvvgalllvlafaggyavaacktvtltvdgtamrvttmksrvidive    50
g2052146    engfsvddrddlypaagvqvhdadtivlrrsrplqisldghdakqvwtta   100
g2052146    stvdealaqlamtdtapaaasrasrvplsgmalpvvsaktvqlndgglvr   150

g2052146    tvhlpapnvagllsaagvpllqsdhvvpaatapivegmqiqvtrnrikkv   200
g2791490    ----------------mpvgwlwrartakgttlknarttliaaaiagt      32

g2440090    ------------------------------------------msesyrkl     8
e1254009    ------------------------------------------msgrhrkpt     9
g2052146    terlplppnarrvedpemnmsrevvedpgvpgtqdvtfavaevngvetgr   250
MlutZ96935  ---------------------------------------mtlfttsat       9
MSGB38COS   ----------------------------------------mpgemldvrklc   12
g2225976    -----------mhplpadhgrsrcnrhpisplsligniSatsgdmssmt     38
g1655671    --------------------------mtpgllttagagrprdrca         19
g2791490    lvttspagianaddaGldpnaaagpdavgfdpnlppapdaapvdtppape    82

g2440090    ttssiivakitftgamldgsialagqaspatdsEWDQVARCESGGNWSIN    58
e1254009    tsnvsvakiaftgavlggggiamaaqataatdgEWDQVARCESGGNWSIN    59
g2052146    lpvanvvvtpaheavvrvgtkpgtevppvidgsIWDAIAGCEAGGNWAIN   300
MlutZ96935  rsrratasivagmtlagaaavgfsapaqaatvdTWDRLAECESNGTWDIN    59
MSGB38COS   klfvksavvsgivtasmalststgmanavprePNWDAVAQCESGRNWRAN    62
g2225976    riakpliksamaaglvtasmslstavahagpsPNWDAVAQCESGGNWAAN    88
g1655671    rivctvfietavvatmfvallglstisskaddIDWDAIAQCESGGNWAAN    69
g2791490    dagfdpnlppplapdflsppaeeappvpvaysVNWDAIAQCESGGNWSIN   132

g2440090    TGNGYLGGLQFSQGTWASHGGGEYAPSAQLATREQQIAVAERVLATQGSG   108
e1254009    TGNGYLGGLQFTQSTWAAHGGGEFAPSAQLASREQQIAVGERVLATQGRG   109
g2052146    TGNGYYGGVQFDQGTWEANGGLRYAPRADLATREEQIAVAEVTRLRQGWG   350
MlutZ96935  TGNGFYGGVQFTLSSWQAVGGEG---YPHQASKAEQIKRAEILQDLQGWG   106
MSGB38COS   TGNGFYGGLQFKPTIWARYGGVG---NPAGASREQQITVANRVLADQGLD   109
g2225976    TGNGKYGGLQFKPATWAAFGGVG---NPAAASREQQIAVANRVLAEQGLD   135
g1655671    TGNGLYGGLQISQATWDSNGGVG---SPAAASPQQQIEVADNIMKTQGPG   116
g2791490    TGNGYYGGLRFTAGTWRANGGSG---SAANASREEQIRVAENVLRSQGIR   179

g2440090    AWPACGHGLSGPSLQEVLPAG--MGAPw----INGAPAPLAPPPPAEPAP   152
e1254009    AWPVCGRGLSNATPREVLPASaaMDAPldaaaVNGEPAPLA-PPPADPAP   158
g2052146    AWPVCAaragar--------------------------------------   362
MlutZ96935  AWPLCSQKLgltqadadagdvdateaapvavertatvqrqsaadeaaaeq   156
MSGB38COS   AWPKCGAASDLPITLWSHPAQGVKQIINDIIqmgdttlaaialngl----   155
g2225976    AWPTCGAASGLPIALWSKPAQGIKQIINEIIwaqiqasipr---------   176
g1655671    AWPKCSscsqgdaplgslthiltflaaetggcsgsrdd------------   154
g2791490    AWPVCGrrg-----------------------------------------   188

g2440090    pqppadnf----------------------PPTPGDVPSPLarp-----   174
e1254009    pvelaandlpaplgeplpaapadpappadlaPPAPADVAPPVelavndlp   208
MlutZ96935  aaaaeqavvaeaetivvksgdslwtlaneyeveggwtalyeankgavsda   206

e1254009    aplgeplpaapadpappadlappapadlappapadlappapadlappvel   258
MlutZ96935  aviyvgqelvlpqa-----------------------------------   220

e1254009    avndlpaplgeplpaapaelappadlapasadlappapadlappapaela   308
e1254009    ppapadlappaavneqtapgdqpatapggpvglatdlelpepdpqpadap   358
e1254009    ppgdvteapaetpqvsniaytkklwqairaqdvcgndaldslaqpyvig-   407
```

Fig. 3. *M. tuberculosis* and *M. leprae* contain genes whose products are similar to Rpf. Multiple sequence alignment of *M. luteus* Rpf (Z96935) with predicted gene products from *M. tuberculosis* g2052146 (MTCI237.26), g2791490 (MTV008.06c), g1655671 (MTCY253.32), g2225976 (MTCY180.34), e1254009 (MTV043.60c), and *M. leprae* g2440090 (MLCB57.05c), MSGB38COS (L01095, nt 12292–12759). Conserved blocks are in upper case, residues conserved or conservatively substituted in five or more sequences are in bold and predicted signal sequences are underlined.

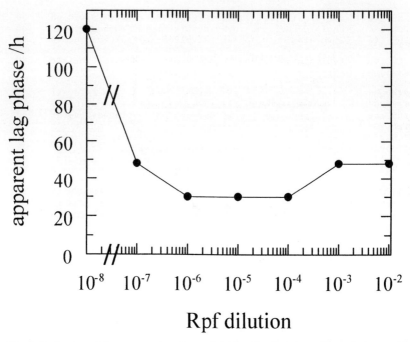

Fig. 4. Reduction of the apparent lag phase of viable cells of *M. luteus* by purifed recombinant Rpf. For experimental details see Mukamolova *et al.* (1998a). A dilution factor of 10^{0} corresponds to 250 µg Rpf ml^{-1}.

similar biological function to that of the *M. luteus* Rpf (evidence to support this assumption is presented below), the following hypothesis can be proposed. One of the Rpf-like gene products (g2052146) is much larger than the others and is predicted to have a membrane anchor at its N-terminus. The Rpf-like domain lies at the extreme C-terminus of the protein, and it is tempting to speculate that this gene product may traverse the bacterial cell wall to present its Rpf-like domain at the cell surface functioning as a signalling molecule over the short distance that separates adherent cells (juxtacrine function). The formation of cell aggregates during growth of *M. tuberculosis* is well known. All the other known Rpf-like gene products (including four in *M. tuberculosis*) are secreted (Fig. 3). There is also an evident and intriguing consonance between the Rpf-like gene products of *M. tuberculosis* and the *Euplotes raikovi* cytokine, where the membrane-bound form serves as a receptor for the soluble pheromone.

Given the significant similarity between *M. luteus* Rpf and the products of other genes found in several Gram-positive bacteria, purified Rpf was tested to see whether it can also be employed as a growth factor for mycobacteria. It is active with both rapidly growing and slowly growing species. Growth of *Mycobacterium bovis* (BCG), *M. tuberculosis*, *Mycobacterium avium* and

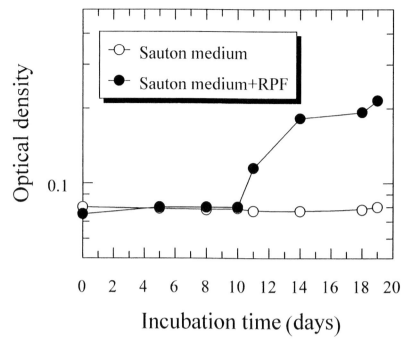

Fig. 5. Effect of *M. luteus* Rpf on the growth of *Mycobacterium bovis* in batch culture. *M. bovis* was grown in Sauton medium to which a 500-fold dilution of RPF was either added (closed circles) or not (open circles). The inoculum was *ca.* 1 × 10⁵ cells ml⁻¹, and the OD shown is the average of ten tubes.

Mycobacterium kansasii in Sauton medium was stimulated by Rpf. Similar results were obtained with *M. smegmatis* growing in either a minimal or a rich medium (Mukamolova *et al.*, 1998*a*). Growth of *M. bovis* (BCG) in Sauton medium was also strongly stimulated by Rpf; growth occurred after 14 days, whereas the control lacking Rpf showed no visible growth after 20 days (Fig. 5). Additional experiments showing that Rpf stimulates the growth of *M. tuberculosis*, *M. avium*, *M. bovis* (BCG) and *M. kansasii* are summarized in Table 2.

These results may have important implications for (i) the detection of mycobacteria in clinical samples and (ii) controlling the progression of mycobacterial infections.

WHEN ONE VIABLE BACTERIAL CELL CANNOT MULTIPLY IN WHAT SEEM NORMALLY APPROPRIATE CONDITIONS; THE DEPENDENCE OF GROWTH ON RPF

In suggesting a crucial role of secreted bacterial cytokines for cell multi-plication it may be expected that there are conditions in which cell growth

Table 2. *Purified* M. luteus *Rpf stimulates growth of mycobacteria*

Organism	Bacterial growth[a]	
	Rpf omitted	Rpf added
M. tuberculosis H37Ra	1.3 ± 1.9 (5)	110 ± 32 (5)
M. tuberculosis H37Rv	1.5 ± 2 (4)	45 ± 28 (4)
M. avium	0 (3)	>300 (3)
M. bovis (BCG)	0 (5)	54 ± 38 (5)
[b]M. smegmatis	0 (8)	225 ± 44 (8)
Mycobacterium kansasii	2.5 ± 2.5 (3)	90 ± 77 (3)

[a] Growth was estimated microscopically (magnification times 600) after 14 days of incubation; *c*. 50 µl of each culture was fixed, stained using Ziehl–Neelsen reagent and counted. Values in the body of the table are average numbers of cells in a microscope field (10–20 fields counted) $\pm$ standard deviation with the number of independent determinations in parentheses. Rpf (after elution from the Mono Q column and dialysis) was used at a concentration of *c*. 40 pMol l^{-1}; activity was lost after either trypsin treatment, heating (autoclaving) or filtration through a 12 kDa cutoff membrane.
[b] Washed cells of *M. smegmatis* were used for these experiments (Mukamolova *et al.*, 1998*a*).

would be arrested (or significantly retarded) without externally added cytokine. Such conditions may be expected to include suboptimal media, a small inoculum size, cellular depletion of endogeneous cytokine, or any combination of these. Similarly, any treatment resulting in the degradation or decreased production of Rpf by cells should influence their growth pattern. Whereas unwashed *M. luteus* cells proliferate normally in liquid lactate minimal medium (LMM) (Kaprelyants & Kell, 1992), the proliferation of a small inoculum of washed cells in this medium appeared to be absolutely dependent on added Rpf over the 160 h duration of the experiment (Mukamolova *et al.*, 1998*a*).

Additional information on the necessity of Rpf for bacterial growth has been obtained from growth experiments employing conditions even less favourable than LMM. To this end, succinate was used instead of lactate in minimal medium with the results shown in Table 3 and Fig. 6. Succinate minimal medium (SMM) does not normally support macroscopically observable *M. luteus* growth when the inoculum is less than 10^5 cells ml^{-1} (cells underwent only a few divisions after which growth stopped; this might be due to carry-over of some Rpf with the unwashed inoculum). However, addition of purified Rpf (4 ng ml^{-1}) resulted in sufficient cell growth to form a turbid suspension (Fig. 6). It should be stressed that washed cells have not lost the ability to make colonies on LMM agar, and normal cells can also grow on agar prepared with SMM without added Rpf. This may be due to the fact that neighbouring cells are in intimate contact during colony development, which should facilitate cell–cell communication by locally accumulated Rpf, as well as juxtacrine signalling by cell surface-associated proteins.

Table 3. *Viable count of* M. luteus *estimated by two methods*

Treatment, medium used	cfu (cells ml^{-1})	MPN (cells ml^{-1})	MPN in presence of Rpf (cells ml^{-1})
Washed cells, LMM (5 times)	8×10^8 (BrothE, LMM)	1.5×10^{6b}	5×10^{8a}
Untreated cells, SMM	3.7×10^9 (BrothE)	4×10^3–7×10^3	5.5×10^{8a}
Stationary phase cells (100 h), LMM	10^8 (BrothE, LMM)	10^5	7×10^8

For experimental details, see text.
[a] Recombinant Rpf (3.3 ng ml^{-1}).
[b] Poor growth.

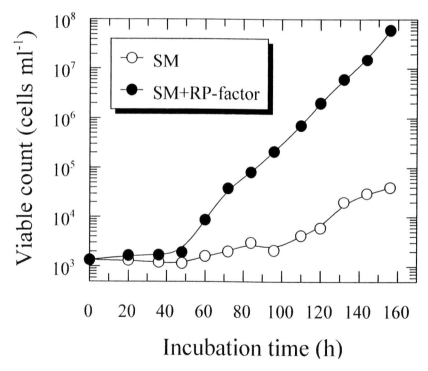

Fig. 6. Effect of *M. luteus* Rpf on the growth of *M. luteus* in batch culture in succinate minimal medium. *M. luteus* was grown in broth E until stationary phase and inoculated into succinate medium (SM) (Mukamolova *et al.*, 1998*a*). Flasks (20 ml) with SM were inoculated with approximately 1000 cells ml^{-1} and growth was monitored by sampling aliquots and plating them out on agar plates supplemented with Lab M nutrient broth and incubated at 30 °C. In some cases (closed circles), Rpf was added to a final concentration of 4 ng ml^{-1}.

M. luteus cells taken from an extended stationary phase grow very poorly in LMM. The culturability of such cells was checked by two methods: (i) plating out on agar plates and (ii) MPN assay. Table 3 shows significant differences between the MPN count and cfu for cells incubated for 100 h in stationary phase (the latter was almost identical for either LMM or rich medium plates (BrothE)). The underestimation of viable cells by MPN indicates that at least 1000 viable cells (as judged by cfu) must be present per tube if they are to produce visible growth. The addition of Rpf to the MPN dilutions led to almost identical numbers of viable cells as estimated by the two methods, demonstrating the ability of one single stationary phase cell to grow in a tube provided Rpf was present (Table 3). A similar effect was obtained when MPN assays were performed either in SMM using untreated cells or in LMM using cells that had been washed several times (Table 3). In both cases the underestimation of viable cells by the MPN method (in comparison with viable counts judged on plates) demonstrated that at least 10^5 'succinate' cells and about 10^2 'washed' cells are required per tube for visible growth. Again, the addition of Rpf resulted in very similar viable counts as judged by the two methods. Similar effects have been observed with other organisms. For example, cultures of *B. subtilis* in poor medium inoculated with fewer than 100 cells per flask failed to initiate growth unless 'shizokinen' were added (Lankford *et al.*, 1966). Cells of *Pasteurella* (now *Francisella*) *tularensis* did not grow to a high optical density from an inoculum of $< 10^5$ cells per flask, unless supernatant taken from a growing culture of the same organism was added (Halman *et al.*, 1967).

These experiments clearly show (i) the dependence of bacterial growth on secreted growth factors or cytokines and (ii) the benefit of studying 'unfavourable' conditions to make this dependence most visible (Christensen *et al.*, 1998; Wheatley *et al.*, 1993). How comparable these phenomena are in other organisms remains to be established, but it is clear that the basic 'one cell-one culture' principle of microbiology may not apply in some circumstances.

The mechanism(s) responsible for 'non-culturability' (in the operational sense) of bacterial cells depleted of exogenous Rpf in the above experiments is/are not yet clear. For such cells to commence multiplication, Rpf must accumulate to a sufficient concentration. The time required will depend on both the initial cell density and the metabolic activity of the cells. A metabolically active cell may have a finite 'lifetime' during which it can survive without division. If held in stationary phase for a period exceeding this lifetime, cell death ensues by a mechanism(s) which remains to be elucidated.

More generally, and as is well established in eukaryotic cells (Raff, 1992), programmed cell death might play a role in bacterial cell autodegradation. To date, 'bacterial apoptosis' (Hochman, 1997) has been described as a mechanism for maintaining plasmids in bacteria, e.g. the 'addiction module' in *E. coli* extrachromosomal elements (Yarmolinsky, 1995), a similar 'module' consisting of two genes *mazE* and *mazF* was recently found in the chromosome of *E. coli* (Aizenman *et al.*, 1996). MazF is a stable protein which is toxic for the cell, while MazE is a less stable protein which protects the cell from MazF. Under starvation conditions, the level of ppGpp increases and inhibits the expression of both genes, which results in a toxic effect of MazF and cell death (Aizenman *et al.*, 1996). This mechanism works under extreme conditions of deep starvation. However, it might also be responsible for the initiation of cell lysis under conditions insufficient for the secretion of growth factor(s), as with the initiation of apoptosis in eukaryotic cells (Raff, 1992).

POSSIBLE SIGNIFICANCE FOR BACTERIAL PATHOGENICITY – NOVEL TARGETS, NEW VACCINES AND DRUGS

The self-promoting mode of bacterial cell growth can have significant implications for medicine and epidemiology. This follows from several phenomena, the most evident of which is an apparent 'non-culturability' of some pathogenic bacteria. There are several unresolved public health problems potentially involving transition to and from a 'non-culturable' state of an infective agent. Principally, these concern aspects of the epidemiology and natural history of infective diseases, which cannot be reconciled with the sample pattern from which the known causal organisms can be isolated. Foremost amongst the epidemiological mysteries are cholera and campylobacteriosis, where the failure to isolate *Vibrio cholerae* and *Campylobacter jejuni* from clearly implicated sources or reservoirs of infection might be accounted for on the basis of their being present in a reversibly non-culturable state. For both these organisms, environmental investigations have provided evidence for the presence of 'non-culturable' cells in appropriate samples (Brayton *et al.*, 1987; Pearson *et al.*, 1993) while *in vitro* studies have demonstrated their capacity to form metabolically active cells, which could not be grown immediately (Rollins & Colwell, 1986; Xu *et al.*, 1982). The list of organisms for which similar phenomena have been claimed (albeit less extensively) is substantial (Kell *et al.*, 1998; Oliver, 1993). From the above discussion it is clear that the involvement of bacterial growth factors in recovery from starvation or resuscitation from dormancy should significantly change the currently accepted methodology for monitoring the environment for biological hazards, and there is evidently a necessity to formulate new protocols for the isolation of bacteria from natural samples, including cultivation of purportedly 'non-culturable' forms in liquid media supplemented by appropriate growth factors.

Further medically significant areas where transition to and from putative 'non-culturable' states have potential relevance include bacterial infections, which have a clinically dormant or latent phase and the effects of antibiotics

(Domingue & Woody, 1997). Tuberculosis (Gangadharam, 1995; Parrish *et al.*, 1998; Wayne, 1994; Young & Duncan, 1995) and melioidosis (Dance, 1991) provide examples of the former. This has especially been suggested for *M. tuberculosis*, which are capable of adapting to dormancy in the tissues of humans and experimental animals, leading to latency of the disease; this is strongly supported by the transition of viable *M. tuberculosis* cells to nonreplicating cells under microaerophilic conditions (Wayne, 1994). Domingue and colleagues demonstrated the presence of bacterial 16S rRNA genes characteristic of Gram-negative bacteria in biopsies of patients with interstitial cystitis, although routine cultures of bacteria from patients' urine were negative. It was suggested that a persistence of such bacteria in a dormant form is involved in the aetiology of this disease (Domingue, 1995; Domingue *et al.*, 1995). In this connection it is worth mentioning the debates on the role in the bacterial persistence of so-called coccoid forms of some pathogenic bacteria. This forms have been described for *Campylobacter jejuni* (Beumer *et al.*, 1992), *Helicobacter pylori* (Cellini *et al.*, 1994), *Mycobacterium tuberculosis* (Khomenko & Golyshevskaya, 1984) and unknown bacteria in patients with cystitis (Domingue *et al.*, 1995) as a result of either prolonged starvation *in vitro*, or persistence in patients *in vivo*. These forms represent a 'nonculturable' state, and it has been suggested they might be infective agents representing dormant forms (Domingue & Woody, 1997). Whilst the experimental evidence for their dormancy and resuscitation to viable bacteria is largely not yet to hand (Kell *et al.*, 1998), reversibly 'nonculturable' or dormant cells of pathogens could provide a straightforward microbiological explanation for latent bacterial infections and for the lack of a clinical response to antimicrobial agents shown to be effective against growing cells *in vitro*. It is very likely that many more diseases will have a microbial aetiology than is currently recognized (Davey & Kell, 1996).

Again, host cytokines can play a key role in ending latency and beginning the development of an active state of such diseases, especially when the initial concentration of infecting bacteria is likely to be very low. For example, although the many and often conflicting roles of the various cytokines here remain to be elucidated (Flynn & Bloom, 1996; Henderson *et al.*, 1996; Rook & Hernandez-Pando, 1996*a,b*; Toossi, 1996), transforming growth factor (TGF β-1) accelerates the growth of *M. tuberculosis* in monocytes and may be important in the pathogenesis of tuberculosis (Hirsch *et al.*, 1994). Incubation of pathogenic mycobacteria (*M. tuberculosis* and *M. avium*) in the presence of epidermal growth factor resulted in the acceleration of bacterial growth within macrophages which might have a role in bacterial multiplication in both granulomatous and necrotic tissues (Bermudez *et al.*, 1996). The decrease of insulin concentration in the blood of diabetics results in a significant proliferation of *Ps. pseudomallei* – the causative agent of melioidosis (Woods *et al.*, 1993). Similarly, transferrin is essential for

Francisella tularensis growth and survival in the acidic vacuole of murine macrophages (Fortier *et al.*, 1995).

Specific signal(s), derived from the host or the invader, may help to resuscitate such dormant forms to active states and promote their growth, just as was found under laboratory conditions (Wai *et al.*, 1996; Whitesides & Oliver, 1997). The resuscitation of non-culturable forms of a pathogen, *Legionella pneumophilia*, in cells of an amoeba (*Acanthamoboe castellanii*) which serves as a host for this bacterium was recently reported (Steinert *et al.*, 1997), although co-cultivation of 'non-culturable' *L. pneumophilia* with another ciliate, *Tetrhymena pyriformis*, failed to recover viable cells (Yamamoto *et al.*, 1996). Possibly intrinsic *bacterial* cytokines can also be involved in the process of transition to and from dormant (latent) state as well as in cell growth and thus the development of infection.

Given the well-known problems of emerging antibiotic resistance (Bloom & Murray, 1992; Davies, 1994; Duncan, 1998; Murray, 1991; WHO, 1997), the recognition that growth factors can control the development and multiplication bacteria opens the possibility of finding new targets for antibacterial agents, which should not be toxic for animals and may be chosen (or otherwise) to be reasonably selective among species. The protein sequence of *M. luteus* Rpf revealed strong similarities only within a fairly narrow family of Gram-positive bacteria, and not with other bacteria and higher organisms of known genome sequence. Equally, appropriate agonists may be of value in decreasing the time of treatment in medically important organisms displaying latency.

Another possibility to exploit bacterial cytokines is to make 'attenuated' bacterial strains suitable for vaccination. Knowing the nature of the growth factor(s) for the particular bacteria it might be possible to make a strain lacking this by constructing a defined knockout mutant either directly in respect to the growth factor if it is a protein, or indirectly by knocking out a step in the biosynthetic pathway. Such strains could be cultivated easily on suitable media in the presence of exogenous growth factor *in vitro* while being unable to grow *in vivo* because the exogenous growth factor will be absent. Finally, antibodies or vaccines targeted against bacterial virulence (Balaban *et al.*, 1998) or growth factors themselves may prove useful therapeutic agents.

'AS YET UNCULTURED' BACTERIA

While there are well-known organisms, which have not yet been cultured axenically, e.g. *M. leprae, Tropheryma whippelii* – the causual agent of Whipple's disease (Relman *et al.*, 1992) – or the agent of human ehrlichiosis (Fredricks & Relman, 1996), as well as many organisms present in environmental samples (Amann *et al.*, 1995), which are clearly the progeny of viable cells, it is now known that the bacterial species which have been cultivated in laboratory conditions may actually represent only a very small fraction (0.01–0.1%) of those occupying our biosphere. These 'as yet uncultured' cells are operationally non-culturable but in many cases they can be recognized by molecular and cytological methods such as rRNA analysis and *in situ* hybridization to be previously undetected and unclassified (Amann *et al.*, 1995; Fredricks & Relman, 1996; Head *et al.*, 1998). The reasons for the purported 'unculturability' of these bacteria are not understood, but numerous attempts at cultivation by using different media, oxygen tensions, etc. leads to the conclusion that the formulation of standard nutritional media alone may be not enough to cultivate them in the laboratory. But it should be stressed that, in many cases, they do not form phylogenetically distinct branches from known, cultured organisms (McVeigh *et al.*, 1996).

It is particularly worth mentioning that the media researchers tend to use have evolved in concert with their recognition of what organisms require in order to grow, such that certain media will inevitably cause slow-growers on such media to be outgrown by fast-growers. One solution to this problem is the exact equivalent of the MPN method, which allows one to culture such organisms by dilution to extinction, as particularly well exemplified by the work of Schut and colleagues (Button *et al.*, 1993; Schut *et al.*, 1993, 1997*a,b*).

In addition, it is entirely plausible, and even likely (Kaprelyants & Kell, 1996) that these 'uncultured microorganisms' actually need some growth factors for their cultivation *in vitro*. The example of the cultivation of *M. luteus* on succinate minimal medium is a good simulation of this situation, when apparently 'uncultured' bacteria (on SMM) became culturable in the presence of a cytokine (Rpf) (Fig. 6). Also, washed *M. smegmatis*, which revealed an absolute requirement for an autocrine growth factor to be cultivated in appropriate medium (Table 2) could be considered as a model for a more 'natural' population of starved bacteria in soil, which might need a cytokine to start growth (and clearly such proteinaceous factors may be degraded over time in nature).

MECHANISMS OF BACTERIAL CELL MULTIPLICATION AND CYTOKINES

While the mechanisms responsible for division of bacterial cells are not fully understood, recent findings on the structure, regulation and functioning of the cell division machinery give some insight into the problem of cytokinesis. Genetic studies have revealed some of the central events of cell division, such as formation of the division septum and initiation of division, which are controlled by several gene products. In particular, the gene cluster *ftsQAZ* is most important for cell septation (Vicente & Errington, 1996). FtsZ, an abundant bacterial protein from this gene cluster, plays a crucial role in bacterial division, and is a homologue of the eukaryotic tubulin, which is able to undergo GTP-driven polymerization to produce long thin sheets of protofilaments. This polymer is evidently a part of the system (motor) which drives septa to the appropriate place of the cell (Erickson, 1997). The expression of the *FtsQAZ* cluster, in turn, is regulated by SdiA protein which stimulates transcription of *Fts* genes via promoter P_2. *SdiA* has strong homology with *luxR* – the key gene in controlling cell-density regulation luminescence of some *Vibrio* spp. via the secretion of low molecular weight inducers belonging to the N-acyl homoserine lactone (HSL) family (Garcia-Lara *et al.*, 1996; Sitnikov *et al.*, 1996). It is obviously possible that SdiA might function similarly to LuxR and (therefore) cell division in bacteria may also be controlled by cell density via a specific inducer.

Thus, Sitnikov *et al.* (1996) reported that conditioned medium of growing *E. coli* stimulates transcription from the SdiA-dependent promoter P_2 during cell growth in early log phase. The nature of the inducer(s) in conditioned medium was not clarified; however, different known inducers from the HSL family (but not non-acylated HSL) also have a positive effect on the same transcription process. The authors interpreted these results in favour of a 'quorum sensing' mechanism for the regulation of *E. coli* growth (Sitnikov *et al.*, 1996). However, in similar experiments, Garcia-Lara *et al.* (1996) did not find stimulation of the same promoter (SdiA-P_2) in early log phase by conditioned medium. In contrast, they reported a down-regulation of this promoter in mid- to late-log phase. The extracellular factor responsible for this regulation was soluble, and resistant to heating, but neither HSL nor OHSL could substitute for the factor from conditioned medium. The evident discrepancy in these results could be based on the use of different media, supporting lower growth rates in the case of the studies reported by Sitnikov *et al.* (1996). In all events, the above results serve to suggest one possible link between the extracellular control of cell multiplication and the molecular mechanisms responsible for the cell division machinery.

CONCLUDING REMARKS

Bacterial cytokines: what for?

Why and under what circumstances should the multiplication of bacterial cells in culture need to be controlled by secreted pheromones or cytokines? While the general role of growth factors and cytokines in controlling the development of differentiated eukaryotic systems is clear, the role of bacterial cytokines is less evident, although the advantages of differentiation in prokaryotic culture are relatively easy to rationalize (Davey & Kell, 1996; Koch, 1987, 1993). In addition, at least teleologically, the advantage of the social behaviour of cells during culture growth could follow from the context of a general strategy of catabolizing substrate (Kell, 1987; Westerhoff *et al.*, 1983) and of increasing biomass as quickly as possible, and in the rapid response to changing nutritional and other circumstances (Kell *et al.*, 1995). In particular, the role of the cytokine may be in 'monitoring' the relevant ecological niche for the presence of substrates appropriate for culture development. In the case of 'rich' environments, 'starter' cells will pass the signal in the form of a cytokine, resulting in an autocatalytic enchancement of multiplication. Hence, bacterial cytokines may play a regulatory role at the level of the population by controlling the balance between multiplication rate and the availability of substrates.

This suggested role of bacterial cytokines may be especially important for non-motile bacteria (e.g. 'branching' bacteria like mycobacteria and strepto-mycetes) which, in contrast to motile bacteria, are deprived of an efficient machinery (taxis) for searching for locations appropriate for multiplication. Starting with random growth, cells in a 'successful' branch (surrounded by enough nutrients) may promote growth in this direction by the secretion of cytokines. Otherwise, diffusing growth factors may specifically target to appropriate substrates, e.g. the surface of an infected host cell, providing bacterial cell multiplication in close vicinity to the host and increasing the probability of invasion. Interestingly, the *M. luteus* Rpf C-terminal domain (and a putative Rpf of *M. tuberculosis*) contains regions with significant similarity to p60, a *Listeria monocytogenes* protein required for this micro-organism's adherence to and invasion of mouse fibroblasts (Bubert *et al.*, 1992). More experimental evidence should be accumulated to verify these hypotheses.

It is also recognized that there may be a selective advantage for an organism to engage in full growth of all cells only if conditions are favourable for a certain period of time. During this 'probing time' a very few cells – 'sentinels' (Postgate, 1995) – might be able to grow, but their growth will provide a growth stimulus in the form of Rpf for the rest of the population. If for any reason (the presence of antibiotics or a change to harmful conditions) growth of the sentinels is stopped, the bacterial cytokines will not be produced to any extent and the cells which had remained in the resting state will stay protected as a result of the stress resistance, which typically accompanies the possession of a stationary phase physiology (Hengge-Aronis, 1996).

Biological unity and evolution

It seems obvious that many or most phenomena thought characteristic of higher eukaryotes are likely to have evolved from older clades, including ancestral prokaryotes and/or archaea (Forterre, 1997*b*), such that phenomena observable in one group may be expected to have recognizable counterparts in others (and this would be expected to be confirmed in the post-genomic era). Thus, it can now be seen how the long-standing recognition of the importance of pheromones in effecting communication between higher organisms (Eisner & Meinwald, 1995) is now being extended to prokaryotes, see Stephens (1986) and many other contributors to this volume.

Regarding the phenomena of growth, cell cycle progression, stasis and dormancy, it is noted that even tumours can enter a state of dormancy, which can be reversed by immunodepression (Wheelock *et al.*, 1981). Possibly the immune system normally keeps the growth factors required by the tumour at a sufficiently low level, and there is also evidence that the addition of appropriate antibodies will induce a long-lived dormant tumour state (Racila *et al.*, 1995; Yefenof *et al.*, 1993), although it appears that here the antibody reagents themselves act as signal transduction agonists.

The authors have therefore sought to stress in this review that, especially in view of their own discovery of bacterial cytokines, and the recent findings that polypeptides with potent activity in promoting cell division may be found in higher plants (Matsubayashi & Sakagami, 1996; Matsubayashi *et al.*, 1997; van de Sande *et al.*, 1996), in multicellular invertebrates (Ottaviani *et al.*, 1996) and even in unicellular eukaryotes such as ciliates (Christensen *et al.*, 1998; Luporini *et al.*, 1995), a non-eukaryotic origin for such activities is most likely. The roles of polypeptide hormones, derived form larger pre-cursors by proteolytic activity, exhibit significant similarities between plants and animals (Bergey *et al.*, 1996; Schaller & Ryan, 1996), and this type of signal processing may thus have occurred rather early in evolution.

Bacterial culture as a social phenomenon

Thus, it is now clear that cells in bacterial cultures (as a socially organized system) are not independent but are talking to each other using specific chemical messages for many processes (Kell *et al.*, 1995; Lenard, 1992), which clearly include multiplication. From this point of view a bacterial culture resembles a tissue cell culture, and a bacterial colony (Shapiro, 1995), or even the entire microbial world (Mathieu & Sonea, 1995, 1996), may be considered as an organism composed of physiologically distinct tissues.

However our knowledge of the control of bacterial cell growth by secreted cytokines is very limited, due to both conceptual/terminological problems (Kell *et al.*, 1998) and more straightforward experimental difficulties. For unstressed (uninjured) bacteria and optimal growth media, the 'self-promoting' mode of culture growth can be masked, owing to the high rate of production of growth factors and the high sensitivity of the cells to these pheromones, which can result in the successful multiplication of one isolated bacterial cell to form a culture in a test tube or a colony on agar-solidified medium. As with unicellular eukaryotes (Christensen *et al.*, 1998), only under unfavourable conditions (poor medium, low cell concentration in the inoculum, starved cells, or their combination) is this behaviour visible. These circumstances should be taken into account for future work aimed at generalizing the idea of bacterial cytokines to other prokaryotes and nominally undifferentiated, unicellular microorganisms.

ACKNOWLEDGEMENTS

We thank the Royal Society, the BBSRC, The Russian Foundation for Basic Research (grant 97-04-49987) and the WHO Global Programme for Vaccines and Immunization for financial support.

REFERENCES

Aizenman, E., Engelberg Kulka, H. & Glaser, G. (1996). An *Escherichia coli* chromosomal addiction module regulated by $3',5'$- bispyrophosphate – a model for programmed bacterial cell death. *Proceedings of the National Academy of Sciences, USA*, **93**, 6059–63.

Alberts, B., Bray, D., Lewis, J., Raff, M., Roberts, K. & Watson, J. D. (1989). *Molecular Biology of the Cell*. 2nd edn. New York: Garland.

Amann, R. I., Ludwig, W. & Schleifer, K. H. (1995). Phylogenetic identification and in *situ* detection of individual microbial cells without cultivation. *Microbiological Reviews*, **59**, 143–69.

Baganz, F., Hayes, A., Marren, D., Gardner, D. C. J. & Oliver, S. G. (1997). Suitability of replacement markers for functional analysis studies in *Saccharomyces cerevisiae*. *Yeast*, **13**, 1563–73.

Balaban, N., Goldkorn, T., Nhan, R. T., Dang, L. B., Scott, S., Ridgley, R. M., Rasooly, A., Wright, S. C., Larrick, J. W., Rasooly, R. & Carlson, J. R. (1998). Autoinducer of virulence as a target for vaccine and therapy against *Staphylococcus aureus*. *Science*, **280**, 438–40.

Barer, M. R. (1997). Viable but non-culturable and dormant bacteria: time to resolve an oxymoron and a misnomer? *Journal of Medical Microbiology*, **46**, 629–31.

Barer, M. R., Gribbon, L. T., Harwood, C. R. & Nwoguh, C. E. (1993). The viable but non-culturable hypothesis and medical microbiology. *Reviews of Medical Microbiology*, **4**, 183–91.

Barer, M. R., Karelyants, A. S., Weichart, D. H., Harwood, C. R. & Kell, D. B. (1998). Microbial stress and culturability: conceptual and operational domains. *Microbiology*, in press.

Batchelor, S. E., Cooper, M., Chhabra, S. R., Glover, L. A., Stewart, G. S. A. B., Williams, P. & Prosser, J. I. (1997). Cell density-regulated recovery of starved biofilm populations of ammonia-oxidizing bacteria. *Applied and Environmental Microbiology*, **63**, 2281–6.

Beck, G. & Habicht, G. S. (1994). Invertebrate cytokines. *Annals of the New York Academy of Sciences*, **712**, 206–12.

Berdicevsky, I. & Mirsky, N. (1994). Effects of insulin and glucose-tolerance factor (GTF) on growth of *Saccharomyces cerevisiae*. *Mycoses*, **37**, 405–10.

Bergey, D. R., Hoi, G. A. & Ryan, C. A. (1996). Polypeptide signaling for plant defensive genes exhibits analogies to defense signaling in animals. *Proceedings of the National Academy of Sciences, USA*, **93**, 12 053–8.

Bermudez, L. E., Petrofsky, M. & Shelton, K. (1996). Epidermal growth factor-binding protein in *Mycobacterium avium* and *Mycobacterium tuberculosis* – a possible role in the mechanism of infection. *Infection and Immunity*, **64**, 2917–22.

Beumer, R. R., Devries, J. & Rombouts, F. M. (1992). *Campylobacter jejuni* nonculturable coccoid cells. *International Journal of Food Microbiology*, **15**, 153–63.

Binnerup, S. J., Jensen, D. F., Thordal-Christensen, H. & Sorgensen, J. (1993). Detection of viable, but non-culturable *Pseudomonas fluorescens* DF57 in soil using a microcolony epifluorescence technique. *FEMS Microbiology and Ecology*, **12**, 97–105.

Bloom, B. R. & Murray, C. J. L. (1992). Tuberculosis – commentary on a reemergent killer. *Science*, **257**, 1055–64.

Bloomfield, S. F., Stewart, G. S. A. B., Dodd, C. E. R., Booth, I. R. & Power, E. G. M. (1998). The viable but non-culturable phenomenon explained? *Microbiology*, **144**, 1–3.

Bloomquist, C. G., Reilly, B. E. & Liljemark, W. F. (1996). Adherence, accumulation, and cell division of a natural adherent bacterial population. *Journal of Bacteriology*, **178**, 1172–7.

Bogosian, G., Morris, P. J. L. & O'Neil, J. P. (1998). A mixed culture recovery method indicates that enteric bacteria do not enter the viable but nonculturable state. *Applied Environmental Microbiology*, **64**, 1736–42.

Brayton, P. R., Tamplin, M. L., Huq, A. & Colwell, R. R. (1987). Enumeration of *Vibrio cholerae* O1 in Bangladesh waters by fluorescent-antibody direct viable count. *Applied and Environmental Microbiology*, **53**, 2862–5.

Brown, J. W. (1998). Metabolic and membrane-altering toxins, molecular differentiation factors, and pheromones in the evolution and operation of endocrine signalling systems. *Hormone and Metabolic Research*, **30**, 66–9.

Bubert, A., Kuhn, M., Goebel, W. & Kohler, S. (1992). Structural and functional properties of the p60 proteins from different *Listeria* species. *Journal of Bacteriology*, **174**, 8166–71.

Bu'lock, J. D. (1961). Intermediary metabolism and antibiotic synthesis. *Advances in Microbial Physiology*, **3**, 293–333.

Button, D. K., Schut, F., Quang, P., Martin, R. & Robertson, B. R. (1993). Viability and isolation of marine bacteria by dilution culture – theory, procedures, and initial results. *Applied and Environmental Microbiology*, **59**, 881–91.

Callard, R. & Gearing, A. (1994). *The Cytokine Facts Book*. London: Academic Press.

Carrell, D. T., Hammond, M. E. & Odell, W. D. (1993). Evidence for an autocrine paracrine function of chorionic gonadotropin in *Xanthomonas maltophila*. *Endocrinology*, **132**, 1085–9.

Carroll, L. (1974, orig. 1871). *Through the Looking Glass*. London: Bodley Head.

Cellini, L., Allocati, N., Angelucci, D., Lezzi, T., Di Campi, E., Marzio, L. & Dainelli, B. (1994). Cocccoid *Helicobacter pylori* not culturable *in vitro* reverts in mice. *Microbiology and Immunology*, **38**, 843–50.

Christensen, S. T., Leick, V., Rasmussen, L. & Wheatley, D. N. (1998). Signaling in unicellular eukaryotes. *International Review of Cytology – A Survey of Cell Biology*, **177**, 181–253.

Christensen, S. T., Wheatley, D. N., Rasmussen, M. I. & Rasmussen, L. (1995). Mechanisms controlling death, survival and proliferation in a model unicellular eukaryote *Tetrahymena thermophila*. *Cell Death Differentiation*, **2**, 301–8.

Clegg, J. S. (1997). Embryos of *Artemia franciscana* survive four years of continuous anoxia: the case for complete metabolic rate depression. *Journal of Experimental Biology*, **200**, 467–75.

Cooper, E. L., Zhang, Z., Raftos, D. A., Habicht, G. S., Beck, G., Connors, V., Cossarizza, A., Franceschi, C., Ottaviani, E., Scapigliati, G. & Parrinello, N. (1994). When did communication in the immune system begin? *International Journal of Immunopathology and Pharmacology*, **7**, 203–17.

Csaba, G. (1994). Outgrowth and ontogeny of chemical signalling: origin and development. *International Reviews of Cytology*, **155**, 1–48.

Dagley, S., Dawes, E. A. & Morrison, G. A. (1950). Factors influencing the early phases of growth of *Aerobacter aerogenes*. *Journal of General Microbiology*, **4**, 437–47.

Dance, D. (1991). Melioidosis – the tip of the iceberg. *Clinical Microbiology Reviews*, **4**, 52–60.

Davey, H. M. & Kell, D. B. (1996). Flow cytometry and cell sorting of heterogeneous microbial populations: the importance of single cell analysis. *Microbiological Reviews*, **60**, 641–96.

Davies, D. G., Parsek, M. R., Pearson, J. P., Iglewski, B. H., Costerton, J. W. & Greenberg, E. P. (1998). The involvement of cell-to-cell signals in the development of a bacterial biofilm. *Science*, **280**, 295–8.

Davies, J. (1994). Inactivation of antibiotics and the dissemination of resistance genes. *Science*, **264**, 375–82.

Denis, M. (1992). Interleukin-6 is used as a growth factor by virulent *Mycobacterium avium*. Presence of specific receptors. *Cellular Immunity*, **141**, 182–8.

Denis, M., Campbell, D. & Gregg, E. O. (1991*a*). Interleukin-2 and granulocyte-macrophage colony-stimulating factor stimulate growth of a virulent strain of *Escherichia coli*. *Infections and Immunity*, **59**, 1853–6.

Denis, M., Campbell, D. & Gregg, E. O. (1991*b*). Cytokine stimulation of parasitic and microbial growth. *Research in Microbiology*, **142**, 979–83.

Dixon, N. M. & Kell, D. B. (1989). The inhibition by CO_2 of the growth and metabolism of microorganisms. *Journal of Applied Bacteriology*, **67**, 109–36.

Domingue, G. J. (1995). Electron dense cytoplasmic particles and chronic infection – a bacterial pleomorphy hypothesis. *Endocytobiosis Cell Research*, **11**, 19–40.

Domingue, G. J. & Woody, H. B. (1997). Bacterial persistence and expression of disease. *Clinical Microbiology Reviews*, **10**, 320.

Domingue, G. J., Ghoniem, G. M., Bost, K. L., Fermin, C. & Human, L. G. (1995). Dormant microbes in interstitial cystitis. *Journal of Urology*, **153**, 1321–6.

Doolittle, W. F. & Logsdon, J. M. (1998). Archaeal genomics: do archaea have a mixed heritage? *Current Biology*, **8**, R209–11.

Duncan, K. (1998). Tuberculosis 1998. *Expert Opinion on Therapeutic Patents*, **8**, 137–42.

Dykhuizen, D. E. (1993). Chemostats used for studying natural-selection and adaptive evolution. *Methods in Enzymology*, **224**, 613–31.

Dykhuizen, D. E. & Hartl, D. L. (1983). Selection in chemostats. *Microbiological Reviews*, **47**, 150–68.

Eisner, T. & Meinwald, J. (1995). *Chemical Ecology: The Chemistry of Biotic Interactions*. Washington, DC: National Academy Press.

Erickson, H. P. (1997). FtsZ, a tubulin homologue in prokaryote cell division. *Trends in Cell Biology*, **7**, 362–7.

Flynn, J. L. & Bloom, B. R. (1996). Role of T1 and T2 cytokines in the response to *Mycobacterium tuberculosis*. *Annals of the New York Academy of Sciences*, **795**, 137–46.

Forterre, P. (1997*a*). Protein versus rRNA: problems in rooting the universal tree of life. *ASM News*, **63**, 89–95.

Forterre, P. (1997*b*). Archaea: what can we learn from their sequences? *Current Opinion in Genetics and Development*, **7**, 764–70.

Fortier, A. H., Leiby, D. A., Narayanan, R. B., Asafoadjei, E., Crawford, R. M., Nacy, C. A. & Meltzer, M. S. (1995). Growth of *Francisella tularensis* LVS in macrophages – the acidic intracellular compartment provides essential iron required for growth. *Infections and Immunity*, **65**, 1478–83.

Fredricks, D. N. & Relman, D. A. (1996). Sequence-based identification of microbial pathogens – a reconsideration of Koch's postulates. *Clinical Microbiology Reviews*, **9**, 18–33.

Fuqua, W. C., Winans, S. C. & Greenberg, E. P. (1994). Quorum sensing in bacteria – the luxR–luxI family of cell density-responsive transcriptional regulators. *Journal of Bacteriology*, **176**, 269–75.

Gangadharam, P. R. J. (1995). Mycobacterial dormancy. *Tubercle and Lung Disease*, **76**, 477–9.

Garcia-Lara, J., Shang, L. H. & Rothfield, L. I. (1996). An extracellular factor regulates expression of SdiA, a transcriptional activator of cell-division genes in *Escherichia coli*. *Journal of Bacteriology*, **178**, 2742–8.

Givskov, M., Denys, R., Manefield, M., Gram, L., Maximilien, R., Eberl, L., Molin, S., Steinberg, P. D. & Kjelleberg, S. (1996). Eukaryotic interference with homo-serine lactone-mediated prokaryotic signaling. *Journal of Bacteriology*, **178**, 6618–22.

Greenberg, E. P., Winans, S. & Fuqua, C. (1996). Quorum sensing by bacteria. *Annual Reviews of Microbiology*, **50**, 727–51.

Greenwood, D. R. S. & Peutherer, J. (1992). *Medical Microbiology*. London: Churchill Livingstone.

Grover, S., Woodward, S. R. & Odell, W. D. (1995). Complete sequence of the gene encoding a chorionic gonadotropin-like protein from *Xanthomonas maltophila*. *Gene*, **156**, 75–8.

Hallmann, A., Godl, K., Wenzl, S. & Sumper, M. (1998). The highly efficient sex-inducing pheromone system of *Volvox*. *Trends in Microbiology*, **6**, 185–9.

Halman, M. & Mager, J. (1967). An endogenously produced substance essential for growth initiation of *Pasteurella tularensis*. *Journal of General Microbiology*, **49**, 461–8.

Halman, M., Benedict, M. & Mager, J. (1967). Nutritional requirements of *Pasteurella tularensis* for growth from small inocula. *Journal of General Microbiology*, **49**, 451–60.

Hardie, D. G. (1991). *Biochemical Messengers: Hormones, Neurotransmitters and Growth Factors*. London: Chapman and Hall.

Head, I. M., Saunders, J. R. & Pickup, R. W. (1998). Microbial evolution, diversity, and ecology: A decade of ribosomal RNA analysis of uncultivated microorganisms. *Microbial Ecology*, **35**, 1–21.

Henderson, B., Poole, S. & Wilson, M. (1996). Bacterial modulins – a novel class of virulence factors which cause host tissue pathology by inducing cytokine synthesis. *Microbiological Reviews*, **60**, 316.

Hengge-Aronis, R. (1996). Regulation of gene expression during entry into stationary phase. In *Escherichia coli and Salmonella typhimurium: Cellular and Molecular Biology*, ed. F. C. Neidhardt, J. L. Ingraham, E. C. C. Lin, K. B. Low & B. Magasanik, pp. 1497–512. Washington, DC: American Society for Micro-biology.

Hinshelwood, C. N. (1946). *The Chemical Kinetics of the Bacterial Cell*. Oxford, UK: The Clarendon Press.

Hirsch, C. S., Yoneda, T., Averill, L., Ellner, J. J. & Toossi, Z. (1994). Enhancement of intracellular growth of *Mycobacterium tuberculosis* in human monocytes by transforming growth-factor-β-1. *Journal of Infectious Diseases*, **170**, 1229–37.

Hochman, A. (1997). Programmed cell death in prokaryotes. *Critical Reviews In Microbiology*, **23**, 207–14.

Hosoya, H., Matsuoka, T., Hosoya, N., Tkahashi, T. & Kosaka, T. (1995). Presence of a *Tetrahymena* growth promoting activity in fetal bovine serum. *Development, Growth and Differentiation*, **37**, 347–53.

Ishizaki, Y., Voyvodic, J. T., Burne, J. & Raff, M. C. (1993). Control of lens cell survival. *Journal of Cell Biology*, **121**, 899–908.

Janssens, P. M. (1988). The evolutionary origin of eukaryotic transmembrane signal transduction. *Comparative Biochemistry and Physiology A*, **90**, 209–23.

Kaiser, D. & Losick, R. (1993). How and why bacteria talk to each other. *Cell*, **73**, 873–85.

Kaprelyants, A. S. & Kell, D. B. (1992). Rapid assessment of bacterial viability and vitality using rhodamine 123 and flow cytometry. *Journal of Applied Bacteriology*, **72**, 410–22.

Kaprelyants, A. S. & Kell, D. B. (1993). Dormancy in stationary-phase cultures of *Micrococcus luteus*: flow cytometric analysis of starvation and resuscitation. *Applied Environmental Microbiology*, **59**, 3187–96.

Kaprelyants, A. S. & Kell, D. B. (1996). Do bacteria need to communicate with each other for growth? *Trends in Microbiology*, **4**, 237–42.

Kaprelyants, A. S., Gottschal, J. C. & Kell, D. B. (1993). Dormancy in non-sporulating bacteria. *FEMS Microbiology Reviews*, **104**, 271–86.

Kaprelyants, A. S., Mukamolova, G. V. & Kell, D. B. (1994). Estimation of dormant *Micrococcus luteus* cells by penicillin lysis and by resuscitation in cell-free spent medium at high dilution. *FEMS Microbiology Letters*, **115**, 347–52.

Kell, D. B. (1987). Forces, fluxes and the control of microbial growth and metabolism. The twelfth Fleming lecture. *Journal of General Microbiology*, **133**, 1651–65.

Kell, D. B. & Sonnleitner, B. (1995). GMP – good modelling practice: an essential component of good manufacturing practice. *Trends in Biotechnology*, **13**, 481–92.

Kell, D. B., Kaprelyants, A. S. & Grafen, A. (1995). On pheromones, social behaviour and the functions of secondary metabolism in bacteria. *Trends in Ecology and Evolution*, **10**, 126–9.

Kell, D. B., Kaprelyants, A. S., Weichart, D. H., Harwood, C. R. & Barer, M. R. (1998). Viability and activity in readily culturable bacteria: a review and discussion of the practical issues. *Antonie van Leeuwenhoek*, in press.

Kell, D. B., Ryder, H. M., Kaprelyants, A. S. & Westerhoff, H. V. (1991). Quantifying heterogeneity – flow cytometry of bacterial cultures. *Antonie Van Leeuwenhoek International Journal of General and Molecular Microbiology*, **60**, 145–58.

Khomenko, A. G. & Golyshevskaya, V. I. (1984). Filtrable forms of *Mycobacterium tuberculosis*. *Z. Erkrank. Atm.-Org*, **162**, 147–54.

Kim, K. S. & Le, M. (1992). IL-1-beta and *Escherichia coli*. *Science*, **258**, 1562.

Kjelleberg, S. (1993). *Starvation in Bacteria*. New York: Plenum Press.

Kjelleberg, S., Steinberg, P., Givskov, M., Gram, L., Manefield, M. & deNys, R. (1997). Do marine natural products interfere with prokaryotic AHL regulatory systems? *Aquatic Microbial Ecology*, **13**, 85–93.

Kleerebezem, M., Quadri, L. E. N., Kuipers, O. P. & deVos, W. M. (1997). Quorum sensing by peptide pheromones and two-component signal-transduction systems in Gram-positive bacteria. *Molecular Microbiology*, **24**, 895–904.

Koch, A. L. (1987). The variability and individuality of the bacterium. In *Escherichia coli and Salmonella typhimurium: Cellular and Molecular Biology*, ed. by F. C. Neidhardt, K. B. Low, B. Magasanik, M. Schaechter & H. E. Umbarger, pp. 1606–14. Washington: American Society for Microbiology.

Koch, A. L. (1993). Genetic response of microbes to extreme challenges. *Journal of Theoretical Biology*, **160**, 1–21.

Kolter, R., Siegele, D. A. & Tormo, A. (1993). The stationary phase of the bacterial life cycle. *Annual Review of Microbiology*, **47**, 855–874.

Koonin, E. V. & Galperin, M. Y. (1997). Prokaryotic genomes: the emerging paradigm of genome-based microbiology. *Current Opinion in Genetics and Development*, **7**, 757–63.

Koonin, E. V., Mushegian, A. R., Galperin, M. Y. & Walker, D. R. (1997). Comparison of archaeal and bacterial genomes: computer analysis of protein sequences predicts novel functions and suggests a chimeric origin for the archaea. *Molecular Microbiology*, **25**, 619–37.

Lankford, C. E., Kustoff, T. Y. & Sergeant, T. P. (1957). Chelating agents in growth initiation of *Bacillus globigii. Journal of Bacteriology*, **74**, 737–48.

Lankford, C. E., Walker, J. R., Reeves, J. B., Nabbut, N. H. & Byers, B. R. (1966). Inoculum-dependent division lag of *Bacillus* cultures and its relation to an endogenous factor(s) ('schizokinen'). *Journal of Bacteriology*, **91**, 1070–9.

Lenard, J. (1992). Mammalian hormones in microbial cells. *Trends in Biochemical Sciences*, **17**, 147–50.

LeRoith, D., Roberts, C., Lesniak, M. A. & Roth, J. (1986). Receptors for intercellular messenger molecules in microbes – similarities to vertebrate receptors and possible implications for diseases in man. *Experientia*, **42**, 782–8.

Levine, J. E. & Prystowsky, M. B. (1995). Polypeptide growth-factors in the nucleus – a review of function and translocation. *Neuroimmunomodulation*, **2**, 290–8.

Lujan, H. D., Nowatt, M. R., Helman, L. J. & Nash, T. E. (1994). Insulin-like growth-factors stimulate growth and L-cysteine uptake by the intestinal parasite *Giardia lamblia. Journal of Biological Chemistry*, **269**, 13 069–72.

Luporini, P., Vallesi, A., Miceli, C. & Bradshaw, R. A. (1994). Ciliate pheromones as early growth factors and cytokines. *Annals of the New York Academy of Sciences*, **712**, 195–205.

Luporini, P., Vallesi, A., Miceli, C. & Bradshaw, R. A. (1995). Chemical signaling in ciliates. *Journal of Eukaryotic Microbiology*, **42**, 208–12.

Lyte, M. (1992). The role of catecholamines in Gram-negative sepsis. *Medical Hypotheses*, **37**, 255–8.

Lyte, M. & Bailey, M. T. (1997). Neuroendocrine-bacterial interactions in a neurotoxin-induced model of trauma. *Journal of Surgical Research*, **70**, 195–201.

Lyte, M. & Ernst, S. (1992). Catecholamine-induced growth of Gram-negative bacteria. *Life Sciences*, **50**, 203–12.

Lyte, M., Arulanandam, B. P. & Frank, C. D. (1996). Production of shiga-like toxins by *Escherichia coli* O157-H7 can be influenced by the neuroendocrine hormone norepinephrine. *Journal of Laboratory and Clinical Medicine*, **128**, 392–8.

Lyte, M., Arulanandam, B., Nguyen, K., Frank, C., Erickson, A. & Francis, D. (1997). Norepinephrine induced growth and expression of virulence associated factors in enterotoxigenic and enterohemorrhagic strains of *Escherichia coli. Advances in Experimental Medicine and Biology*, **412**, 331–9.

MacDonell, M. & Hood, M. (1982). Isolation and characterization of ultramicrobacteria from a gulf coast estuary. *Applied and Environmental Microbiology*, **43**, 566–71.

McKenzie, M. A., Fawell, S. E., Cha, M. & Lenard, J. (1988). Effects of mammalian insulin on metabolism, growth and morphology of a wall-less strain of *Neurospora crassa. Endocrinology*, **122**, 511–17.

McVeigh, H. P., Munro, J. & Embley, T. M. (1996). Molecular evidence for the presence of novel actinomycete lineages in a temperate forest soil. *Journal of Industrial Microbiology*, **17**, 197–204.

Massague, J. & Pandiella, A. (1993). Membrane-anchored growth factors. *Annual Review of Biochemistry*, **62**, 515–41.

Mathieu, L. G. & Sonea, S. (1995). A powerful bacterial world. *Endeavour*, **19**, 112–17.

Mathieu, L. G. & Sonea, S. (1996). Review of the unique mode of evolution of bacteria: an opinion. *Symbiosis*, **21**, 199–207.

Matin, A. (1991). The molecular basis of carbon-starvation-induced general resistance in *Escherichia coli*. *Molecular Microbiology*, **5**, 3–10.

Matin, A. (1994). Starvation promoters of *Escherichia coli* – their function, regulation, and use in bioprocessing and bioremediation. *Annals of the New York Academy of Sciences*, **721**, 277–91.

Matsubayashi, Y. & Sakagami, Y. (1996). Phytosulfokine, sulfated peptides that induce the proliferation of single mesophyll-cells of *Asparagus-officinalis* L. *Proceedings of the National Academy of Sciences, USA*, **93**, 7623–7.

Matsubayashi, Y., Takagi, L. & Sakagami, Y. (1997). Phytosulfokine-α, a sulfated pentapeptide, stimulates the proliferation of rice cells by means of specific high- and low-affinity binding sites. *Proceedings of the National Academy of Sciences, USA*, **94**, 13 357–62.

Morita, R. Y. (1988). Bioavailability of energy and its relationship to growth and starvation survival in nature. *Canadian Journal of Microbiology*, **34**, 346–441.

Mukamolova, G. V., Kaprelyants, A. S. & Kell, D. B. (1995). Secretion of an antibacterial factor during resuscitation of dormant cells in *Micrococcus luteus* cultures held in an extended stationary phase. *Antonie Van Leeuwenhoek International Journal of General and Molecular Microbiology*, **67**, 289–95.

Mukamolova, G. V., Kaprelyants, A. S., Young, D. I., Young, M. & Kell, D. B. (1998a). A bacterial cytokine. *Proceedings of the National Academy of Sciences, USA*, **95**, 8916–21.

Mukamolova, G. V., Yanopolskaya, N. D., Kell, D. B. & Kaprelyants, A. S. (1998b). On resuscitation from the dormant state of *Micrococcus luteus*. *Antonie Van Leeuwenhoek International Journal of General and Molecular Microbiology*, **73**, 237–43.

Mullis, K. B., Pollack, J. R. & Neilands, J. B. (1971). Structure of schizokinen, an iron-transport compound from *Bacillus megaterium*. *Biochemistry*, **10**, 4894–8.

Murray, B. E. (1991). New aspects of antimicrobial resistance and the resulting therapeutic dilemmas. *Journal of Infectious Diseases*, **163**, 1185–94.

Oliver, J. D. (1993). Formation of viable but nonculturable cells. In *Starvation in Bacteria*, ed. S. Kjelleberg, pp. 239–272. New York: Plenum.

Ottaviani, E., Franchini, A., Kletsas, D. & Franceschi, C. (1996). Presence and role of cytokines and growth factors in invertebrates. *Italian Journal of Zoology*, **63**, 317–23.

Overgaard, A. K., Friis, J., Christensen, L., Christensen, H. & Rasmussen, L. (1995). Effects of glucose, terapyrroles and protein kinase C activators on cell proliferation in cultures of *Saccharomyces cerevisiae*. *FEMS Microbiology Letters*, **132**, 159–63.

Pancholi, V. & Fischetti, V. A. (1992). A major surface protein on group A streptococci is a glyceraldehyde-3-phosphate-dehydrogenase with multiple binding activities. *Journal of Experimental Medicine*, **176**, 415–26.

Parrish, N. M., Dick, J. D. & Bishai, W. R. (1998). Mechanisms of latency in *Mycobacterium tuberculosis*. *Trends in Microbiology*, **6**, 107–12.

Pearson, A. D., Greenwood, M., Healing, T. D., Rollins, D., Shahamat, M., Donaldson, J. & Colwell, R. R. (1993). Colonization of broiler chickens by waterborne *Campylobacter jejuni*. *Applied and Environmental Microbiology*, **59**, 987–96.

Penfold, W. J. (1914). On the nature of bacterial lag. *Journal of Hygiene*, **14**, 215–41.

Pertseva, M. (1991). The evolution of hormonal signalling systems. *Comparative Biochemistry and Physiology*, **100A**, 775–87.

Pignolo, R. J., Rotenberg, M. O. & Cristofalo, V. J. (1994). Alterations in contact and density-dependent arrest state in senescent WI-38 cells. *In vitro – Cellular Developmental Biology – Animal*, **30A**, 471–6.

Porat, R., Clark, B. D., Wolff, S. M. & Dinarello, C. A. (1991). Enhancement of growth of virulent strains of *Escherichia coli* by interleukin-1. *Science*, **254**, 430–2.

Porat, R., Clark, B. D., Wolff, S. M. & Dinarello, C. A. (1992). IL-1-beta and *Escherichia coli* – reply. *Science*, **258**, 1562–3.

Postgate, J. (1967). Viability measurements and the survival of microbes under minimum stress. In *Advances in Microbial Physiology*, ed. A. H. Rose & J. Wilkinson, pp. 1–21. London: Academic Press.

Postgate, J. R. (1969). Viable counts and viability. *Methods in Microbiology*, **1**, 611–28.

Postgate, J. R. (1976). Death in microbes and macrobes. In *In The Survival of Vegetative Microbes*, ed. T. R. G. Gray & J. R. Postgate, pp. 1–19. Cambridge: Cambridge University Press.

Postgate, J. R. (1995). Danger of sleeping bacteria. *The (London) Times* November 13th issue, 19.

Primas, H. (1981). *Chemistry, Quantum Mechanics and Reductionism*. Berlin: Springer.

Racila, E., Scheuermann, R. H., Picker, L. J., Yefenof, E., Tucker, T., Chang, W., Marches, R., Street, N. E., Vitetta, E. S. & Uhr, J. W. (1995). Tumor dormancy and cell signalling. II. Antibody as an agonist in inducing dormancy of a B cell lymphoma in SCID mice. *Journal of Experimental Medicine*, **181**, 1539–50.

Raff, M. C. (1992). Social controls on cell survival and cell death. *Nature*, **356**, 397–400.

Relman, D. A., Schmidt, T. M., Macdermott, R. P. & Falkow, S. (1992). Identification of the uncultured bacillus of Whipple's disease. *New England Journal of Medicine*, **327**, 293–301.

Rollins, D. M. & Colwell, R. R. (1986). Viable but nonculturable stage of *Campylobacter jejuni* and its role in survival in the natural aquatic environment. *Applied and Environmental Microbiology*, **52**, 531–8.

Rook, G. A. W. & Hernandez-Pando, R. (1996a). Cellular immune responses in tuberculosis: protection and immunopathology. *Médécine et Maladies Infectieuses*, **26**, 904–10.

Rook, G. A. W. & Hernandez-Pando, R. (1996b). The pathogenesis of tuberculosis. *Annual Review of Microbiology*, **50**, 259–84.

Roszak, D. & Colwell, R. R. (1985). Viable but non-culturable bacteria in the aquatic environment. *Journal of Applied Bacteriology*, **59**, R 9.

Roszak, D. B. & Colwell, R. R. (1987). Survival strategies of bacteria in the natural environment. *Microbiology Reviews*, **51**, 365–79.

Roszak, D. B., Grimes, D. J. & Colwell, R. R. (1984). Viable but nonrecoverable stage of *Salmonella enteritidis* in aquatic systems. *Canadian Journal of Microbiology*, **30**, 334–8.

Roth, J., LeRoith, D., Collier, E. S., Watkinson, A. & Lesniak, M. A. (1986). The evolutionary origins of intercellular communication and the Maginot lines of the mind. *Annals of the New York Academy of Sciences*, **463**, 1–11.

Salmond, G. P. C., Bycroft, B. W., Stewart, G. S. A. B. & Williams, P. (1995). The bacterial enigma – cracking the code of cell–cell communication. *Molecular Microbiology*, **16**, 615–24.

Schaller, A. & Ryan, C. A. (1996). Systemin: a polypeptide defense signal in plants. *Bioessays*, **18**, 27–33.

Schut, F., Devries, E. J., Gottschal, J. C., Robertson, B. R., Harder, W., Prins, R. A. & Button, D. K. (1993). Isolation of typical marine bacteria by dilution culture –

growth, maintenance, and characteristics of isolates under laboratory conditions. *Applied and Environmental Microbiology*, **59**, 2150–60.

Schut, F., Gottschal, J. C. & Prins, R. A. (1997*a*). Isolation and characterisation of the marine ultramicrobacterium *Sphingomonas* sp. strain RB2256. *FEMS Microbiology Reviews*, **20**, 363–9.

Schut, F., Prins, R. A. & Gottschal, J. C. (1997*b*). Oligotrophy and pelagic marine bacteria: facts and fiction. *Aquatic Microbial Ecology*, **12**, 177–202.

Shapiro, J. A. (1995). The significances of bacterial colony patterns. *Bioessays*, **17**, 597–607.

Shida, T., Mitsugi, K. & Komagata, K. (1977). Reduction of lag time in bacterial growth. 3. Effect of inoculum size and growth phases of seed cultures. *Journal of General Applied Microbiology*, **23**, 187–200.

Sitnikov, D. M., Schineller, J. B. & Baldwin, T. O. (1996). Control of cell division in *Escherichia coli* – regulation of transcription of *ftsQA* involves both rpoS and sdiA-mediated autoinduction. *Proceedings of the National Academy of Sciences, USA*, **93**, 336–41.

Srinivasan, S., Ostling, J., Charlton, T., DeNys, R., Takayama, K. & Kjelleberg, S. (1998). Extracellular signal molecule(s) involved in the carbon starvation response of marine *Vibrio* sp. strain S14. *Journal of Bacteriology*, **180**, 201–9.

Steinert, M., Emody, L., Amann, R. & Hacker, J. (1997). Resuscitation of viable but nonculturable *Legionella pneumophila* Philadelphia JR32 by *Acanthamoeba castellanii*. *Applied and Environmental Microbiology*, **63**, 2047–53.

Stephens, K. (1986). Pheromones among the prokaryotes. *CRC Critical Reviews in Microbiology*, **13**, 309–34.

Strakhovskaya, M. G., Ivanonva, E. V. & Fraikin, G. Y. (1993). Stimulation of growth of *Candida guillermondii* and *Streptococcus faecalis* by serotonin. *Microbiology (Russ.)*, **62**, 32–4.

Sussman, S. & Halvorson, H. (1966). *Spores, Their Dormancy and Germination*. New York: Harper and Row.

Swift, S., Bainton, N. J. & Winson, M. K. (1994). Gram-negative bacterial communication by N-acyl homoserine lactones: a universal language? *Trends in Microbiology*, **2**, 193–8.

Tanabe, H., Nishi, N., Takagi, Y., Wada, F., Akamatsu, I. & Kaji, K. (1990). Purification and identification of a growth factor produced by *Paramecium tetraurelia*. *Biochemical and Biophysical Research Communications*, **170**, 786–92.

Thatcher, J. W., Shaw, J. M. & Dickinson, W. J. (1998). Marginal fitness contributions of nonessential genes in yeast. *Proceedings of the National Academy of Sciences, USA*, **95**, 253–7.

Tokusumi, Y., Nishi, N. & Takagi, Y. (1996). A substance secreted from *Tetrahymena* and mammalian sera act as mitogens on *Paramecium tetraurelia*. *Zoological Science*, **13**, 89–96.

Toossi, Z. (1996). Cytokine circuits in tuberculosis. *Infectious Agents and Disease – Reviews Issues and Commentary*, **5**, 98–107.

Torrella, F. & Morita, R. (1981). Microcultural study of bacterial size changes and microcolony and ultramicrocolony formation by heterotrophic bacteria in seawater. *Applied Environmental Microbiology*, **41**, 518–27.

Vallesi, A., Giuli, G., Bradshaw, R. A. & Luporini, P. (1995). Autocrine mitogenic activity of pheromones produced by the protozoan ciliate *Euplotes raikovi*. *Nature*, **376**, 372–4.

van de Sande, K., Pawlowski, K., Czaja, I., Wieneke, U., Schell, J., Schmidt, J., Walden, R., Matvienko, M., Wellink, J., Vankammen, A., Franssen, H. & Bisseling, T. (1996). Modification of phytohormone response by a peptide encoded by ENOD40 of legumes and a nonlegume. *Science*, **273**, 370–3.

Vicente, M. & Errington, J. (1996). Structure, function and controls in microbial division. *Molecular Microbiology*, **20**, 1–7.

Votyakova, T. V., Kaprelyants, A. S. & Kell, D. B. (1994). Influence of viable cells on the resuscitation of dormant cells in *Micrococcus luteus* cultures held in an extended stationary phase – the population effect. *Applied and Environmental Microbiology*, **60**, 3284–91.

Wai, S. N., Moriya, T., Kondo, K., Misumi, H. & Amako, K. (1996). Resuscitation of *Vibrio cholerae* O1 Strain Tsi-4 from a viable but nonculturable state by heat-shock. *FEMS Microbiology Letters*, **136**, 187–91.

Watson, L. (1987). *The Biology of Death* (previously published as *The Romeo Error*). London: Sceptre Books.

Wayne, L. G. (1994). Dormancy of *Mycobacterium tuberculosis* and latency of disease. *European Journal of Clinical Microbiology and Infectious Diseases*, **13**, 908–14.

Weiss, M. S., Anderson, D. H., Raffioni, S., Bradshaw, R. A., Ortenzi, C., Luporini, P. & Eisenberg, D. (1995). A cooperative model for receptor recognition and cell-adhesion – evidence from the molecular packing in the 1.6-ångstrom crystal structure of the pheromone Er-1 from the ciliated protozoan *Euplotes raikovi*. *Proceedings of the National Academy of Sciences, USA*, **92**, 10172–6.

Westerhoff, H. V., Hellingwerf, K. J. & van Dam, K. (1983). Thermodynamic efficiency of microbial growth is low but optimal for maximal growth rate. *Proceedings of the National Academy of Sciences, USA*, **80**, 305–9.

Wheatley, D. N., Christensen, S. T., Schousboe, P. & Rasmussen, P. (1993). Signalling in cell growth and death. Adequate nutrition alone may not be sufficient for ciliates – a minireview. *Cell Biology International*, **17**, 817–23.

Wheelock, E. F., Weinhold, K. J. & Levich, J. (1981). The tumor dormant state. *Advances in Cancer Research*, **34**, 107–40.

Whitesides, M. D. & Oliver, J. D. (1997). Resuscitation of *Vibrio vulnificus* from the viable but nonculturable state. *Applied and Environmental Microbiology*, **63**, 1002–5.

WHO (1997). *WHO Report on the Tuberculosis Epidemic 1997* (see also http:www.who.ch/gtb/publications/dritw/index.htm). Geneva: World Health Organization.

Winding, A., Binnerup, S. J. & Sorensen, J. (1994). Viability of indigenous soil bacteria assayed by respiratory activity and growth. *Applied and Environmental Microbiology*, **60**, 2869–75.

Woods, D. E., Jones, A. L. & Hill, P. J. (1993). Interaction of insulin with *Pseudomonas pseudomallei*. *Infections and Immunity*, **61**, 4045–50.

Xu, H. S., Roberts, N., Singleton, F. L., Attwell, R. W., Grimes, D. J. & Colwell, R. R. (1982). Survival and viability of nonculturable *Escherichia coli* and *Vibrio cholerae* in the estuarine and marine environment. *Microbial Ecology*, **8**, 313–23.

Yamamoto, H., Hashimoto, Y. & Ezaki, T. (1996). Study of nonculturable *Legionella pneumophila* cells during multiple-nutrient starvation. *FEMS Microbiology and Ecology*, **20**, 149–54.

Yarmolinsky, M. B. (1995). Programmed cell death in bacterial populations. *Science*, **267**, 836–7.

Yefenof, E., Picker, L. J., Scheuermann, R. H., Tucker, T. F., Vitetta, E. S. & Uhr, J. W. (1993). Cancer dormancy: isolation and characterization of dormant lymphoma cells. *Proceedings of the National Academy of Sciences, USA*, **90**, 1829–33.

Young, D. B. & Duncan, K. (1995). Prospects for new interventions in the treatment and prevention of mycobacterial disease. *Annual Review of Microbiology*, **49**, 641–73.

QUORUM SENSING IN GRAM-NEGATIVE BACTERIA: ACYLHOMOSERINE LACTONE SIGNALLING AND CELL–CELL COMMUNICATION

E. PETER GREENBERG

Department of Microbiology, University of Iowa, Iowa City, IA 52240, USA

INTRODUCTION

This chapter represents an overview of the rapidly developing area of acylhomoserine lactone signalling and cell density-dependent control of gene expression in Gram-negative bacteria. Although it will not be covered in this chapter, it seems clear that bacterial cell-to-cell signalling and cell density-dependent gene expression is not always mediated by acylhomoserine lactones and other signalling systems. For example, recent evidence indicates that a novel non-acylhomoserine lactone-mediated cell-to-cell communication system is relevant to the pathogenesis of *Staphylococcus aureus* (Balaban *et al.*, 1998). The ability to communicate with one another and to organize into communal groups with characteristics not exhibited by individual cells is prevalent in the bacterial world. Acylhomoserine lactone signalling is perhaps the most well-studied communication mechanism but, even with this type of signalling, many important questions remain to be answered. It seems of critical importance that this emerging area of microbial cell-to-cell signalling continue its rapid expansion as a scientific endeavour.

Over the past several years there has been an increasing appreciation among microbiologists that bacteria can sense other bacteria and, in response, they can differentially express specific sets of genes. This capability is often important in the colonization of animal and plant hosts by symbiotic or pathogenic bacterial species. Although different bacterial groups have different mechanisms for monitoring their own abundance in a local environment, one mechanism that has emerged as common in many Gram-negative bacteria is acylhomoserine-mediated quorum sensing (for recent reviews, see Fuqua & Greenberg, 1998; Fuqua *et al.*, 1996).

The basic framework for quorum sensing was established in the early 1970s by Nealson *et al.* (1970). They showed that the marine luminescent bacteria, *Vibrio fischeri* and *Vibrio harveyi*, produce diffusible compounds, termed autoinducers, that accumulate in the medium during growth. These autoinducer signals can accumulate to sufficient concentrations only when there is a critical mass of cells in a confined environment. The signals from *V.*

fischeri and *V. harveyi* do not cross-react. Thus there is species specificity. *V. fischeri* is a specific symbiont in light organs of certain fish where it is found at very high density (10^{10}–10^{11} cells ml^{-1}) and it also can be found free in seawater where it occurs at much lower densities (perhaps 5 cells ml^{-1}). Thus the autoinducer system allows *V. fischeri* to sense its presence in the light organ and express the luminescence system there, where it is required for the symbiosis, but not in seawater where luminescence, which is energetically expensive, would be frivolous.

This concept that bacteria produce signals and communicate with one another was not readily accepted by the scientific community in the 1970s. Evidence in favour of autoinduction began to build in the late 1970s and early 1980s, first with a careful chemostat study confirming that luminescence required high cell density (Rosson & Nealson, 1981). Next, the structure of the *V. fischeri* autoinducer signal was solved, *N*-3-(oxohexanoyl)homoserine lactone (Eberhard *et al.*, 1981). This molecule was shown to move out of, and into, cells by passive diffusion (Kaplan & Greenberg, 1985). The genes for luminescence were cloned from *V. fischeri* into *E. coli* (Engebrecht *et al.*, 1983). Fortunately, the genes for autoinduction are linked to the luminescence structural genes, and *E. coli* containing this *lux* gene cluster produce light in a cell density-dependent fashion (Fig. 1). Thus quorum sensing could now be analysed with the tools of *E. coli* genetics.

The regulatory region that enables autoinduction of luminescence consists of two genes, *lux*R, which encodes an autoinducer-responsive transcriptional activator, and *lux*I, which encodes a protein required for autoinducer synthesis (Engebrecht *et al.*, 1983). The region between *lux*R and *I* contains the regulated *lux* promoter elements. Of note, *lux*I is positively autoregulated, so that basal levels of luminescence operon transcription lead to low rates of autoinducer production, and quite high densities of cells are necessary for activation of the luminescence genes. Once activation has occurred, the rate of autoinducer synthesis is more rapid and the cell density must drop considerably before the rate of transcription of the luminescence operon returns to the basal level. Also, not surprisingly, autoinduction is just one of the regulatory systems that comes to bear on luminescence gene expression. It is known that *lux*R requires activation by cAMP and the cAMP receptor protein (Dunlap & Greenberg, 1988), iron can influence expression of luminescence (Dunlap, 1992; Haygood & Nealson, 1985), FNR seems to exert an effect on *lux*R (Jekosch & Winkler, 1996), and there appear to be other cellular regulatory elements that may exert effects on expression of the luminescence genes of *V. fischeri* (Nealson & Hastings, 1979).

As mentioned above, the *V. fischeri* autoinducer is free to diffuse out of, and into, cells. Thus the cellular concentration and the environmental concentration of this signal are equivalent (Kaplan & Greenberg, 1985). For this reason, the transcriptional activator, LuxR, which is located on the cytoplasmic side of the cell membrane (Kolibachuk & Greenberg, 1993) can

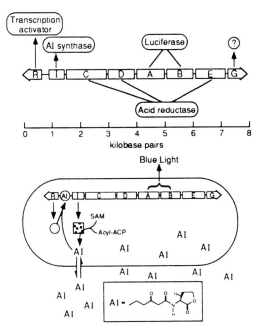

Fig. 1. Quorum sensing in *Vibrio fischeri*. (*Above*) The *lux* gene cluster. The *luxR* gene encodes an autoinducer-dependent transcriptional activator of the *luxI-G* operon. The *luxI* product is the autoinducer synthase, *luxC*, *D*, and *E* form a complex responsible for generation of one of the substrates for the luciferase reaction, the long chain fatty aldehyde, *luxA*, and *B* encode the two subunits of luciferase and the function of *luxG* remains unknown. (*Below*) Cartoon of a *V. fischeri* cell producing the diffusible autoinducer signal. At low cell densities the luminescence operon is transcribed at a basal level. At high cell densities the autoinducer signal can reach a sufficient concentration and bind to the cellular LuxR protein, which will then activate transcription of the luminescence operon.

respond to the environmental concentration of the autoinducer. The environmental concentration of the autoinducer increases with *V. fischeri* cell density (Dunlap, 1992).

A considerable amount is now known about LuxR and the regulatory DNA with which it interacts to activate expression of the luminescence operon (for a recent review, see Sitnikov *et al.*, 1995). LuxR is a 250-amino acid polypeptide that consists of two domains and functions as a homomultimer (Choi & Greenberg, 1991, 1992*a,b*; Hanzelka & Greenberg, 1995), probably a dimer, with $\sigma70$-RNA polymerase to activate *lux* gene expression (Stevens *et al.*, 1994; Stevens & Greenberg, 1997). It is a member of the LuxR superfamily of transcription factors, all of which contain somewhat similar H-T-H motifs in their DNA binding regions (Fuqua *et al.*, 1994). This family includes true LuxR homologs (see below) and also, MalT, GerE, NarL, and others. The N-terminal 160 amino acids or so constitute an autoinducer-binding, regulatory domain, which in the absence of sufficient autoinducer

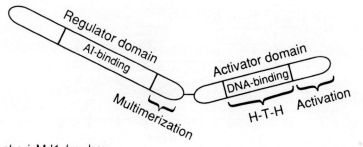

V. fischeri MJ1 *lux* box   A C C T G T A G G A T C G T A C A G G T
lux box-like consensus sequence   R N S T G V A X G A T N X T R C A S R T

Fig. 2. Elements of autoinducible *lux* gene expression. (*Above*) Key regions of LuxR, the activator of luminescence gene transcription. The polypeptide consists of two domains. There is a C-terminal helix–turn–helix (H-T-H) containing activator domain extending from about residue 160 to the C-terminal residue, 250. This domain interacts with the transcription initiation complex. The region from residue 230 to 250 is thought to be required for transcriptional activation but not for DNA binding. There is an N-terminal regulator domain extending to about residue 160. A region of this domain is involved with autoinducer binding, residues 79–127, and a region is involved in multimer formation, around 120 to 160. In the absence of autoinducer the regulator domain interferes with the activity of the activator domain. (*Below*) The *lux* box from *V. fischeri* strain MJ1 and a consensus sequence for *lux*-box like elements found in promoter regions of acylhomoserine lactone-regulated genes from other bacterial species. The 20-bp *lux* box is centred at -42.5 from the start of *luxI* transcription. Consensus sequence abbreviations: N = A, T, C, or G: R = A or G; S = C or G; Y = T or C; X = N or a gap in the sequence.

interferes with the C-terminal domain, the last 90 or so amino acids, binds to RNA polymerase, and the *lux* regulatory DNA, and activates transcription of the luminescence operon (Fig. 2). There is a 20-bp inverted repeat at -42.5 from the start of transcription of the luminescence operon (Fig. 2), and this genetic element is required for autoinduction of luminescence (Devine *et al.*, 1989). *In vitro* studies of LuxR have been difficult and slow, but from such studies we believe LuxR and σ70-RNA polymerase are the only transcription factors required for activation of the *luxI* promoter, and that these two factors bind synergistically to the promoter region (Stevens *et al.*, 1994; Stevens & Greenberg, 1997). Many autoinducer analogues with alterations in the acyl side chain can bind to LuxR, and some can serve weakly as autoinducers. It is perhaps more important to note that many can inhibit the activity of the natural autoinducer presumably by competition for the autoinducer binding site (Schaefer *et al.*, 1996).

More recently a clearer understanding of the mechanism by which the *luxI* gene directs the synthesis of the autoinducer signal has emerged (Hanzelka & Greenberg, 1996; Hanzelka *et al.*, 1997; Schaefer *et al.*, 1996). It is now known that LuxI and LuxI homologues are autoinducer synthases that catalyse the formation of an amide bond between two substrates, a 6 carbon fatty acyl-ACP and *S*-adenosylmethionine (Fig. 3). Although it has been

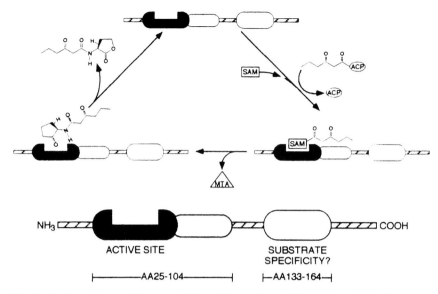

ACTIVE SITE   SUBSTRATE
                     SPECIFICITY?

|————AA25-104————| |—AA133-164—|

Fig. 3. A scheme for autoinducer synthesis and key regions of the LuxI protein. (*Above*) LuxI binds an acyl-ACP and SAM. The acyl group is transferred from the ACP to the enzyme forming amide bond with the SAM. The acyl-SAM is converted to acylhomoserine lactone with release of 5′-methylthioadenosine (MTA) and release of the acylhomoserine lactone. (*Below*) LuxI is 193 amino acids in length. There is a region extending from about residue 25 to somewhere between residue 70 and 104 that appears to represent the active site for amide bond formation. There is limited evidence to suggest that a region between residue 133 and 164 is involved in selection of the appropriate acyl-ACP from the cellular pools.

suggested that the acyl group forms a covalent bond with an active-site cysteine in LuxI, recent studies of cysteine substitution mutants indicate that this is not the case (Hanzelka *et al.*, 1997). Through studies of LuxI mutants and mutant forms of a related protein from *Pseudomonas aeruginosa*, RhlI (Parsek *et al.*, 1997) a view of the protein has been developed that indicates the active site in which amide bond formation is catalysed is roughly in the region of residues 25 to 104, and a region in the C-terminus may be involved in selection of the appropriate acyl-ACP from those existing in the cellular pools (Fig. 3).

THE DISCOVERY OF LUXR–LUXI TYPE QUORUM SENSING SYSTEMS IN OTHER BACTERIA

Within the last 10 years, several groups have made key discoveries that have led to the current view that quorum sensing is common to many Gram-negative bacterial species. First, LuxR homologues were discovered in *P. aeruginosa* (Gambello & Iglewski, 1991) and *Agrobacterium tumefaciens* (Piper *et al.*, 1993), and several bacterial species were shown to produce

N-3-(oxohexanoyl)homoserine lactone (Bainton *et al.*, 1992). Then it was found that the *A. tumefaciens* and *P. aeruginosa* autoinducers are analogues of the *V. fischeri* autoinducer. For *A. tumefaciens* the autoinducer is *N*-3-(oxooctanoyl)homoserine lactone (Zhang *et al.*, 1993) and *P. aeruginosa* has at least two quorum sensing systems: one that uses *N*-3-(oxododecanoyl)-homoserine lactone (Pearson *et al.*, 1994) and one that uses *N*-butyrylhomoserine lactone (Latifi *et al.*, 1995; Ochsner *et al.*, 1994; Ochsner & Reiser, 1995; Pearson *et al.*, 1995; Winson *et al.*, 1995). The genes responsible for autoinducer production were sequenced and their products are homologous to LuxI. Thus the two systems have been termed *lasI–lasR* and the *rhlI–rhlR*.

There are now over 15 LuxI homologues and 15 LuxR homologues in the protein sequence data bases. Furthermore, *lux* box-like sequences can be found in the promoter regions of many of the genes regulated by LuxR homologues in bacteria other than *V. fischeri*. The LuxI homologues direct the synthesis of acylhomoserine lactones with saturated or unsaturated acyl chains of 4 to 14 carbons, with either a hydroxyl group, a carbonyl group, or a hydrogen on the third carbon from the amide bond. The constant is the homoserine lactone, the acyl group provides signal specificity. Different LuxI homologues produce different autoinducers and the cognate LuxR homologues respond best to the appropriate autoinducer (for example, see Schaefer *et al.*, 1996). Table 1 provides a partial list of bacteria that are known to produce acyl homoserine lactones.

What does autoinduction control in different bacteria? As discussed, *V. fischeri* uses quorum sensing to regulate transcription of the luminescence genes so that they are expressed in the light organ symbiosis. *A. tumefaciens* controls conjugal transfer genes by quorum sensing (Fuqua & Winans, 1994; Piper *et al.*, 1993). This is thought to ensure that this bacterial species possesses its catabolic Ti plasmid when present at high density in a crown gall tumour. *P. aeruginosa* is an opportunistic human pathogen, and quorum

Table 1. *A partial list of bacteria that produce acyl homoserine lactone signals*

Vibrio fischeri	*Rhizobium meliloti*
Vibrio harveyi	*Aeromonas hydrophila*
Pseudomonas aeruginosa	*Aeromonas salmonicida*
Agrobacterium tumefaciens	*Burkholderia cepacia*
Erwinia carotovora	*Citrobacter freundii*
Erwinia herbicola	*Enterobacter agglomerans*
Erwinia stewartii	*Obesumbacterium*
Chromobacterium violaceum	*Proteus mirabilis*
Rhizobium leguminosarum	*Rhodobacter sphaeroides*
Vibrio anguillarum	*Serratia liquefaciens*
Pseudomonas aurefociens	*Yersinia enterocolitica*
Pseudomonas solanacearum	*Hafnia*
	Pseudomonas fluorescens

sensing is used to regulate expression of a battery of extracellular virulence genes including enzymes and exotoxins (for a recent review, see Fuqua *et al.*, 1996). Autoinduction of extracellular enzymes is a common theme. One can envision that, at low density, production of extracellular enzymes would be of no value. The enzymes would diffuse away from the cell, convert relatively little substrate to product, and because the environmental concentration of the product would not change appreciably, the bacterial cells would not benefit. When the bacteria have achieved a high enough density, production of an extracellular enzyme could have an impact on the environment. What about exotoxins? Here the analogy is to an invading army. The bacterial pathogen first masses the troops, but it does not reveal its weapons until they can be deployed in sufficient quantity to overwhelm the opposition. By not producing exotoxins at low cell densities and waiting until the host defences can be overwhelmed, *P. aeruginosa* deprives the host of the chance to respond immunologically. *P. aeruginosa* quorum sensing mutants are capable of initial colonization in a mouse lung model, but the progression of the disease is impaired, and unlike the wild type, infection with the quorum sensing mutants does not lead to death (Tang *et al.*, 1996). Another easily understood example of LuxR–LuxI-type quorum sensing is control, not only of extracellular enzymes in *Erwinia carotovora*, but also of carbepenem antibiotic synthesis see pp. 161–71, this volume (Bainton *et al.*, 1992; Pirhonnen *et al.*, 1993). The significance of some quorum sensing systems is more difficult to picture, for example, the autoinduction of a set of genes in *Rhizobium leguminosarum* that is expressed just prior to root hair penetration, together with the expression of functions that lead to stationary phase (Gray *et al.*, 1996). It would not be surprising to find that quorum sensing in *R. leguminosarum* is more elaborate than what is revealed by currently available information. A general theme that has emerged is that the bacteria that exhibit this type of cell-density dependent gene regulation experience a plant or animal host association as part of their lifestyle. However, there are recently described examples that might provide an exception to this rule. For example, the photosynthetic bacterium *Rhodobacter sphaeroides* has a quorum sensing system and, although one cannot rule out an involvement with a eukaryotic host, such an association has not been described (Puskas *et al.*, 1997).

Although the divergently transcribed *luxI* and *luxR* genes in *V. fischeri* are linked to each other and to the genes they regulate (Fig. 1), this is not always the case. Every sort of arrangement imaginable has been reported. The *R* and *I* genes can regulate unlinked genes and in some cases are not even linked to each other. There can be multiple LuxR and LuxI homologues in a single bacterium, for example, *P. aeruginosa* (Latifi *et al.*, 1995; Ochsner *et al.*, 1994; Ochsner & Reiser, 1995). There is now some evidence that at least some strains of *Burkholderia cepacia*, an opportunistic pathogen that can colonize lungs of cystic fibrosis patients, may sense and respond to the density of another bacterial species infecting the cystic fibrosis lung, *P. aeruginosa* (McKenney *et al.*, 1995). It appears that the elements of the LuxR–LuxI system evolved in Gram-negative bacteria early, or have moved from species to species by gene transfer, and that each species has adapted these elements to its own needs.

RECENTLY DISCOVERED ROLES FOR ACYLHOMOSERINE LACTONE SIGNALLING IN BIOFILM DIFFERENTIATION AND DISPERSAL FROM COMMUNITY STRUCTURES

Biofilms of mixed bacterial communities, and of individual species such as *P. aeruginosa*, form thick layers consisting of differentiated mushroom and pillar-like structures that consist primarily of an extracellular polysaccharide matrix in which the bacterial cells are embedded. The development of a biofilm involves several steps. First, individual bacteria must adhere to a surface, they must then proliferate, and at an appropriate time there must be a differentiation or morphogenesis into a mature, structured biofilm. This, taken together with the knowledge that *P. aeruginosa* produces extracellular signals involved in cell-to-cell communication and cell density-dependent expression of many secreted virulence factors, suggests cell-to-cell signalling could be involved in the differentiation of *P. aeruginosa* into the mature biofilm form. Because quorum sensing requires a sufficient density of bacteria, *P. aeruginosa* quorum sensing signals would not be expected to be involved in the initial attachment stage of biofilm formation. However, quorum sensing may be involved in biofilm differentiation.

The wild-type *P. aeruginosa* forms structured biofilms with stalked mushroom-shaped aggregates approximately 120 µm in thickness with water-filled spaces in-between. Biofilm bacteria are hundreds of times more resistant to antibiotics than are bacteria growing in broth culture. This makes biofilms of organisms like *P. aeruginosa* clinically relevant (Davies *et al.*, 1998). One signal generator mutant, a *lasI* mutant, but not the other, a *rhlI* mutant, produces thin, unstructured biofilms about 10–20 µm in thickness. These thin biofilms are sensitive to environmental challenges that do not affect the wild-type biofilms. These challenges result in dispersal of the bacteria from the glass surface to which they are attached (Davies *et al.*, 1998). When the biofilms of the *lasI* mutant are provided with 3-oxododecanoylhomoserine lactone, the LasI-generated signal, they appear identical to the wild type.

The control of biofilm differentiation and integrity by quorum sensing has important implications in medicine. *P. aeruginosa* can colonize devices such as catheters and it colonizes the lungs of most cystic fibrosis patients (Govan & Deretic, 1996; Pollack, 1990). Because of their innate resistance to antibiotics and other biocides, biofilms in these environments are difficult if not impossible to eradicate. Bacterial biofilms also present other problems of significant economic importance in both industry and medicine. The connection between biofilm differentiation and a quorum sensing signal suggests that inhibition of these cell-to-cell signals could aid in the treatment of biofilms.

At least for *P. aeruginosa*, quorum sensing signals are required for conversion to a mature biofilm. Are there signals for dispersal of biofilm cells back into the planktonic community? This is an area that merits intensive investigation. Although much more study is required, the case of quorum sensing in *Rhodobacter sphaeroides* (Puskas *et al.*, 1997) brings evidence to bear on this question. It has been found that *R. sphaeroides* contains *luxI–luxR* homologues called *cerI–cerR*. The *cerI* gene directs the synthesis of the quorum sensing signal 7,8-*cis*-tetradecenoylhomoserine lactone. In a broth medium, the wild type grows as individual cells in suspension. However, a *cerI* mutant grown under identical conditions occurs as a single large mucilaginous clump. Addition of the signal, 7,8-*cis*-tetradecenoylhomoserine lactone to a clumped mutant causes cells to disperse from the clump and grow like the wild type. The ecological significance of this observation is unknown, but it does provide some evidence that there may be acylhomoserine lactone signals involved in dispersal of biofilms. The gene designation *cer* is for community escape response (Puskas *et al.*, 1997).

A CASE OF CONVERGENT EVOLUTION

Understanding autoinduction of luminescence in *V. harveyi* has come more slowly than our understanding of autoinduction in *V. fischeri*. This is, in large part, because the *V. harveyi* system is more complicated. There are two signalling systems that can function independently of each other (Bassler & Silverman, 1995). One of the systems involves an acylhomoserine lactone, *N*-(3-hydroxybutyryl)homoserine lactone (Cao & Meighen, 1993). The structure of the signal for the other system remains unknown and it does not appear to be an acylhomoserine lactone (Surette & Bassler, 1998). LuxR homologues have not been identified in *V. harveyi*. Rather, the signal sensors are complex proteins with sequence similarities to both components of two component regulatory proteins. Two genes, *luxL* and *luxM*, are required for synthesis of *N*-(3-hydroxybutyryl)homoserine lactone (Bassler & Silverman, 1995). Neither of these genes encodes a protein that show similarity to the *V. fischeri* LuxI protein or any of its homologues. It was surprising when Kuo *et al.* (1994) discovered that *V. fischeri luxI* mutants produce octanoylhomoserine lactone, which serves as a very poor substitute for the LuxI-produced 3-oxohexanoylhomoserine lactone in luminescence gene activation. The gene required for octanoylhomoserine lactone synthesis was cloned, sequenced, and although its product is not a LuxI homologue there is a 38% sequence identity of its amino-terminal region and the *V. harveyi* LuxM. This suggests that there is a second family of acylhomoserine lactone synthesizing enzymes

(Gilson *et al.*, 1995). The mechanism of acylhomoserine lactone synthesis by this family has not yet been investigated.

CONCLUDING REMARKS

Thus LuxR–LuxI-type quorum sensing plays a role not only in the curious light organ relationship between *V. fischeri* and its marine animal hosts but also in the virulence of certain human and plant pathogens. The research field is young, and there are many important areas that merit investigation. It is clear that LuxI and LuxR homologues of pathogens are targets for development of novel antimicrobial factors. More needs to be known about how they function and inhibitors need to be identified. With respect to inhibitors, at least one marine algal species produces a furanone compound that can inhibit autoinduction (Givskov *et al.*, 1996). This may provide an explanation as to why luminescent marine bacteria, which can be isolated from a variety of marine habitats, are not found on the surface of algae. Little is known about quorum sensing in natural environments and in the biofilms in which bacteria often grow. Is communication between bacterial species in complex natural environments common or important? Why do bacteria have multiple quorum sensing systems and how many can be found in an individual strain? What is the significance of the 'second family' of auto-inducer synthetic enzymes? Finally, one might expect that, with the intimate associations known to exist between mutualistic and pathogenic quorum sensing bacteria and their plant and animal hosts, the hosts may have evolved systems that can sense and respond to acyl homoserine lactone signals. It has been reported that one of the *P. aeruginosa* autoinducers stimulates epithelial cell production of interleukin-8 (DiMango *et al.*, 1995); other acylhomoserine lactones also appear to affect production of immunomodulators and cytokines (Telford *et al.*, 1998), and the authors have shown that, at concentrations as low as 5 nM, butyrylhomoserine lactone, which is produced by *Pseudomonas aeruginosa*, stimulates mouse spleen cells to produce interferon-γ. In general, host detection and response to autoinducers is as yet an untapped avenue of investigation.

ACKNOWLEDGEMENTS

Research on quorum sensing in the authors laboratory is supported by the National Science Foundation and the Cystic Fibrosis Foundation. Similar accounts of quorum sensing will appear in other symposium volumes concurrent with this chapter.

REFERENCES

Bainton, N. J., Stead, P., Chhabra, S. R., Bycroft, B. W., Salmond, G. P. C., Stewart, G. S. A. B. & Williams, P. (1992). A general role for the *lux* autoinducer in bacterial cell signalling: control of antibiotic synthesis in *Erwinia. Gene*, **116**, 87–91.

Balaban, N., Goldkorn, T., Nhan, R. T., Dang, L. B., Scott, S., Ridgley, R. M., Rasooly, A., Wright, S. C., Larrick, J. W., Rasooly, R. & Carlson, J. R. (1998). Autoinducer of virulence as a target for vaccine and therapy against *Staphylococcus aureus. Science*, **280**, 438–41.

Bassler, B. L. & Silverman, M. R. (1995). Intercellular communication in marine Vibrio species: density-dependent regulation of the expression of bioluminescence. In *Two-Component Signal Transduction*, ed J. A. Hoch & T. J. Silhavy, pp. 431–5. Washington, DC: ASM Press.

Cao, J. & Meighen, E. A. (1993). Biosynthesis and stereochemistry of the autoinducer controlling luminescence in *Vibrio harveyi. Journal of Bacteriology*, **175**, 3856–62.

Choi, S. H. & Greenberg, E. P. (1991). The C-terminal region of the *Vibrio fischeri* LuxR protein contains an inducer-independent lux gene activating domain. *Proceedings of the National Academy of Sciences, USA*, **88**, 11 115–19.

Choi, S. H. & Greenberg, E. P. (1992*a*). Genetic dissection of the DNA binding and luminescence gene activation by the *Vibrio fischeri* LuxR protein. *Journal of Bacteriology*, **174**, 4064–9.

Choi, S. H. & Greenberg, E. P. (1992*b*). Genetic evidence for multimerization of LuxR, the transcriptional activator of *Vibrio fischeri* luminescence. *Molecular Marine Biology and Biotechnology*, **1**, 408–13.

Davies, D. G., Parsek, M. R., Pearson, J. A., Iglewski, B. H., Costerton, J. W. & Greenberg, E. P. (1998). The involvement of cell-to-cell signals in the development of a bacterial biofilm. *Science*, **280**, 295–8.

Devine, J. H., Shadel, G. S. & Baldwin, T. O. (1989). Identification of the operator of the *lux* regulon from *Vibrio fischeri* ATCC7744. *Proceedings of the National Academy of Sciences, USA*, **86**, 5688–92.

DiMango, E., Zar, H. J., Bryan, R. & Prince, A. (1995). Diverse *Pseudomonas aeruginosa* gene products stimulate respiratory epithelial cells to produce interleukin-8. *Journal of Clinical Investigation*, **96**, 2204–10.

Dunlap, P. V. (1992). Mechanism for iron control of the *Vibrio fischeri* luminescence system: involvement of cyclic AMP and cyclic AMP receptor protein and modulation of DNA level. *Journal of Bioluminescence and Chemiluminescence*, **7**, 203–14.

Dunlap, P. V. & Greenberg, E. P. (1988). Analysis of the mechanism of *Vibrio fischeri* luminescence gene regulation by cyclic AMP and cyclic AMP receptor protein in *Escherichia coli. Journal of Bacteriology*, **170**, 4040–6.

Eberhard, A., Burlingame, A. L., Eberhard, C., Kenyon, G. L., Nealson, K. H. & Oppenheimer, N. J. (1981). Structural identification of autoinducer of *Photobacterium fischeri* luciferase. *Biochemistry*, **20**, 2444–9.

Engebrecht, J., Nealson, K. H. & Silverman, M. (1983). Bacterial bioluminescence: isolation and genetic analysis of the functions from *Vibrio fischeri. Cell*, **32**, 773–81.

Fuqua, C. & Greenberg, E. P. (1998). Self perception in bacteria: quorum sensing with acylated homoserine lactones. *Current Opinions in Microbiology*, **1**, 183–9.

Fuqua, W. C. & Winans, S. C. (1994). A LuxR–LuxI type regulatory system activates *Agrobacterium* Ti plasmid conjugal transfer in the presence of a plant tumor metabolite. *Journal of Bacteriology*, **176**, 2796–806.

Fuqua, W. C., Winans, S. C. & Greenberg, E. P. (1994). Quorum sensing in bacteria: the LuxR–LuxI family of cell density-responsive transcriptional regulators. *Journal of Bacteriology*, **176**, 269–75.

Fuqua, W. C., Winans, S. C. & Greenberg, E. P. (1996). Census and consensus in bacterial ecosystems: the LuxR–LuxI family of quorum-sensing transcriptional regulators. *Annual Review of Microbiology*, **50**, 727–51.

Gambello, M. J. & Iglewski, B. H. (1991). Cloning and characterization of the *Pseudomonas aeruginosa lasR* gene, a transcriptional activator of elastase expression. *Journal of Bacteriology*, **173**, 3000–9.

Gilson, L., Kuo, A. & Dunlap, P. V. (1995). AinS and a new family of autoinducer synthesis proteins. *Journal of Bacteriology*, **177**, 6946–51.

Givskov, M., deNys, R., Manefield, M., Gram, L., Maximilien, R., Eberl, L., Steinberg, P. D. & Kjelleberg, S. (1996). Eukaryotic interference with homoserine lactone-mediated prokaryotic signalling. *Journal of Bacteriology*, **178**, 6618–22.

Govan, J. R. W. & Deretic, V. (1996). Microbial pathogenesis in cystic fibrosis: mucoid *Pseudomonas aeruginosa* and *Burkholderia cepacia*. *Microbiology Reviews*, **60**, 539–74.

Gray, K. M., Pearson, J. P., Downie, J. A., Boboye, B. E. A. & Greenberg, E. P. (1996). Cell-to-cell signaling in the symbiotic nitrogen-fixing bacterium *Rhizobium leguminosarum*: autoinduction of a stationary phase and rhizosphere-expressed genes. *Journal of Bacteriology*, **178**, 372–6.

Hanzelka, B. L. & Greenberg, E. P. (1995). Evidence that the N-terminal region of the *Vibrio fischeri* LuxR protein constitutes an autoinducer-binding domain. *Journal of Bacteriology*, **177**, 815–17.

Hanzelka, B. L. & Greenberg, E. P. (1996). Quorum sensing in *Vibrio fischeri*: evidence that *S*-adenosylmethionine is the amino acid substrate for autoinducer synthesis. *Journal of Bacteriology*, **178**, 5291–4.

Hanzelka, B. L., Stevens, A. M., Parsek, M. R., Crone, T. J. & Greenberg, E. P. (1997). Mutational analysis of the *Vibrio fischeri* LuxI polypeptide: critical regions of an autoinducer synthase. *Journal of Bacteriology*, **179**, 4882–7.

Haygood, M. G. & Nealson, K. H. (1985). Mechanisms of iron regulation of luminescence in *Vibrio fischeri*. *Journal of Bacteriology*, **162**, 209–16.

Jekosch, K. & Winkler, U. K. (1996). Anaerobic expression of the *Photobacterium fischeri lux* regulon requires the FNR protein which acts upon the left operon. In *Proceedings of the 9th International Symposium on Bioluminescence and Chemiluminescence*, ed J. W. Hastings, L. J. Kricka & P. E. Stanley, pp. 93–6. New York: John Wiley.

Kaplan, H. B. & Greenberg, E. P. (1985). Diffusion of autoinducer is involved in regulation of the *Vibrio fischeri* luminescence system. *Journal of Bacteriology*, **163**, 1210–14.

Kolibachuk, D. & Greenberg, E. P. (1993). The *Vibrio fischeri* luminescence gene activator LuxR is a membrane-associated protein. *Journal of Bacteriology*, **175**, 7307–12.

Kuo, A., Blough, N. V. & Dunlap, P. V. (1994). Multiple *N*-acyl-L-homoserine lactone autoinducers of luminescence in the marine symbiotic bacterium *Vibrio fischeri*. *Journal of Bacteriology*, **176**, 7558–65.

Latifi, A., Winson, K. M., Foglino, M., Bycroft, B. W., Stewart, G. S. A. B., Lazdunski, A. & Williams, P. (1995). Multiple homologues of LuxR and LuxI control expression of virulence determinants and secondary metabolites through quorum sensing in *Pseudomonas aeruginosa* PAO1. *Molecular Microbiology*, **17**, 333–44.

McKenney, D., Brown, K. E. & Allison, D. G. (1995). Influence of *Pseudomonas aeruginosa* exoproducts on virulence factor production in *Burkholderia cepacia*: evidence of interspecies communication. *Journal of Bacteriology*, **177**, 6989–92.

Nealson, K. H. & Hastings, J. W. (1979). Bacterial bioluminescence: its control and ecological significance. *Microbiological Reviews*, **43**, 496–518.

Nealson, K. H., Platt, T. & Hastings, J. W. (1970). Cellular control of the synthesis and activity of the bacterial luminescence system. *Journal of Bacteriology*, **104**, 313–22.

Ochsner, U. A. & Reiser, J. (1995). Autoinducer-mediated regulation of rhamnolipid biosurfactant synthesis in *Pseudomonas aeruginosa*. *Proceedings of the National Academy of Sciences, USA*, **92**, 6424–8.

Ochsner, U. A., Koch, A. K., Fiechter, A. & Reiser, J. (1994). Isolation and characterization of a regulatory gene affecting rhamnolipid biosurfactant synthesis in *Pseudomonas aeruginosa*. *Journal of Bacteriology*, **176**, 2044–54.

Parsek, M. R., Schaefer, A. L. & Greenberg, E. P. (1997). Analysis of random and site-directed mutations in *rhlI*, a *Pseudomonas aeruginosa* gene encoding an acylhomoserine lactone synthase. *Molecular Microbiology*, **26**, 301–10.

Pearson, J. P., Gray, K. M., Passador, L., Tucker, K. D., Eberhard, A., Iglewski, B. H. & Greenberg, E. P. (1994). Structure of the autoinducer required for expression of *Pseudomonas aeruginosa* virulence genes. *Proceedings of the National Academy of Sciences, USA*, **91**, 197–201.

Pearson, J. P., Passador, L., Iglewski, B. H. & Greenberg, E. P. (1995). A second *N*-acylhomoserine lactone signal produced by *Pseudomonas aeruginosa*. *Proceedings of the National Academy of Sciences, USA*, **92**, 1490–4.

Piper, K. R., Bodman, S. B. V. & Farrand, S. K. (1993). Conjugation factor of *Agrobacterium tumefaciens* regulates Ti plasmid transfer by autoinduction. *Nature (London)*, **362**, 448–50.

Pirhonnen, M., Flego, D., Heikiheimo, R. & Palva, E. T. (1993). A small diffusible signal molecule is responsible for the global control of virulence and exoenzyme production in the plant pathogen *Erwinia carotovora*. *EMBO Journal*, **12**, 2467–76.

Pollack, M. (1990). *Pseudomonas aeruginosa*. In *Principles and Practices of Infectious Diseases*, ed G. L. Mandell, R. G. Douglas & J. E. Bennett, pp. 1673–91. Edinburgh: Churchill Livingstone.

Puskas, A., Greenberg, E. P., Kaplan, S. & Schaefer, A. L. (1997). A quorum sensing system in the free-living photosynthetic bacterium *Rhodobacter sphaeroides*. *Journal of Bacteriology*, **179**, 7530–37.

Rosson, R. A. & Nealson, K. H. (1981). Autoinduction of bacterial luminescence in a carbon limited chemostat. *Archives of Microbiology*, **159**, 160–7.

Schaefer, A. L., Hanzelka, B. L., Eberhard, A. & Greenberg, E. P. (1996). Quorum sensing in *Vibrio fischeri*: autoinducer–LuxR interactions with autoinducer analogues. *Journal of Bacteriology*, **178**, 2897–901.

Schaefer, A. L., Val, D. L., Hanzelka, B. L., Cronan, J. E., Jr. & Greenberg, E. P. (1996b). Generation of cell-to-cell signals in quorum sensing: acyl homoserine lactone synthase activity of a purified *Vibrio fischeri* LuxI protein. *Proceedings of the National Academy of Sciences, USA*, **93**, 9505–9.

Sitnikov, D. M., Schineller, J. B. & Baldwin, T. O. (1995). Transcriptional regulation of bioluminescence genes from *Vibrio fischeri*. *Molecular Microbiology*, **17**, 801–12.

Stevens, A. M. & Greenberg, E. P. (1997). Quorum sensing in *Vibrio fischeri*: essential elements for activation of the luminescence genes. *Journal of Bacteriology*, **179**, 557–62.

Stevens, A. M., Dolan, K. M. & Greenberg, E. P. (1994). Synergistic binding of the *Vibrio fischeri* LuxR transcriptional activator domain and RNA polymerase to the *lux* promoter region. *Proceedings of the National Academy of Sciences, USA*, **91**, 12 619–23.

Surette, M. G. & Bassler, B. L. (1998). Quorum sensing in *Escherichia coli* and *Salmonella typhimurium*. *Proceedings of the National Academy of Sciences, USA*, **95**, 7046–50.

Tang, H. B., Dimango, E., Bryan, R., Gambello, M. J., Iglewski, B. H., Goldberg, J. B. & Prince, A. (1996). Contribution of specific *Pseudomonas aeruginosa* virulence factors to pathogenesis of pneumonia in a neonatal mouse model of infection. *Infection and Immunity*, **64**, 37–43.

Telford, G., Wheeler, D., Williams, P., Tompkins, P. T., Appleby, P., Sewell, H., Stewart, G. S., Bycroft, B. W. & Pritchard, D. I. (1998). The *Pseudomonas aeruginosa* quorum-sensing signal molecule *N*-(3-oxododecanoyl)-L-homoserine lactone has immunomodulatory activity. *Infections and Immunity*, **66**, 36–42.

Winson, M. K., Camara, M., Latifi, A., Foglino, M., Chhabra, S. R., Daykin, M., Bally, M., Chapon, V., Salmond, G. P. C., Bycroft, B. W., Lazdunski, A., Stewart, G. S. A. B. & Williams, P. (1995). Multiple *N*-acyl-L-homoserine lactone signal molecules regulate production of virulence determinants and secondary metabolites in *Pseudomonas aeruginosa*. *Proceedings of the National Academy of Sciences, USA*, **92**, 9427–31.

Zhang, L., Murphy, P. J., Kerr, A. & Tate, M. E. (1993). *Agrobacterium* conjugation and gene regulation by *N*-acyl-L-homoserine lactones. *Nature (London)*, **362**, 446–8.

QUORUM SENSING IN *AEROMONAS* AND *YERSINIA*

SIMON SWIFT[1], KAREN E. ISHERWOOD[2],
STEVE ATKINSON[1], PETRA C. F. OYSTON[2] AND
GORDON S. A. B. STEWART[3]

[1] *Institute of Infections and Immunity, Queen's Medical Centre, University of Nottingham, Nottingham NG7 2UH, UK;*
[2] *DERA, CBD, Porton Down, Salisbury, Wiltshire SP4 0JQ, UK;*
[3] *School of Pharmaceutical Sciences, University of Nottingham, University Park, Nottingham NG7 2RD, UK*

INTRODUCTION

The elegant control of bacterial gene expression by small *N*-acyl homoserine lactone (AHL) signalling molecules (often termed quorum sensing) seemed, until recently, to reside exclusively in the control of bioluminescence in marine *Vibrios* (Sitnikov *et al.*, 1995). Following on from the discoveries between 1992 and 1993 that AHLs regulate multiple genetic loci in *Erwinia* (Bainton *et al.*, 1992*a,b*; Williams *et al.*, 1992; Chhabra *et al.*, 1993; Jones *et al.*, 1993; Pirhonen *et al.*, 1993; McGowan & Salmond, this volume), *Pseudomonas* (Jones *et al.*, 1993; Passador *et al.*, 1993; Pesci & Iglewski, this volume) and *Agrobacterium* (Piper *et al.*, 1993; Zhang *et al.*, 1993), however, the realization that quorum sensing is widespread in Gram-negative bacteria became accepted.

From the above, and with the parallel emergence of biosensors to search for the presence of AHLs in bacterial culture supernatants (Bainton *et al.*, 1992*a*), it was observed that many diverse Gram-negative bacteria were sample positive. Included in this group were pathogenic members of the species *Aeromonas* and *Yersinia*. Since a role for quorum sensing control had already been demonstrated in the regulation of pathogenesis in *Erwinia*, *Pseudomonas* and *Agrobacterium*, there was encouragement to initiate the studies that are now presented below.

AEROMONAS

Aeromonas spp. are important and emerging pathogens of fish and man. *A. salmonicida* is the causative agent of furunculosis in salmonid fish, a disease of considerable importance to the fish farming industry (Fryer & Bartholomew, 1996). *A. hydrophila* causes disease in both fish (motile aeromonad septicemia; Fryer & Bartholomew, 1996) and man (gastroenteritis and septicaemia; Merino *et al.*, 1995; Thornley *et al.*, 1997). Other species, e.g. *A. caviae, A. veronii*, have also been implicated in human disease (Thornley *et al.*, 1997). The virulence of *Aeromonas* is multifactorial, involving a number of secreted and surface-associated factors (Merino *et al.*, 1995; Noonan & Trust, 1997; Pemberton *et al.*, 1997; Thornley *et al.*, 1997). Importantly, the study of secreted enzymes in the laboratory has demonstrated that they are expressed at high cell densities in the stationary phase of growth (Anguita *et al.*, 1993; Ljung, Wretland & Molby, 1981; MacIntyre & Buckley, 1978; MacIntyre *et al.*, 1979). Furthermore, it appears that the combined effect of these exoenzymes is the important factor, as the simple creation of mutations in a given factor has little effect on overall virulence (Wong *et al.*, 1998; Vipond *et al.*, 1998). It was hypothesized therefore, that the co-regulation of these factors may be an important feature of virulence and that, as expression was confined to high cell density, AHL-mediated quorum sensing could be responsible for their coordinated activation. An understanding of quorum sensing in this organism would therefore allow a better understanding of virulence and thus, perhaps, the development of improved treatments.

AEROMONAS QUORUM SENSING CIRCUITRY

The activation of AHL-biosensors *Chromobacterium violaceum* CV026 (McClean *et al.*, 1997) and *Escherichia coli* JM109 (pSB401) (Winson *et al.*, 1998) by a number of *Aeromonas* spp. in T-streak (biosensor and test organism are streaked at 90° to one another in a 'T' shape on an agar plate, allowing a few millimetres gap between the two. Activation of the biosensor by AHLs diffusing across the gap from the test organism can then be visualized) and spent supernatant assays demonstrated that members of the genus produced AHLs (Swift *et al.*, 1997). To characterize AHL production further, the broad host range (IncP) *luxRI'::luxCDABE*-containing AHL-reported plasmid was conjugated into *A. hydrophila* to investigate AHL production throughout the growth phase. As expected, and as has been observed for other Gram-negative bacteria (Swift *et al.*, 1993), bioluminescence due to AHL-production was induced at high cell density (Swift *et al.*, 1997).

Genomic libraries of *A. hydrophila* strain A1 and *A. salmonicida* strain NCIMB 1102 were screened for the ability to complement the AHL biosensors *C. violaceum* CV026 and *E. coli* JM109 (pSB401). AHL-producing clones were identified on cross-hybridizing DNA fragments of *Pst*I (2.8 kbp), *Eco*RI (10 kbp) and *Hin*dIII (11 kbp) clones derived from *A. hydrophila*. Subsequent subcloning and DNA sequencing experiments identified the *luxR* and *luxI* homologues *ahyR* and *ahyI* as a divergently transcribed unit. An AHL-producing clone was also identified from an *A. salmonicida* cosmid. Subcloning and DNA sequencing was then used to identify the *luxR* and *luxI* homologues *asaR* and *asaI* (Swift *et al.*, 1997).

In excess of 20 000 independent clones from three *A. hydrophila* libraries were screened for AHL production. Positive clones from the *Pst*I (2), *Hind*III (1) and *Eco*RI (1) libraries all contained *ahyI*, strongly suggesting that this is the only synthase for C4-C8 AHLs present in *A. hydrophila*. Additionally, the presence of any longer chain AHLs was not detected using the *C. violaceum* CV026 reverse assay (McClean *et al.*, 1997) or with the *lasRI'::luxCDABE* biosensor *E. coli* JM109 [pSB1075] (Winson *et al.*, 1998). It is likely therefore, that a single AHL synthase exists in *Aeromonas*.

SIGNAL MOLECULES

To identify the AHL signal molecules produced by AhyI and AsaI, dichloromethane extracts of the respective *Aeromonas* strain and corresponding *E. coli* clone were analysed (Swift *et al.*, 1997). High performance liquid chromatography (HPLC) and thin layer chromatography (TLC) fractionation tentatively identified two activating molecules with HPLC retention times and TLC R_f values corresponding to C4-HSL and C6-HSL (*N*-(butanoyl)-L-homoserine lactone (BHL) and *N*-(hexanoyl)-L-homoserine lactone (HHL) respectively). The identification was confirmed by fast atom bombardment mass spectrometry (FAB-MS) of the TLC-purified compounds with tandem mass spectrometry (MS–MS) of specific molecular ion peaks and in comparison with the synthetic molecule. Quantification of each molecule in spent medium demonstrated that C4-HSL was predominant, at approximately 100 times the concentration of C6-HSL (Swift *et al.*, 1997).

To gauge the importance of quorum sensing during an *Aeromonas* infection, it was important to demonstrate the presence of AHLs actually at the site of infection. In an *A. hydrophila* infection of fingerling rainbow trout (*Oncorhyncus mykiss*), where a high density of bacteria is often localized in the kidney (Ellis *et al.*, 1988). Kidneys from both infected and uninfected fish were therefore analysed for the presence of AHLs. The AHL-biosensor *E. coli* JM109 (pSB406) (Winson *et al.*, 1998), responsive to C4-HSL and C6-HSL, was able to detect activating molecules in dichloromethane extracts of homogenized infected samples, but not from uninfected controls (L. Fish, D. L. Milton, S. Swift, P. Williams & G. S. A. B. Stewart, unpublished observations).

GENE REGULATION BY QUORUM SENSING IN *AEROMONAS*

Positive feedback to produce a rapid amplification of the AHL signal is a feature of many quorum-sensing circuits (Engebrecht *et al.*, 1983; Hwang *et al.*, 1994; Chan *et al.*, 1995; Seed *et al.*, 1995; Winson 1995). For example, in *Vibrio fischeri*, LuxR and the LuxI biosynthetic produce, 3-oxo-C6-HSL (*N*-(3-oxohexanoyl)-L-homoserine lactone (OHHL)), activates *luxI* transcription (Engebrecht *et al.*, 1983; see also Greenberg, this volume). Hence, positive feedback and signal amplification are seen, with more 3-oxo-C6-HSL activating more *luxI* transcription to give more LuxI and thus more 3-oxo-C6-HSL.

To test if such a positive feedback loop exists in *Aeromonas*, a plasmid borne *ahyRI'::luxCDABE* fusion was constructed (Fig. 1) and transformed into *E. coli*. The recombinant was unable to direct the synthesis of AHLs; however the addition of exogenous C4-HSL induced expression from P_{ahyI} giving a 100-fold increase in reporter bioluminescence (Swift *et al.*, 1997). Signal amplification of this type may be important in pathogenesis, where a rapid expression of virulence determinants could be required to overcome the host defences.

To investigate the contribution of quorum sensing to the phenotype of *A. hydrophila*, a null mutation in *ahyI* was constructed by the insertion of a chloramphenicol resistance cassette into *ahyI* and crossing this onto the chromosome of strain A1N. The resultant *A. hydrophila* A1N *ahyI'::cat* mutant was unable to produce detectable levels of C4-HSL, C6-HSL or any other known AHLs (S. Swift, L. Fish, M. J. Lynch, D. F. Kirke, P. Williams & G. S. A. B. Stewart, unpublished observations). The ability of this mutant to produce a selection of virulence factors was assayed, and although the elaboration of a number of traits (lipase, nuclease, amylase and motility) was unaffected, a key virulence determinant, exoprotease was abolished (S. Swift, L. Fish, M. J. Lynch, D. F. Kirke, P. Williams & G. S. A. B. Stewart, unpublished observations).

PROTEASES

Aeromonas hydrophila A1N produces both a serine protease (PMSF inhibited) and a metalloprotease (EDTA inhibited) activity, which are absent from the *A. hydrophila* A1N *ahyI'::cat* mutant. The provision of an exogenous C4-HSL signal at 1 µM and 10 µM (not shown) restores both activities in a dose responsive manner (Fig. 2). The mutant therefore, is inactive against common proteolysis substrates, e.g. gelatin, casein and azocasein, unless provided with exogenous C4-HSL. A comparison of the proteins secreted by *A. hydrophila* A1N and *A. hydrophila* A1N *ahyI'::cat* by SDS–PAGE has identified two polypeptides absent in the mutant and restorable with 1 and 10 µM C4-HSL (S. Swift, L. Fish, M. J. Lynch, D. F. Kirke, P. Williams & G. S. A. B. Stewart, unpublished observations) that correspond to the 70 kDa serine protease (Leung & Stevenson, 1988; Rivero *et al.*, 1991) and the 35 kDa metalloprotease (Leung & Stevenson, 1988; Rivero *et al.*, 1990) described previously for *A. hydrophila*.

An important role in the pathogenicity of *Aeromonas* has been established for the exoproteases. Protease null mutants of both *A. hydrophila* and *A. salmonicida* show a reduced virulence in animal models (Leung & Stevenson, 1988; Sakai, 1985) and protection against *Aeromonas* proteases

i

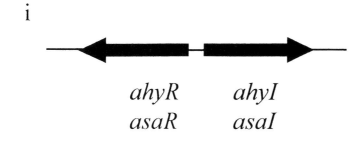

ahyR   *ahyI*

asaR   *asaI*

ii

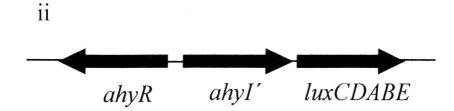

ahyR   *ahyI′*   *luxCDABE*

iii

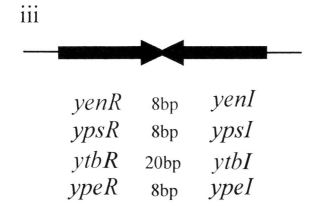

yenR	8bp	*yenI*
ypsR	8bp	*ypsI*
ytbR	20bp	*ytbI*
ypeR	8bp	*ypeI*

Fig. 1. Schematic representation of the *luxRI* homologues from *Aeromonas* and *Yersinia*. Gene organization in: (i) *A. hydrophila* (*ahyRI*) and *A. salmonicida* (*asaRI*); (ii) the *ahyRI*::*lux* construct with *luxCDABE* cloned into the *Bam*HI site at the 3′ end of *ahyI*; (iii) *Y. enterocolitica* (*yenRI*), *Y. pseudotuberculosis* (*ypsRI* and *ytbRI*) and *Y. pestis* (*ypeRI*). The *luxRI* homologues from *Aeromonas* are divergently transcribed and contain an intergenic region of dyad symmetry. The *luxRI* homologues from *Yersinia* are convergently transcribed and overlap by between 8 and 20 bp.

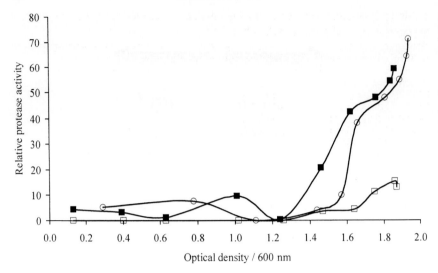

Fig. 2. The role of AhyI and its product C4-HSL in protease production by *Aeromonas hydrophila* A1N. The variation of protease activity in cell free supernatants from overnight cultures of *A. hydrophila* A1N (open circle); *A. hydrophila* A1N *ahyI*::*cat* (open square); and *A. hydrophila* A1N *ahyI'*::*cat* (closed square) incubated with 1 μM C4-HSL throughout the growth phase is indicated. Protease activity was assayed according to the method of Braun and Schmitz (1980). The mean value is plotted, *n* = 3.

is demonstrably effective against *Aeromonas* infection. For example, resistance to furunculosis has been correlated with circulating levels of a protease inhibitor α-macroglobulin inhibitors (Ellis & Stapleton, 1988) and the serine protease of *A. salmonicida* has been successfully used in fish vaccination trials (Coleman *et al.*, 1993). Quorum sensing control of important virulence determinants offers an obvious advantage to a pathogen. It is likely that expression of disease-related proteins by low numbers of bacteria would serve to alert host defences without causing much damage. Thus, only when sufficient bacteria are present would the pathogen be able to mount a strong enough challenge to overcome the host.

An emerging trend within quorum sensing controlled virulence factors is the required involvement of additional gene regulators. For example, in *Erwinia*, *P. aeruginosa* and *Aeromonas* it is not possible to advance the expression of the quorum sensing controlled exoprotease activity simply by the addition of exogenous AHL signal to low cell density cultures. In *Erwinia carotovora* a repressor that fulfils this role, RsmA (Cui *et al.*, 1995; Chatterjee *et al.*, 1995), has been identified (the repressor Rsa has also been recently identified in *P. aeruginosa*; Pesci & Iglewski, this volume). RsmA homologues are found in many bacteria including *Aeromonas* spp., *Serratia* spp. and *Yersinia* spp (Swift *et al.*, 1997). In *P. aeruginosa* a second mechanism is

Table 1. *The effect of quorum sensing agonists and antagonists on protease production in* Aeromonas hydrophila

AHL added/concentration	Relative protease activity
None added	176 $\pm$ 1.2
C10-HSL/10 μM	16 $\pm$ 1.5
C12-HSL/10 μM	62 $\pm$ 2.3
3-oxo-C10-HSL/10 μM	106 $\pm$ 1.7
3-oxo-C12-HSL/10 μM	1 $\pm$ 0.6
3-oxo-C14-HSL/10 μM	zero

being elucidated where expression only occurs in the presence of activating AHL and an appropriate alternative sigma factor (Latifi *et al.*, 1996; Pearson *et al.*, 1997; Pesci & Iglewski, this volume).

QUORUM SENSING BLOCKERS

Empirical studies with AHL analogues of the natural ligands for LuxR, LasR and CarR have shown certain structures to be antagonistic (Chhabra *et al.*, 1993; Schaefer *et al.*, 1996; Passador *et al.*, 1996). The application of this antagonism to the induction of pigment by *C. violaceum* CV026 by C6-HSL has been used as an assay for long chain (C > 8) AHLs (McClean *et al.*, 1997) and was used to detect 3-oxo-C10-HSL as a signal molecule produced by *Vibrio anguillarum* (Milton *et al.*, 1997). As an extension of this approach, 3-oxo-C10-HSL was demonstrated to antagonize both the time of induction and final level of exoprotease expressed by *A. salmonicida* (Swift *et al.*, 1997). The same inhibitory effect on exoprotease production has been shown for 3-oxo-C10-HSL by *A. hydrophila* (S. Swift, L. Fish, M. J. Lynch, D. F. Kirke, P. Williams & G. S. A. B. Stewart, unpublished observations). Furthermore, analysis of the antagonistic effects of a range of C4-HSL analogues towards exoprotease expression by *A. hydrophila* has consistently shown that AHLs with acyl chains of C10, 12 or 14, at 10 μM antagonize protease expression. In the parent strain, 3-oxo-C12-HSL and 3-oxo-C14-HSL almost totally inhibited protease expression (Table 1; S. Swift, L. Fish, M. J. Lynch, D. F. Kirke, P. Williams & G. S. A. B. Stewart, unpublished observations).

In nature a significant role for quorum sensing blocking may exist. Furanones produced by the seaweed *Delisea pulchra* inhibit the swarming of *Serratia liquefaciens* by blocking quorum sensing, thus providing an example of quorum sensing blockers in the natural world (Givskov *et al.*, 1996). It can now be speculated that *V. anguillarum*, a pathogen of fish species also infected by *A. hydrophila* and *A. salmonicida*, produces 3-oxo-C10-HSL to gain an advantage over *Aeromonas* spp. by blocking the quorum sensing regulated traits of exoprotease production.

AEROMONAS CONCLUDING REMARKS

The study of quorum sensing in *Aeromonas* has given new insights into the regulation of both virulence and multicellularity in this genus. The use of quorum sensing blocking now offers the potential to develop new drugs to control these phenomena. One potential target for the design of quorum sensing blockers is the AHL synthase (LuxI homologue). Over-expression of both AhyI and AsaI in *E. coli* has allowed the purification of the proteins for biochemical study. In agreement with results obtained with TraI (*Agrobacterium tumefaciens*; Moré *et al.*, 1996) and LuxI (*Vibrio fischeri*; Hanzelka *et al.*, 1997) the substrates required derive from fatty acid synthesis (acyl-ACP) and amino acid synthesis (*S*-adenosyl methionine). Both AhyI and AsaI differed from RhlI (*Pseudomonas aeruginosa*) as they were unable to utilize acyl-coenzyme A (and so presumably the products of fatty acid degradation; Jiang *et al.*, 1998). Initial structure/function studies using an amber suppression strategy (S. Swift, K. H. McClean, L. Fish, P. Williams & G. S. A. B. Stewart, unpublished observations) has determined that conserved residues at arginine-24, 68, 103 and aspartate-68 are intolerant to replacements. The results were consistent with these residues being essential to the activity of the proteins, as only the conservative change of arginine-103 to lysine showed an activity profile similar to the native protein.

YERSINIA

The genus *Yersinia* belongs to the family *Enterobacteriacea* and comprises 11 species, seven of which are non-pathogenic (*Yersinia aldovae, Yersinia bercovieri, Yersinia frederiksenii, Yersinia intermedia, Yersinia kristensenii, Yersinia mollaretii*, and *Yersinia rohdei*). Of the pathogenic species, *Yersinia ruckeri* is an economically important fish pathogen, causing Red Mouth disease in many species, particularly trout. The remaining three pathogenic yersiniae, *Yersinia enterocolitica, Yersinia pseudotuberculosis* and *Yersinia pestis*, infect a wide range of mammals, including man; *Y. enterocolitica* and *Y. pseudotuberculosis* causing an enteric disease known as yersiniosis, and *Y. pestis* causing plague (Johnson, 1992; Salyers & Whitt, 1994).

The enteropathogenic yersiniae are ubiquitous in the environment and are transmitted from mammal to mammal via the ingestion of contaminated food or water to cause yersiniosis. *Y. enterocolitica* is the most common cause of human yersiniosis, the symptoms of which range from a mild diarrhoea and abdominal pain through to cases of fever and severe abdominal pain that is often mistaken for appendicitis (Baert *et al.*, 1994; Bottone, 1997).

It was during the third plague pandemic, in Hong Kong in 1894, that Alexandre Yersin (Butler, 1994; Solomon, 1995) and Professor Kitasato (Kitasato, 1894), working independently of each other, first isolated the plague bacillus. *Y. pestis* is now described as a Gram-negative, non-spore forming rod. Unlike *Y. enterocolitica* and *Y. pseudotuberculosis, Y. pestis* is non-motile and is defective in many biosynthetic pathways.

There are three main forms of human plague, bubonic, septicaemic and pneumonic. Bubonic plague results from the bite of an infected flea and got its name from the characteristic swelling of regional lymph nodes, termed buboes, which can sometimes appear black due to haemorrhaging. Septicaemic plague occurs when *Y. pestis* cells do not become localized to the lymph nodes and buboes are not observed. In bubonic and septicaemic cases of plague, *Y. pestis* may spread to the lungs, resulting in a secondary pneumonia, which is highly infectious due to the carriage of bacteria in aerosols generated by coughing. Pneumonic plague is a major public health concern, since it can be transmitted from person to person via the airborne route and is fatal within 2 days in over 90% of cases if antibiotic therapy is delayed after the onset of symptoms. Plague is now listed by the World Health Organisation (WHO) as a notifiable disease and, as such, all cases of human infection must be reported to the WHO via public health authorities.

VIRULENCE GENES

Although plague and yersiniosis have very different clinical presentations, many virulence determinants are shared by *Y. pestis, Y. enterocolitica* and *Y. pseudotuberculosis*, the genes for which are located on either the chromosome or on the shared plasmid, pYV (Salyers & Whitt, 1994). In addition, *Y. pestis* harbours two unique plasmids, which also carry virulence genes (see Table 2).

Y. pseudotuberculosis and *Y. enterocolitica* possess several virulence factors that are not expressed in *Y. pestis*. First, two invasion factors, invasin and the attachment-invasion locus (Ail) are encoded on the chromosome by the *inv* and *ail* genes, respectively (Miller *et al.*, 1989; Miller & Falkow, 1988). A pYV-encoded adhesin, known as YadA, is also involved in invasion of mammalian cells by the enteric yersiniae. This fibrillar structure is capable of mediating the clumping effect associated with these organisms, adherence and bacterial entry into eukaryotic cells (Bliska *et al.*, 1993; Yang & Isberg, 1993). Invasin is a surface-exposed protein, which is regulated by temperature and cell-density, being maximally produced at approximately 22 °C when cells are in stationary phase. The *inv* gene is present in *Y. pestis* but is non-functional due to the insertion of an IS*200*-like sequence in the central region of the gene (Simonet *et al.*, 1996). Ail and YadA differ from invasin in that they are maximally expressed in *Y. pseudotuberculosis* and *Y. enterocolitica* at a temperature of 37 °C. A frame shift mutation in the *yadA* gene of *Y. pestis* accounts for the absence of functional YadA protein. Finally, both *Y. pseudotuberculosis* and *Y. enterocolitica* are motile at 28 °C, whereas *Y. pestis* is non-motile at all tested temperatures.

The regulation of virulence genes by the pathogenic yersiniae is complex and relies on environmental fluctuations such as changes in pH, iron levels

Table 2. *Expression of virulence factors of the pathogenic* Yersinia

Location	Virulence determinant	*Y. pestis*	*Y. pseudo-tuberculosis*	*Y. entero-colitica*	Role
Chromosome	Ail	−	+	+	Invasion
	hms locus	+	?	+	Storage of haemin
	Invasin	−	+	+	Invasion
	Iron uptake mechanisms	+	+	+	Acquisition of iron
	pH6 antigen	+	+	+ (Myf)	Adhesin; intracellular survival
pYV	V antigen	+	+	+	Protective antigen; translocation of Yops
	Effector Yops	+	+	+	Act against eukaryotic cell (YopE, YopH)
	Translocation mechanism	+	+	+	Translocation of effector Yops (YopB, YopD)
	Ysc system	+	+	+	Type III secretion system
	YopN	+	+	+	Control protein
	Syc	+	+	+	Molecular chaperones to Yops
	YadA	−	+	+	Adherence; invasion
pTox	Murine toxin	+	−	−	Functions in flea infection. Lethal to rodents.
	Fraction 1	+	−	−	Capsular antigen
pPst	Pesticin	+	−	−	Bacteriocin
	Pla	+	−	−	Coagulase/fibrinolytic properties

and temperature. However, the fact that these organisms sense alterations in their environment and regulate gene expression accordingly, implies that a variety of regulatory systems may be utilized. Two-component systems of gene regulation have previously been identified in the yersiniae (Wren *et al.*, 1995) and, although these systems play a role in virulence, they do not appear to regulate any of the recognised virulence factors.

QUORUM SENSING CIRCUITS

Throup *et al.* (1995) were the first group to evaluate the yersiniae for the presence of acyl-homoserine lactones (AHLs) in culture supernatants by the use of sensitive bioassays. These assays have subsequently been shown to be universally positive for the pathogenic yersiniae by the inclusion of *Y. pestis*. Fifteen different strains of *Y. pestis* were found to activate violacein production in *C. violaceum* CVO26 (McClean *et al.*, 1997) indicating that AHLs are also produced by the plague bacillus (K. E. Isherwood, P. C. F. Oyston, S. Atkinson, P. Williams, G. S. A. B. Stewart & R. W. Titball, unpublished observations). Given the cloning of the *luxRI* homologues *yenR* and *yenI* from *Y. enterocolitica* by biosensor activation (Throup *et al.*, 1995), a similar programme was initiated for *Y. pseudotuberculosis* (S. Atkinson,

J. P. Throup, P. Williams & G. S. A. B. Stewart, unpublished observations). A combination of the close genetic similarity between *Y. pseudotuberculosis* and *Y. pestis* and the biological containment required for *Y. pestis*, determined that a PCR-based approach to screen for these homologues in *Y. pestis* was initiated (K. E. Isherwood, P. C. F. Oyston, S. Atkinson, P. Williams, G. S. A. B. Stewart & R. W. Titball, unpublished observations).

CLONING *LUXRI* HOMOLOGUES FROM *Y. PSEUDOTUBERCULOSIS* AND *Y. PESTIS*

Complementation of the truncated *luxI* gene in the AHL-biosensor *E. coli* JM109 [pSB401] (Winson *et al.*, 1998) by a chromosomal bank of *Y. pseudotuberculosis* DNA, was used to identify clones with a bioluminescent phenotype. Among these clones, one was characterized further and *luxRI* homologues, designated *ypsR* and *ypsI*, were identified (S. Atkinson, J. P. Throup, P. Williams & G. S. A. B. Stewart, unpublished observations). PCR primers designed from *ypsR* and *ypsI* were used to obtain the corresponding homologues, designated *ypeR* and *ypeI*, from *Y. pestis* (K. E. Isherwood, P. C. F. Oyston, S. Atkinson, P. Williams, G. S. A. B. Stewart & R. W. Titball, unpublished observations). Both sets of genes were found to be convergently transcribed and overlapping and the *ypsRI* genes shared approximately 99% homology with the *ypeRI* genes. Phylogenetic analyses of the amino acid sequences of LuxR and LuxI homologues from a variety of Gram-negative bacteria are given in Fig. 3. This reveals that LuxR and LuxI homologues from the yersiniae have the highest identity with the other Enterobacteriaceae, such as the members of the genus *Erwinia*, and the least similarity to the pseudomonads. Nevertheless, in both the LuxR homologues and in the LuxI homologues, there are several amino acid residues that are conserved between all of the tested strains.

SIGNAL MOLECULES

Y. pseudotuberculosis

Dichloromethane extracts of cell-free *E. coli* culture supernatants of *ypsI* and *ypeI* clones were purified by HPLC. Assays using the *C. violaceum* CV026 biosensor (McClean *et al.*, 1997) indicated a positive result in several fractions, which were then purified further using HPLC, TLC overlaid with *C. violaceum* CV026 of the active HPLC sub-fractions revealed that the AHLs produced by the clones had identical R_f values to the synthetic standards *N*-3-(oxohexanoyl)-L-homoserine lactone (3-oxo-C6-HSL; OHHL), *N*-(hexanoyl)-L-homoserine lactone (C6-HSL; HHL) and *N*-(octanoyl)-L-homoserine lactone (C8-HSL; OHL). The identity was confirmed by liquid chromatography mass spectrometry (LCMS).

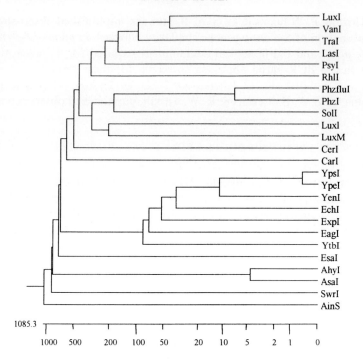

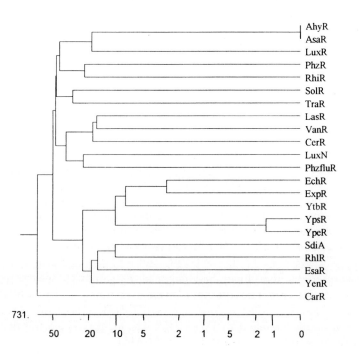

TLC analysis (McClean *et al.*, 1997) of AHLs extracted from culture supernatants of *E. coli* containing the cloned *ypsI* gene produced two spots consistent with the R_f values for 3-oxo-C6-HSL and C6-HSL. The absence of any spot consistent with C8-HSL should be noted in the context of the subsequent identification of a second *luxI* homologue (see section entitled A Second Quorum Sensing System).

Y. pestis

Dichloromethane extracts of culture supernatants of *E. coli* containing the cloned *ypeI* gene were purified by HPLC. Assays using the *C. violaceum* CV026 biosensor indicated a positive result in several fractions, which were then further purified using HPLC. TLC analysis of the active HPLC-subfractions revealed that the AHLs produced by YpeI had identical R_f values to the synthetic standards 3-oxo-C6-HSL and C6-HSL, an allocation confirmed by mass spectrometry.

FINDING A ROLE FOR QUORUM SENSING IN *YERSINIA*

The role of quorum sensing in *Y. enterocolitica* remains elusive and so a role in either *Y. pestis* or *Y. pseudotuberculosis* was sought to provide a model. To explore the role of the *yps* system in *Y. pseudotuberculosis*, insertion/deletion mutations were constructed for *ypsR* and *ypsI*, using the method described previously for *yenI* in *Y. enterocolitica* (Throup *et al.*, 1995). This involved the introduction of a kanamycin cassette replacing an internal fragment of approximately 200 base pairs of each *yps* gene. The constructs were transformed into *Y. pseudotuberculosis* and, following allelic replacement, mutants were selected by growth on appropriately supplemented agar. TLC analysis of supernatant extracts from the *ypsI* mutant of *Y. pseudotuberculosis*, the wild-type strain and the *E. coli* clone containing the functional *ypsI* gene grown at 37 °C, indicated that the AHL profile of the *ypsI* mutant was identical to that of the wild type, whereas the *E. coli* clone extracts were lacking 3-oxo-C6-HSL. However, at 28 °C it was shown that the *ypsI* mutant failed to produce 3-oxo-C6-HSL and at 22 °C the *ypsR* mutant lacked C8-HSL (S. Atkinson, J. P. Throup, P. Williams & G. S. A. B. Stewart,

---

Fig. 3. Phylogenetic tree of LuxR and LuxI homologes and their relatedness to *Aeromonas* spp. and *Yersinia* spp. Constructed using the procedures of Higgins and Sharp (1989). All accession numbers are Genbank unless otherwise stated: LuxR/I (M19039, M96844, M25752), VanR/I (U69677), TraR/I (L22207, L17024), LasR/I (SwissProt P33883, M59425), PsyI (U39802), RhlR/I (L08962, U11811, U15644), PhzfluR/I (L48616), PhzR/I (L33724, L32729), SolR/I (AFO21840), LuxL/M/N (L13940), CerR/I (AFO16298), CarR/I (U17224, X72891, X74299, X80475), YenR/I (X76082), EchR/I (U45854), ExpR/I (X96440), EagI (X74300), EsaR/I (L32184, L32183), AhyR/I (X89469), AsaR/I (U65741), SwrI (U22823), RhiR (M98835), AinS (L37404), SdiA (X03691).

unpublished observations). These results suggest that temperature may prove to be an important factor in cell density-dependent regulation. This is interesting, considering that the expression of many *Yersinia* virulence factors is also regulated by temperature (Isberg *et al.*, 1988; Mikulskis *et al.*, 1994; Badger & Miller, 1998).

YERSINIA PSEUDOTUBERCULOSIS

When examined under the light microscope, liquid cultures of the *Y. pseudotuberculosis ypsR* mutant grown at 28 °C and 37 °C exhibit an altered morphology and appear to clump together when compared to the wild-type or the *ypsI* strain. Surface protein extracts were prepared from the *ypsR* and *ypsI* mutants and the wild-type strain. Although the *ypsI* mutant extracts appeared the same as the wild-type in sodium-dodecyl-sulphate polyacrylamide gel electrophoresis (SDS-PAGE) gels, there were differences in the profile observed for the *ypsR* mutant. *N*-terminal sequence analysis of the major upregulated protein produced by the *ypsR* mutant has shown it to be identical to *fleA* of *Y. enterocolitica* encoding the flagellin subunit. Motility studies suggest that the *ypsR* mutant becomes motile earlier in the growth phase when compared to the wild-type, which would support the suggestion of a role of the quorum sensing system in the regulation of flagella production in *Y. pseudotuberculosis* (S. Atkinson, J. P. Throup, P. Williams & G. S. A. B. Stewart, unpublished observations). This result poses an interesting question with respect to the role of the homologous quorum sensing system in *Y. pestis*, since this species is non-motile.

YERSINIA PESTIS

To explore the role of the *ype* system, an isogenic mutant of *Y. pestis* GB was made by homologous recombination between the *Y. pestis* chromosome and the insertionally inactivated *ypsR* gene construct, as described above for *Y. pseudotuberculosis* (K. E. Isherwood, P. C. F. Oyston, S. Atkinson, P. Williams, G. S. A. B. Stewart & R. W. Titball, unpublished observations). The effects of this mutation on *Y. pestis* were analysed by comparing whole-cell protein profiles of the wild-type and the *ypeR* mutant by SDS–PAGE. No significant differences between the wild-type and the mutant were observed when bacteria were cultured at either 28 °C or 37 °C. The expression of virulence genes by *Y. pestis* GB and *Y. pestis ypeR* at the two different temperatures were analysed by Western blotting using antibodies raised against several of the main virulence factors of *Y. pestis*. The V antigen was weakly expressed by both the wild type and the mutant at 28 °C but was strongly up-regulated in both strains at 37 °C. the pH6 antigen was not produced by either strain when the bacteria were cultured at 28 °C at an acidic pH6, but an increase in temperature to 37 °C at acidic pH resulted in the production of the PsaA subunit by both strains indicating that a mutation in the *ypeR* gene does not affect the regulation of expression of pH6 antigen by temperature. Comparable levels of expression of Pla were observed when wild-type and mutant were grown at either 28 °C and 37 °C, suggesting that the *ypeR* gene is not involved in the expression of the fibrinolytic and coagulase properties of *Y. pestis*. Lipopolysaccharide (LPS) profiles from *Y. pestis* GB and *Y. pestis ypeR* were compared on silver stained SDS–PAGE gels and were both found to produce the characteristic smear of the lipid A component at the bottom of the gel; the usual LPS ladder is not observed for *Y. pestis* since its LPS is rough.

Therefore, as yet it has not been possible to identify a virulence factor of *Y. pestis* whose regulation involves quorum sensing (K. E. Isherwood, P. C. F. Oyston, S. Atkinson, P. Williams, G. S. A. B. Stewart & R. W. Titball, unpublished observations). It was necessary to determine whether the *ypsR* mutation affected the virulence of *Y. pestis in vivo*. To study this, groups of Balb/c mice were challenged subcutaneously with either wild type *Y. pestis* GB or the *ypeR* mutant. The MLD was determined for each strain and found to be 2.5 for the wild-type and < 1.4 for the mutant, indicating that the *ypeR* mutant is no less virulent than the wild-type strain of *Y. pestis* GB. An increase in time to death of mice infected with the mutant when compared to the wild-type was observed, particularly at the higher challenge doses. For example, at a challenge dose of approximately 10^4 cfu, the mean time to death increased from 5.4 ± 0.29 days for mice challenged with *Y. pestis* GB to 6.7 ± 0.3 days for mice challenged with the *ypeR* mutant (K. E. Isherwood, P. C. F. Oyston, S. Atkinson, P. Williams, G. S. A. B. Stewart & R. W. Titball, unpublished observations). These results indicate that, although the mutation does not significantly alter the virulence of the bacteria, there is an increase in time to death implying that the *ypeR* gene has a functional role in *Y. pestis* pathogenicity.

A SECOND QUORUM SENSING SYSTEM

Recently, a second quorum sensing system of *luxR* and *luxI* homologues has been identified in *Y. pseudotuberculosis*, designated *ytbR* and *ytbI*. The genes have been cloned and sequenced as previously described, and, as seen in the first system, show a very high degree of homology to each other (Fig. 3). The presence of a second system suggests the possibility of a hierarchical cascade of control such as that seen in *Pseudomonas aeruginosa* (Pearson *et al.*, 1995; Latifi *et al.*, 1996; Pesci & Iglewski, this volume). Understanding this hierarchy may help explain some of the observations made concerning the *ypsR* mutant.

CONCLUSIONS

It is obvious that regulation of gene expression in the pathogenic *Yersinia* is highly complex and still poorly understood. *Y. pestis* and *Y. pseudotuberculosis* are two very closely related organisms, possessing similar regulatory systems, yet producing very different diseases. A study of quorum sensing in *Y. enterocolitica* and *Y. pestis* has so far failed to unlock the door to the role of quorum sensing in these species. *Y. pseudotuberculosis*, however, appears to be letting us glimpse at the hidden repertoire for quorum sensing in yersiniae and, as such, it may be an excellent candidate for a model from which address the role(s) of quorum sensing through future research.

ACKNOWLEDGEMENTS

S.S. had a BBSRC grant to study Quorum Sensing in *Aeromonas*, K.E.I. was funded as a University of Nottingham registered postgraduate student by CBD, Porton Down, S.A. had a BBSRC research assistant post from the Wellcome Trust.

REFERENCES

Anguita, J., Rodriguez-Aparicio, L. B. & Naharro, G. (1993). Purification, gene cloning, amino acid sequence analysis, and expression of an extracellular lipase from an *Aeromonas hydrophila* human isolate. *Applied and Environmental Microbiology*, **59**, 2411–17.

Badger, J. L. & Miller, V. L. (1998). Expression of invasin and motility are coordinately regulated in *Yersinia enterocolitica*. *Journal of Bacteriology*, **180**, 793–800.

Baert, F., Peetermans, W. & Knockaert, D. (1994). Yersiniosis: the clinical spectrum. *Acta Clinica Belgica*, **49**, 76–85.

Bainton, N. J., Bycroft, B. W., Chhabra, S. R., Stead, P., Gledhill, L., Hill, P. J., Rees, C. E. D., Winson, M. K., Salmond, G. P. C., Stewart, G. S. A. B. & Williams, P. (1992*a*). A general role for the *lux* autoinducer in bacterial cell signalling: control of antibiotic synthesis in *Erwinia*. *Gene*, **116**, 87–91.

Bainton, N. J., Stead, P., Chhabra, S. R., Bycroft, B. W., Salmond, G. P. C., Stewart, G. S. A. B. & Williams, P. (1992*b*). *N*-(3-oxohexanoyl)-L-homoserine lactone regulated carbapenem antibiotic production in *Erwinia carotovora*. *Biochemical Journal*, **288**, 997–1004.

Bliska, J. B., Copass, M. C. & Falkow, S. (1993). The *Yersinia pseudotuberculosis* adhesin YadA mediates intimate bacterial attachment to and entry into Hep-2 cells. *Infection and Immunity*, **61**, 3914–21.

Bottone, E. J. (1997). *Yersinia enterocolitica*: The charisma continues. *Clinical Microbiology Reviews*, **10**, 257–76.

Braun, V. & Schmitz, G. (1980). Excretion of a protease by *Serratia marcescens*. *Archives of Microbiology*, **124**, 55–61.

Butler, T. (1994). Yersinia infections: centennial of the discovery of the plague bacillus. *Clinical Infectious Diseases*, **19**, 655–63.

Chan, P. F., Bainton, N. J., Daykin, M. M., Winson, M. K., Chhabra, S. R., Stewart, G. S. A. B., Salmond, G. P. C., Bycroft, B. W. & Williams, P. (1995). Small molecule mediated autoinduction of antibiotic biosynthesis in the plant pathogen *Erwinia carotovora*. *Biochemical Society Transactions*, **23**, S127.

Chatterjee, A., Cui, Y., Liu, Y., Dumenyo, C. K. & Chatterjee, A. K. (1995). Inactivation of *rsmA* leads to overproduction of extracellular pectinases, cellulases and proteases in *Erwinia carotovora* subsp. *carotovora* in the absence of the starvation cell density-signal, *N*-(3-oxohexanoyl)-L-homoserine lactone. *Applied and Environmental Microbiology*, **61**, 1959–67.

Chhabra, S. R., Stead, P., Bainton, N. J., Salmond, G. P. C., Williams, P., Stewart, G. S. A. B. & Bycroft, B. W. (1993). Autoregulation of carbapenem biosynthesis in *Erwinia carotovora* ATCC 39048 by analogues of *N*-(3-oxohexanoyl)-L-homoserine lactone. *Journal of Antibiotics*, **46**, 441–54.

Coleman, G., Bennett, A. J., Whitby, P. W. & Bricknell, I. R. (1993). A 70 kDa *Aeromonas salmonicida* serine protease-β-galactosidase hybrid protein as an antigen and its protective effect on Atlantic salmon (*Salmo salar L.*) against a virulent *A. salmonicida* challenge. *Biochemical Society Transactions*, **21**, 49S.

Cui, Y., Chatterjee, A., Liu, Y., Dumenyo, C. K. & Chatterjee, A. K. (1995). Identification of a global repressor gene, *rsmA* of *Erwinia carotovora* subsp. *carotovora* that controls extracellular enzymes, *N*-(3-oxohexanoyl)-L-homoserine lactone, and pathogenicity in soft-rotting *Erwinia* spp. *Journal of Bacteriology*, **177**, 5108–15.

Ellis, A. E., Burrows, A. S. & Stapleton, K. J. (1988). Lack of relationship between virulence of *Aeromonas salmonicida* and the putative virulence factors: A-layer, extracellular proteases, and extracellular haemolysins. *Journal of Fish Diseases*, **11**, 309–23.

Ellis, A. E. & Stapleton, K. J. (1988). Differential susceptibility of salmonid fishes to furunculosis correlates with differential serum enhancement of *Aeromonas salmonicida* extracellular protease activity. *Microbial Pathogenicity*, **4**, 299–304.

Engebrecht, J., Nealson, K. & Silverman, M. (1983). Bacterial bioluminescence: isolation and genetic analysis of functions from *Vibrio fischeri*. *Cell*, **32**, 773–81.

Fryer, J. L. & Bartholomew, J. L. (1996). Established and emerging infectious diseases of fish – as fish move, infections move with them. *American Society for Microbiology News*, **62**, 592–4.

Givskov, M., de Nys, R., Manefield, M., Gram, L., Maximilien, R., Eberl, L., Molin, S., Steinberg, P. D. & Kjelleberg, S. (1996). Eukaryotic interference with homoserine lactone mediated prokaryotic signalling. *Journal of Bacteriology*, **178**, 6618–22.

Hanzelka, B. L., Stevens, A. M., Parsek, M. R., Crone, T. J. & Greenberg, E. P. (1997). Mutational analysis of the *Vibrio fischeri* LuxI polypeptide: critical regions of an autoinducer synthase. *Journal of Bacteriology*, **179**, 4882–7.

Higgins, D. G. & Sharp, P. M. (1989). Fast and sensitive multiple sequence alignments on a microcomputer. *CAMBIOS*, **5**, 151–3.

Hwang, I., Pei-Li, L., Zhang, L., Piper, K. R., Cook, D. M., Tate, M. E. & Farrand, S. K. (1994). TraI, a LuxI homologue, is responsible for production of conjugation factor, the Ti plasmid *N*-acyl-homoserine lactone autoinducer. *Proceedings of the National Academy of Sciences, USA*, **91**, 4639–43.

Isberg, R. R., Swain, A. & Falkow, S. (1988). Analysis of expression and thermoregulation of the *Yersinia pseudotuberculosis inv* gene with hybrid proteins. *Infection and Immunity*, **56**, 2133–8.

Jiang, Y., Camara, M., Chhabra, S. R., Hardie, K. R., Bycroft, B. W., Lazdunski, A., Salmond, G. P. C., Stewart, G. S. A. B. & Williams, P. (1998). *In vitro* biosynthesis of the *Pseudomonas aeruginosa* quorum-sensing signal molecule *N*-butanoyl-L-homoserine lactone. *Molecular Microbiology*, **28** 192–203.

Johnson, R. H. (1992). Yersinia infections. *Current Opinion in Infectious Diseases*, **5**, 654–8.

Jones, S., Yu, B., Bainton, N. J., Birdsall, M., Bycroft, B. W., Chhabra, S. R., Cox, A. J. R., Golby, P., Reeves, P. J., Stephens, S., Winson, M. K., Salmond, G. P. C., Stewart, G. S. A. B. & Williams, P. (1993). The *lux* autoinducer regulates the production of exoenzyme virulence determinants in *Erwinia carotovora* and *Pseudomonas aeruginosa*. *EMBO Journal*, **12**, 2477–82.

Kitasato, S. (1894). The bacillus of bubonic plague. *The Lancet*, **25**, 428–30.

Latifi, A., Foglino, M., Tanaka, T., Williams, P. & Lazdunski, A. (1996). A hierarchical quorum sensing cascade in *Pseudomonas aeruginosa* links the transcriptional activators LasR and VsmR to expression of the stationary phase sigma factor RpoS. *Molecular Microbiology*, **21**, 1137–46.

Leung, K. Y. & Stevenson, R. M. (1988). Tn*5-induced protease-deficient strains of Aeromonas hydrophila* with reduced virulence for fish. *Infection and Immunity*, **56**, 2639–44.

Ljung, A., Wretlind, B. & Molby, R. (1981). Separation and characterisation of enterotoxin and two haemolysins from *Aeromonas hydrophila*. *Acta Pathology Microbiology Scandinavian Section B*, **89**, 387–97.

MacIntyre, S. & Buckley, J. T. (1978). Presence of glycerophospholipid cholesterol acyltransferase and phospholipase in culture supernatants of *Aeromonas hydrophila*. *Journal of Bacteriology*, **135**, 402–7.

MacIntyre, S., Trust, T. J. & Buckley, J. T. (1979). Distribution of glycerophospholipid cholesterol acyltransferase in selected bacterial species. *Journal of Bacteriology*, **139**, 132–6.

McClean, K. H., Winson, M. K., Fish, L., Taylor, A., Chhabra, S. R., Camara, M., Daykin, M., Lamb, J. H., Swift, S., Bycroft, B. W., Stewart, G. S. A. B. & Williams, P. (1997). Quorum sensing and *Chromobacterium violaceum*: exploitation and violacein production and inhibition for the detection of *N*-acyl-homoserine lactones. *Microbiology*, **143**, 3703–11.

Merino, S., Rubires, X., Knøchel, S. & Tomas, J. M. (1995). Emerging pathogens: *Aeromonas* spp. *International Journal of Food Microbiology*, **28**, 157–68.

Mikulskis, A. V., Delor, I., Thi, V. H. & Cornelis, G. R. (1994). Regulation of the *Yersinia enterocolitica* enterotoxin Yst gene. Influence of growth phase, temperature, osmolarity, pH and bacterial host factors. *Molecular Microbiology*, **14**, 905–15.

Miller, V. L. & Falkow, S. (1988). Evidence for two genetic loci in *Yersinia enterocolitica* that can promote invasion of epithelial cells. *Infection and Immunity*, **56**, 1242–8.

Miller, V. L., Farmer, III, J. J., Hill, W. E. & Falkow, S. (1989). The *ail* locus is found uniquely in *Yersinia enterocolitica* serotypes commonly associated with disease. *Infection and Immunity*, **57**, 121–31.

Milton, D. L., Hardman, A., Camara, M., Chhabra, S. R., Bycroft, B. W., Stewart, G. S. A. B. & Williams, P. (1997). Quorum sensing in *Vibrio anguillarum*: characterization of the *vanI/R* locus and identification of the autoinducer *N*-(3-oxododecanoyl)-L-homoserine lactone. *Journal of Bacteriology*, **179**, 3004–12.

Moré, M. I., Finger, L. D., Stryker, J. L., Fuqua, C., Eberhard, A. & Winans, S. C. (1996). Enzymatic synthesis of a quorum-sensing autoinducer through use of defined substrates. *Science*, **272**, 1655–8.

Noonan, B. & Trust, T. J. (1997). The synthesis, secretion and role in virulence of the paracrystalline surface protein layers of *Aeromonas salmonicida* and *A. hydrophila*. *FEMS Microbiology Letters*, **154**, 1–7.

Passador, L., Cook, J. M., Gambello, M. J., Rust, L. & Iglewski, B. H. (1993). Expression of *Pseudomonas aeruginosa* virulence genes requires cell-to-cell communication. *Science*, **260**, 1127–30.

Passador, L., Tucker, K. D., Guertin, K. R., Journet, M. P., Kende, A. S. & Iglewski, B. H. (1996). Functional analysis of the *Pseudomonas aeruginosa* autoinducer PAI. *Journal of Bacteriology*, **178**, 5995–6000.

Pearson, J. P., Passador, L., Iglewski, B. H. & Greenberg, E. P. (1995). A 2nd *N*-acyl-homoserine lactone signal produced by *Pseudomonas aeruginosa*. *Proceedings of the National Academy of Sciences, USA*, **92**, 1490–4.

Pearson, J. P., Pesci, E. C. & Iglewski, B. H. (1997). Roles of *Pseudomonas aeruginosa las* and *rhl* quorum-sensing systems in control of elastase and rhamnolipid biosynthesis genes. *Journal of Bacteriology*, **179**, 5756–67.

Pemberton, J. M., Kidd, S. P. & Schmidt, R. (1997). Secreted enzymes of *Aeromonas*. *FEMS Microbiology Letters*, **152**, 1–10.

Piper, K., von Bodman, S. B. & Farrand, S. (1993). Conjugation factor of *Agrobacterium tumefaciens* regulates Ti plasmid transfer by autoinduction. *Nature*, **362**, 448–50.

Pirhonen, M., Flego, D., Heikinheimo, R. & Palva, E. T. (1993). A small diffusible signal molecule is responsible for the global control of virulence and exoenzyme production by the plant pathogen *Erwinia carotovora*. *EMBO Journal*, **17**, 2467–76.

Rivero, O., Anguita, J., Paniagua, C. & Naharro, G. (1990). Molecular cloning and characterization of an extracellular protease gene from *Aeromonas hydrophila*. *Journal of Bacteriology*, **172**, 3905–8.

Rivero, O., Anguita, J., Mateos, D., Paniagua, C. & Naharro, G. (1991). Cloning and characterization of an extracellular temperature-labile serine protease gene from *Aeromonas hydrophila*. *FEMS Microbiology Letters*, **65**, 1–7.

Sakai, D. K. (1985). Loss of virulence in a protease-deficient mutant of *Aeromonas salmonicida*. *Infection and Immunity*, **48**, 146–52.

Salyers, A. A. & Whitt, D. D. (1994). *Yersinia* infections. In *Bacterial Pathogenesis: A Molecular Approach*, ed. A. A. Salyers & D. D. Whitt, pp. 213–28. Washington: ASM Press.

Schaefer, A. L., Hanzelka, B. L., Eberhard, A. & Greenberg, E. P. (1996). Quorum sensing in *Vibrio fischeri*: probing autoinducer-LuxR interactions with autoinducer analogs. *Journal of Bacteriology*, **178**, 2897–901.

Seed, P. C., Passador, L. & Iglewski, B. H. (1995). Activation of the *Pseudomonas aeruginosa lasI* gene by LasR and the *Pseudomonas* autoinducer PAI: an autoinduction regulatory hierarchy. *Journal of Bacteriology*, **177**, 654–9.

Simonet, M., Riot, B., Fortineau, N. & Berche, P. (1996). Invasin production by *Yersinia pestis* is abolished by insertion of an IS*200*-like element within the *inv* gene. *Infection and Immunity*, **64**, 375–9.

Sitnikov, D. M., Schineller, J. B. & Baldwin, T. O. (1995). Transcriptional regulation of bioluminescence genes from *Vibrio fischeri*. *Molecular Microbiology*, **17**, 801–12.

Solomon, T. (1995). Alexandre Yersin and the plague bacillus. *Journal of Tropical Medicine and Hygiene*, **98**, 209–12.

Swift, S., Winson, M. K., Chan, P. F., Bainton, N. J., Birdsall, M., Reeves, P. J., Rees, C. E. D., Chhabra, S. R., Hill, P. J., Throup, J. P., Bycroft, B. W., Salmond, G. P. C., Williams, P. & Stewart, G. S. A. B. (1993). A novel strategy for the isolation of *luxI* homologues: evidence for the widespread distribution of a LuxR:LuxI superfamily in enteric bacteria. *Molecular Microbiology*, **10**, 511–20.

Swift, S., Karlyshev, A. V., Durant, E. L., Winson, M. K., Chhabra, S. R., Williams, P., Macintyre, S. & Stewart, G. S. A. B. (1997). Quorum sensing in *Aeromonas hydrophila* and *Aeromonas salmonicida*: Identification of the LuxRI homologues

AhyRI and AsaRI and their cognate signal molecules. *Journal of Bacteriology*, **179**, 5271–81.

Thornley, J. P., Shaw, J. G., Gryllos, I. A. & Eley, A. (1997). Virulence properties of clinically significant *Aeromonas* species: evidence for pathogenicity. *Reviews in Medical Microbiology*, **8**, 61–72.

Throup, J. P, Camara, M., Briggs, G. S., Winson, M. K., Chhabra, S. R., Bycroft, B. W., Williams, P. & Stewart, G. S. A. B. (1995). Characterisation of the *yenI/yenR* locus from *Yersinia enterocolitica* mediating the synthesis of two *N*-acyl-homoserine lactone signal molecules. *Molecular Microbiology*, **17**, 345–56.

Vipond, R., Bricknell, I. R., Durant, E., Bowden, T. J., Ellis, A. A., Smith, M. & MacIntyre, S. (1998). Defined deletion mutants demonstrate that the major secreted toxins are not essential for the virulence of *Aeromonas salmonicida*. *Infection and Immunity*, **66**, 1990–8.

Williams, P., Bainton, N. J., Swift, S., Winson, M. K., Chhabra, S. R., Stewart, G. S. A. B., Salmond, G. P. C. & Bycroft, B. W. (1992). Small molecule mediated density dependent control of gene expression in prokaryotes: bioluminescence and the biosynthesis of carbapenem antibiotics. *FEMS Microbiology Letters*, **100**, 161–8.

Winson, M. K., Camara, M., Latifi, A., Foglino, M., Chhabra, S. R., Daykin, M., Bally, M., Chapon, V., Salmond, G. P. C., Bycroft, B. W., Lazdunski, A., Stewart, G. S. A. B. & Williams, P. (1995). Multiple *N*-acyl-L-homoserine signal molecules regulate production of virulence determinants and secondary metabolites in *Pseudomonas aeruginosa*. *Proceedings of the National Academy of Sciences, USA*, **92**, 9427–31.

Winson, M. K., Swift, S., Fish, L., Throup, J. P., Jørgensen, F., Chhabra, S. R., Bycroft, B. W., Williams, P. & Stewart, G. S. A. B. (1998). Construction and analysis of *luxCDABE* based plasmid sensors for investigating *N*-acyl-homoserine lactone mediated quorum sensing. *FEMS Microbiology Letters*, **163**, 185–92.

Wong, C. Y., Heuzenroeder, M. W. & Flower, R. L. (1998). Inactivation of two haemolytic toxin genes in *Aeromonas hydrophila* attenuates virulence in a suckling mouse model. *Microbiology UK*, **144**, 291–8.

Wren, B. W., Olsen, A. L., Stabler, R. & Li, S. (1995). Genetic analysis of *phoP* and *ompR* from *Yersinia enterocolitica, Yersinia pseudotuberculosis* and *Yersinia pestis*. In *Yersiniosis: Present and future*, ed. G. Ravagnan & C. Chiesa. *Microbiology and Immunology*, vol. **13**, pp. 318–20.

Yang, Y. & Isberg, R. R. (1993). Cellular internalization in the absence of invasin expression is promoted by the *Yersinia pseudotuberculosis yadA* product. *Infection and Immunity*, **61**, 3907–13.

Zhang, L. H., Murphy, P. J., Kerr, A. & Tate, M. E. (1993). *Agrobacterium* conjugation and gene regulation by *N*-acyl-L-homoserine lactones. *Nature*, **362**, 446–8.

SIGNALLING IN *PSEUDOMONAS AERUGINOSA*

EVERETT C. PESCI[2] AND BARBARA M. IGLEWSKI[1]

[1] *Department of Microbiology and Immunology, University of Rochester School of Medicine and Dentistry, Rochester, NY 14642*
[2] *Department of Microbiology and Immunology, East Carolina University School of Medicine, Greenville, NC 27858-4354, USA*

Historically, bacterial populations have been considered to be nothing more than large groups of individuals independently progressing towards a common goal. This notion has become all but extinct with the discovery of the intercellular signalling mechanism known as quorum sensing. Quorum sensing systems allow bacteria within a population to communicate in order to coordinate the expression of specific genes in response to bacterial cell density. The general model for how quorum sensing works is that, at low cell densities, a basal level of a signal molecule (called autoinducer) is produced by each cell in a population. Autoinducer concentration then increases with cell density until a threshold concentration is reached. At this critical level, autoinducer binds to, and thereby activates, a transcriptional activator protein ('R-protein') which can then induce specific genes. In this manner, a quorum sensing system elegantly controls the expression of genes in response to cell density. This form of bacterial cell–cell signalling was first characterized in the marine symbiont, *Vibrio fischeri*, and has since been discovered in numerous species of bacteria (see Fuqua *et al.*, 1996 for review and Williams *et al.* this volume), including *Pseudomonas aeruginosa*.

P. aeruginosa is found throughout our environment and causes severe, opportunistic infections in humans. This opportunist is a major source of nosocomial infections and is a very serious problem in cystic fibrosis patients (see Govan & Deretic, 1996, and Van Delden & Iglewski, 1998 for reviews). The ability of this bacterium to grow and thrive in a wide range of settings suggests that it must continually adapt to changes in its environment. The focus of this chapter will be to discuss how *P. aeruginosa* responds to its environment through the use of multiple quorum sensing systems that control specific genes in response to cell density.

P. AERUGINOSA CONTAINS TWO COMPLETE AND SEPARATE QUORUM SENSING SYSTEM

It was first discovered that *P. aeruginosa* utilized a quorum sensing system to control gene expression when Gambello and Iglewski (1991) complemented an elastase deficient *P. aeruginosa* strain with the *lasR* gene. This gene, which encodes a LuxR homolog, acts as the transcriptional activator protein of the *las* quorum sensing system (Gambello & Iglewski, 1991). The *las* quorum sensing system consists of both LasR and the autoinducer molecule, N-(3-oxodocecanoyl)-L-homoserine lactone (3-oxo-C_{12}-HSL) (Pearson *et al.*, 1994). LasR is a typical member of the LuxR family of proteins in that it contains a putative autoinducer-binding region in the amino two-thirds of the protein (36% identity to the LuxR autoinducer-binding region), and a putative DNA-binding region (including a helix–turn–helix motif) in the carboxyl third of the protein (53% identity to the LuxR DNA-binding region) (Gambello & Iglewski, 1991). Deletion analysis of the 239 amino acid LasR protein suggested that 3-oxo-C_{12}-HSL interacted with the protein somewhere between amino acids 3 and 155, and that transcriptional activation by LasR required only amino acids 160 to 239 (L. Passador & B. H. Iglewski, unpublished observations). In addition, You *et al.* (1996) have demonstrated that in the presence of 3-oxo-C_{12}-HSL, LasR binds to the promoter region of *lasB* (encodes elastase), which is controlled by the *las* quorum sensing system (Gambello & Iglewski, 1991).

The *las* autoinducer molecule, 3-oxo-C_{12}-HSL, is a typical Gram negative autoinducer molecule that is composed of a homoserine lactone ring with an acyl side chain. The synthesis of 3-oxo-C_{12}-HSL is directed by the auto-inducer synthase, LasI, a member of the LuxI family of autoinducer synthases (Passador *et al.*, 1993). LasI is encoded by the *lasI* gene which was found directly downstream from *lasR* by Passador *et al.* (1993). The *lasR* and *lasI* genes are arranged in a regulon consisting of two independently transcribed genes oriented in the same direction (Gambello & Iglewski, 1991; Seed *et al.*, 1995). Both LasR and 3-oxo-c_{12}-HSL are required for the *las* quorum sensing system to be functional. It has been shown that LasR is only active in the presence of 3-oxo-C_{12}-HSL, and that 3-oxo-C_{12}-HSL specifically binds to *E. coli* cells over-expressing LasR (Passador *et al.*, 1996). Passador *et al.* (1996) also showed that 3-oxo-C_{12}-HSL analogues with shorter acyl side chain length had a decreased ability to activate or bind to LasR, indicating that the LasR/3-oxo-C_{12}-HSL interaction was highly specific. (*Note:* Jones *et al.* (1993) found that, at high concentrations, 3-oxo-C_6-HSL could partially activate elastase production in a non-defined *P. aeruginosa* mutant, indicating that R-protein specificity will decrease in the presence of a non-specific autoinducer at a high concentration.)

The interaction of LasR and 3-oxo-C_{12}-HSL forms a powerful activator complex that induces the expression of numerous *P. aeruginosa* genes. It has been reported that the *las* quorum sensing system controls several virulence factor genes including *lasB*, *lasA* and *toxA* (Gambello & Iglewski, 1991; Gambello *et al.*, 1993; Toder *et al.*, 1991; Pearson *et al.*, 1994). The *las* system has also been shown to control genes of the general secretory pathway (*xcpP* and *xcpR*), indicating that LasR/3-oxo-C_{12}-HSL acts as a global regulator in *P. aeruginosa* (Chapon-Herve *et al.*, 1997). LasR and 3-oxo-C_{12}-HSL are required and sufficient for the expression of the 3-oxo-C_{12}-HSL synthase gene, *lasI* (Seed *et al.*, 1995). This regulation creates a positive autoregulatory loop in which the production of 3-oxo-C_{12}-HSL leads to the production of more 3-oxo-C_{12}-HSL (Seed *et al.*, 1995). It is also interesting to note that Seed *et al.* (1995) reported that ten-fold more 3-oxo-C_{12}-HSL was required for LasR to activate *lasB* as compared to the amount needed to activate *lasI*. This indicated that a hierarchy existed in which *lasI* could be activated before *lasB*, allowing the concentration of 3-oxo-C_{12}-HSL to increase so that *lasB* and other LasR/3-oxo-C_{12}-HSL controlled genes would subsequently be activated.

Shortly after the discovery of the *las* quorum sensing system, there was another finding that made *P. aeruginosa* quorum sensing much more complex. Ochsner *et al.* (1994*b*) found that *P. aeruginosa* contained a second LuxR homologue (31% and 23% identical to LasR and LuxR, respectively), which was named RhlR (encoded by *rhlR*) because of its regulation of the biosurfactant/haemolysin, rhamnolipid. (*Note*: This gene was subsequently discovered by Winson *et al.* (1995) and by Brint & Ohman (1995). Winson *et al.* (1995) named the gene *vsmR* and Brint & Ohman used the original name, *rhlR*.) The autoinducer component of the *rhl* quorum sensing system was then found when Pearson *et al.* (1995) reported the existence of a second *P. aeruginosa* autoinducer molecule, *N*-butyryl-L-homoserine lactone (C_4-HSL). Subsequently, it was shown that a LuxI homologue was encoded by *rhlI*, which lies directly downstream from *rhlR* (Ochsner & Reiser, 1995). The *rhlI* gene has been shown to direct the synthesis of C_3-HSL (Winson *et al.*, 1995) and RhlI was shown to be the C_4-HSL synthase by Jiang *et al.* (1998). Studies on the *rhl* system have indicated that C_4-HSL binds specifically to *E. coli* cells over-expressing RhlR (Pearson *et al.*, 1997). However, unlike the binding of 3-oxo-C_{12}-HSL to LasR-expressing cells seen by Passador *et al.* (1996), the binding of C_4-HSL to RhlR-expressing cells is significantly enhanced by over-expression of the chaperone GroESL. The reason for this effect is not apparent and further studies will be required to understand its significance.

The *rhl* system was discovered because RhlR/C_4-HSL induces the *rhlAB* operon, which encodes for a rhamnosyltransferase required for rhamnolipid production (Ochsner *et al.*, 1994*a*; Pearson *et al.*, 1997). In addition, this system has also been shown to regulate the production of multiple virulence factors including elastase, LasA protease, pyocyanin, and alkaline protease (Ochsner *et al.*, 1994*b*; Brint & Ohman, 1995; Latifi *et al.*, 1995; Ochsner & Reiser, 1995; Pearson *et al.*, 1995; Winson *et al.*, 1995). Winson *et al.* (1995) showed that expression of *rhlI* was positively autoregulated by RhlR/C_4-HSL so that C_4-HSL production activated the production of more C_4-HSL. Finally, Latifi *et al.* (1996) provided evidence for the global regulatory ability of RhlR/C_4-HSL when they showed it induced *rpoS*, which encodes a stationary phase sigma factor.

Quorum sensing controlled genes usually have a sequence in their promoter region that resembles the operator, or *lux* box, of the *V. fischeri lux* operon (Fuqua *et al.*, 1996). The *lasB* promoter region actually has two *lux*-box-like operator sequences that are important for the regulation of *lasB* by quorum sensing (Rust *et al.*, 1996). The presence of these two sequences in the *lasB* promoter region allows one to speculate that they could be the reason that *lasB* is controlled by both the *las* and *rhl* quorum sensing systems as discussed above. Several other *P. aeruginosa* quorum sensing-controlled genes, including *rhlI*, *rhlA*, *lasI*, *lasA*, *rhlR*, and *lasR* have been shown to contain a sequence similar to OP1 operator of *lasB* (ACCTGCCAGTTCTGG-CAGGT) in their promoter region (Freck-O'Donnell & Darzins, 1993; Latifi *et al.*, 1995; Pesci *et al.*, 1997; Albus *et al.*, 1997; Pearson *et al.*, 1997; Pesci & Iglewski, 1998). While the presence of these sequences is quite interesting, further work will be required to determine what, if any, role they play in quorum sensing-dependent gene regulation.

THE HIERARCHY OF *P. AERUGINOSA* QUORUM SENSING

The fact that *P. aeruginosa* contained two separate, complete quorum-sensing systems raised the question of whether these systems were communicating with each other. Pearson *et al.* (1995) suggested that the *las* and *rhl* systems were linked when they presented data that showed the production of C_4-HSL depended on LasR. Interchangeability tests with these two systems showed that 3-oxo-C_{12}-HSL does not activate Rh1R and C_4-HSL does not activate LasR, indicating that the R proteins were specific with regard to the autoinducer that they required for activation (Pearson *et al.*, 1997). Pearson *et al.* (1997) also found that the *las* system preferentially activated *lasB* over *rhlA*, and the *rhl* system preferentially activated *rhlA* over *lasB*, indicating some specificity with regard to the quorum sensing controlled promoters that these systems can activate. The connection between the *las* and *rhl* systems was finally made when Latifi *et al.* (1996) showed that LasR/3-oxo-C_{12}-HSL controlled the transcription of *rhlR*. This meant that the *las* system was atop a hierarchy in which the *rhl* system was only activated after the *las* system became activated. Pesci *et al.* (1997) also demonstrated that the *las* system controlled *rhlR* transcription, and, in addition, discovered that the *las* system controlled the *rhl* system at a post-translational level. They showed that the *las* autoinducer, 3-oxo-C_{12}-HSL, blocked the *rhl* auto-inducer, C_4-HSL, from binding to *E. coli* cells over-expressing RhlR (Pesci *et al.*, 1997). This effect was quite intriguing because it had been shown that C_4-HSL will not block 3-oxo-C_{12}-HSL from binding to cells expressing LasR, and 3-oxo-C_{12}-HSL will not activate RhlR (Passador *et al.*, 1996; Pearson *et al.*, 1997). Pesci *et al.* (1997) also showed that 3-oxo-C_{12}-HSL inhibited the ability of *RhlR* in the presence of C_4-HSL, suggesting that 3-oxo-C_{12}-HSL post-translationally controlled RhlR. The significance of this effect is unknown, but it is speculated that it could allow *P. aeruginosa* to tightly regulate the activation of the *rhl* system by only allowing RhlR to be activated after enough C_4-HSL was present to overcome the blocking effect of 3-oxo-C_{12}-HSL. In support of this theory is the fact that the *las* system must be activated before *rhlR* is induced (Latifi *et al.*, 1996; Pesci *et al.*, 1997), suggesting that the concentration of 3-oxo-C_{12}-HSL will be greater than that of C_4-HSL early in the progression of the quorum sensing signal cascade.

With so much riding on the activation of the *las* system, it becomes apparent that the regulator(s) of *lasR* will control the entire *P. aeruginosa* quorum sensing network. Studies on *lasR* regulation showed that the expression of *lasR* in *P. aeruginosa* was at a basal level until the gene was induced in the last half of log phase growth (Pesci *et al.*, 1997). Pesci *et al.* (1997) also found that *lasR* expression was mildly up-regulated by 3-oxo-C_{12}-HSL, suggesting the possibility of positive autoregulation of *lasR*. Together, these data indicated that *lasR* expression was dependent on cell density. However, Latifi *et al.* (1996) reported that *lasR* was constitutively expressed in *P. aeruginosa*, and that in *E. coli*, LasR negatively autoregulates *lasR* transcription. There is no obvious explanation as to why these well documented reports have conflicting data, but the complex nature of *lasR* regulation (discussed below) could account for the observed differences in *lasR* expression.

Albus *et al.* (1997) discovered that *lasR* was transcribed from two start sites with promoters separated by 30 base pairs, and that *lasR* transcription was positively regulated by a cyclic AMP receptor protein (CRP) consensus sequence located in the *lasR* promoter region. In addition, the *P. aeruginosa* Vfr protein (a homolog CRP) was shown to bind to the *lasR* promoter region, and was absolutely required for even low level *lasR* expression in *P. aeruginosa* (Albus *et al.*, 1997). A second global regulator, GacA, has also been found to have a partial positive effect on both *lasR* and *rhlR* expression (Reimmann *et al.*, 1997). A *P. aeruginosa gacA* mutant produced a decreased amount of C_4-HSL (15% of wild type level), and a wild-type level of 3-oxo-C_{12}-HSL indicating that the effect of *gacA* on *P. aeruginosa* quorum sensing may be complicated and will require further study (Reimann *et al.*, 1997). It is also worthy to note that a *lux* box-like operator sequence has been identified in the promoter region of *lasR* (Albus *et al.*, 1997), which could allow for *lasR* autoregulation as suggested by Pesci *et al.* (1997). Taken together, the above data indicate that multiple factors are involved in a complex scheme that controls *lasR* regulation, and it is speculated that additional, undiscovered *lasR* regulatory mechanisms still exist.

IMPORTANCE OF *P. AERUGINOSA* QUORUM SENSING

P. aeruginosa produces numerous virulence factors, and as discussed above, some of these factors (i.e. elastase, LasA protease, alkaline protease, exotoxin A, and rhamnolipid) are controlled by quorum sensing. Considering this along with the global effectors ($xcpP$, $xcpR$, and $rpoS$) that are regulated by the las and rhl quorum sensing systems, it is not surprising that a $lasR$ mutant has been reported to be less virulent than the wild-type strain in a neonatal mouse model of $P.$ $aeruginosa$ infection (Tang et $al.$, 1996). Another interesting study suggested that the las quorum sensing system was controlling the expression of virulence factors in the lungs of cystic fibrosis patients, providing evidence that quorum sensing occurs in $vivo$ (Storey et $al.$, 1998). It has also been shown that, aside from its effects on virulence factors, 3-oxo-C_{12}-HSL acts as a modulator of the immune response, suggesting a direct role in virulence for this autoinducer molecule (DiMango et $al.$, 1995; Telford et $al.$, 1998).

It is understandable that, with information such as that described above, $P.$ $aeruginosa$ quorum sensing has been viewed in the context of bacterial virulence. Arguments for the in $vivo$ usefulness of quorum sensing often centre on the advantage of delaying virulence factor production early in infection (at low cell densities) until sufficient bacterial numbers are present to overcome the host response that would be elicited by the delayed virulence factors. However, recent data suggest that these bacteria can also use quorum sensing during an environmental lifestyle. $P.$ $aeruginosa$ will form a biofilm on solid surfaces that are exposed to a continuous liquid flow. The complex structures formed by these bacteria implied that perhaps cell–cell communication was involved. This has been confirmed by Davies et $al.$ (1998) who showed that that 3-oxo-C_{12}-HSL was necessary for the formation of a normal biofilm. This research also showed that a $P.$ $aeruginosa$ $lasI$ mutant formed an abnormal biofilm that was much more sensitive to the biocide, sodium dodecyl sulphate, than the wild-type strain (Davies et $al.$, 1998). These studies indicate that a form of multicellular behaviour by $P.$ $aeruginosa$ (biofilm formation) requires the use of intercellular signals that allow individual bacteria to communicate with one another. Such exciting news adds immense importance to the role of $P.$ $aeruginosa$ quorum sensing both in $vivo$ and in $nitro$.

WORK IN PROGRESS

Some very interesting aspects of $P.$ $aeruginosa$ quorum sensing are currently being explored. It has been discovered that a negative regulator is intertwined in the las quorum sensing system (T. deKievit, P. C. Seed & B. H. Iglewski, unpublished observations). This regulator, termed RsaL, is encoded in the intergenic region between $lasR$ and $lasI$ and is transcribed in the opposite direction of these two genes. Transcription of $rsaL$ is dependent on LasR/3-oxo-C_{12}-HSL and RsaL represses the transcription of $lasI$. This repression probably ensures that the low level of LasR and 3-oxo-C_{12}-HSL present at low cell densities cannot activate $lasI$ to cause the las quorum sensing system to be induced before the bacterial population has grown to the desired level. In other work, it was shown that a back-up virulence factor activation system exists in *P. aeruginosa* (Van Delden *et al.*, 1998). Through a secondary mutation, this system allows *P. aeruginosa* to partially suppress a quorum sensing mutation (in *lasR*) and thereby induce some quorum sensing controlled virulence factors genes. And finally, there is a continuing effort on the part of *P. aeruginosa* quorum sensing researchers to discover novel ways to interfere with quorum sensing for the long-term goal of developing improved antimicrobial therapies for this highly resistant pathogen.

CONCLUSIONS

As more is learned about *P. aeruginosa* quorum sensing, it becomes increasingly complex and even confusing to those not directly familiar with it. In order to clarify our view of how quorum sensing works in *P. aeruginosa*, a model is presented that assimilates the current data into a schematic form (Fig. 1). The *P. aeruginosa* quorum sensing cascade probably begins when *lasR* is induced by Vfr (and GacA?) in the presence of enough 3-oxo-C_{12}-HSL (present when cultures become more dense) to overcome the repression of *lasI* by RsaL. The LasR/3-oxo-C_{12}-HSL complex induces *lasI*, leading to a rapid increase in the concentration of 3-oxo-C_{12}-HSL. The cascade will then continue with LasR/3-oxo-C_{12}-HSL activating all *las* controlled genes, including *rhlR*. Once enough RhlR and C_4-HSL are present to overcome the blocking effect of 3-oxo-C_{12}-HSL (which is speculated to be at a much higher concentration than C_4-HSL before the *rhl* system is activated), the RhlR/C_4–HSL complex will form and activate all *rhl* controlled genes.

In summary, the *las* and *rhl* quorum sensing systems are part of a finely tuned network that allows *P. aeruginosa* to regulate the production of numerous genes in response to cell density. While it is apparent from the research presented above that great progress has been made toward understanding this exquisite cell–cell signalling strategy, there is still a lot to be learned. The continuing study of *P. aeruginosa* quorum sensing in the years to come provides hope that a greater understanding of this form of bacterial communication will produce improved treatments for infections caused not only by *P. aeruginosa* but by all Gram-negative bacteria that communicate through quorum sensing.

ACKNOWLEDGEMENTS

This work was supported by National Institutes of Health research grant R01A133713-04, and E. C. Pesci was also supported by Research Fellowship Grant PESCI96FO from the Cystic Fibrosis Foundation.

We thank T. deKievit, and C. S. Pesci for help in manuscript preparation.

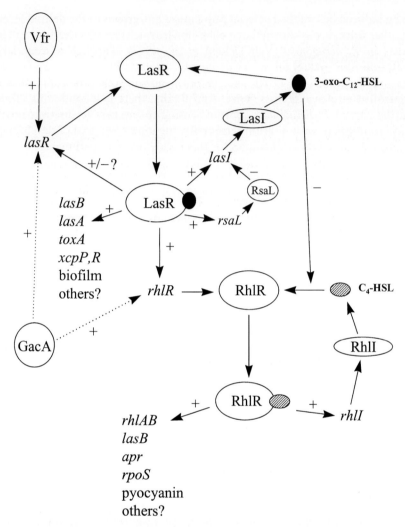

Fig. 1. *P. aeruginosa* quorum sensing. The quorum sensing signalling cascade begins at higher cell densities when *lasR* is induced in the presence of an activating amount of 3-oxo-C_{12}-HSL. It is speculated that the repression of *lasI* by RsaL holds back the quorum sensing cascade until enough LasR/3-oxo-C_{12}-HSL complexes form to override the repression of *lasI*. Once the positive feedback loop of *lasI* regulation begins, the concentration of 3-oxo-C_{12}-HSL would increase and the remaining *las* controlled genes, including *rhlR*, would be induced. Again, it is speculated that 3-oxo-C_{12}-HSL blocks the interaction of RhlR and C_4-HSL until enough RhlR and C_4-HSL are present to overcome the blocking effect of 3-oxo-C_{12}-HSL. At that point, the RhlR/C_4-HSL complex would form and all *rhl* controlled genes would be induced. The reason that the *rhl* system is delayed until after the *las* system is activated is not obvious, but could be due to the stationary phase sigma factor gene (*rpoS*) controlled by *rhl* quorum sensing (Pesci & Iglewski, 1997). Transcriptional activation or repression of a gene or product at the end of an arrow is indicated by plus (+) or minus (−) symbols, respectively. The type of *lasR* autoregulation that occurs is unclear and is therefore indicated by a '+/−?' symbol. The partial effect of GacA on both *lasR* and *rhlR* is indicated by a dotted line.

REFERENCES

Albus, A. M., Pesci, E. C., Runyen-Janecky, L. J., West, S. E. H. & Iglewski, N. H. (1997). Vfr controls quorum sensing in *Pseudomonas aeruginosa*. *Journal of Bacteriology*, **179**, 3928–35.

Brint, J. M. & Ohman, D. E. (1995). Synthesis of multiple exoproducts in *Pseudomonas aeruginosa* is under the control of RhlR-RhlI, another set of regulators in strain PAO1 with homology to the autoinducer-responsive LuxR-LuxI family. *Journal of Bacteriology*, **177**, 7155–63.

Chapon-Herve, V., Akrim, M., Latifi, A., Williams, P., Lazdunski, A. & Bally, M. (1997). Regulation of the *xcp* secretion pathway by multiple quorum-sensing modulons in *Pseudomonas aeruginosa*. *Molecular Microbiology*, **24**, 1169–78.

Davies, D. G., Parsek, M. R., Pearson, J. P., Iglewski, B. H., Costerton, J. W. & Greenberg, E. P. (1998). The involvement of cell-to-cell signals in the development of a bacterial biofilm. *Science*, **280**, 295–8.

DiMango, E., Zar, H. J., Bryan, R. & Prince, A. (1995). Diverse *Pseudomonas aeruginosa* gene products stimulate respiratory epithelial cells to produce interleukin-8. *Journal of Clinical Investigation* **96**, 2204–10.

Freck-O'Donnell, L. C. & Darzins, A. (1993). *Pseudomonas aeruginosa lasA* gene: determination of the transcription start point and analysis of the promoter/regulatory region. *Gene*, **129**, 113–17.

Fuqua, W. C., Winans, S. C. & Greenberg, E. P. (1996). Census and consensus in bacterial ecosystems: the LuxR-LuxI family of quorum-sensing transcriptional regulators. *Annual Review of Microbiology*, **50**, 727–51.

Gambello, M. J. & Iglewski, B. H. (1991). Cloning and characterization of the *Pseudomonas aeruginosa lasR* gene: a transcriptional activator of elastase expression, *Journal of Bacteriology*, **173**, 3000–9.

Gambello, M. J., Kaye, S. & Iglewski, B. H. (1993). LasR of *Pseudomonas aeruginosa* is a transcriptional activator of the alkaline protease gene (*apr*) and an enhancer of exotoxin A expression. *Infection and Immunity*, **61**, 1180–4.

Govan, J. R. W. & Deretic, V. (1996). Microbial pathogenesis in cystic fibrosis: mucoid *Pseudomonas aeruginosa* and *Burkholderia cepacia*. *Microbiological Reviews*, **60**, 539–74.

Jiang, Y., Camara, M., Chhabra, S. R., Hardie, K. R., Bycroft, B. W., Lazdunski, A., Salmond, G. P. C., Stewart, G. S. A. B. & Williams, P. (1998). In vitro biosynthesis of the *Pseudomonas aeruginosa* quorum-sensing signal molecule *N*-butanoyl-L-homoserine lactone. *Molecular Microbiology*, **28**, 193–203.

Jones, S., Yu, B., Bainton, N. J., Birdsall, M., Bycroft, B. W., Chhabra, S. R., Cox, A. J. R., Golby, P., Reeves, P. J., Stephens S., Winson, M. K., Salmond, G. P. C., Stewart, G. S. A. B. & Williams, P. (1993). The *lux* autoinducer regulates the production of exoenzyme virulence determinants in *Erwinia carotovora* and *Pseudomonas aeruginosa*. *The EMBO Journal*, **12**, 2477–82.

Latifi, A., Winson, M. K., Foglino, M., Bycroft, B. W., Stewart, G. S. A. B., Lazdunski, A. & Williams, P. (1995). Multiple homologues of LuxR and LuxI control expression of virulence determinants and secondary metabolites through quorum sensing in *Pseudomonas aeruginosa* PAO1. *Molecular Microbiology*. **17**, 333–43.

Latifi, A., Foglino, M., Tanaka, K., Williams, P. & Lazdunski, A. (1996). A hierarchical quorum-sensing cascade in *Pseudomonas aeruginosa* links the transcriptional activators LasR and RhlR to expression of the stationary-phase sigma factor RpoS. *Molecular Microbiology*, **21**, 1137–46.

Ochsner, U. A., Koch, A. K., Fiechter, A. & Reiser, J. (1994a). Isolation, characterization, and expression in *Escherichia coli* of the *Pseudomonas aeruginosa rhlAB* genes encoding a rhamnosyltransferase involved in rhamnolipid biosurfactant synthesis. *Journal of Biological Chemistry*, **269**, 19 787–95.

Ochsner, U. A., Koch, A. K., Fiechter, A. & Reiser, J. (1994*b*). Isolation and characterization of a regulatory gene affecting rhamnolipid biosurfactant synthesis in *Pseudomonas aeruginosa. Journal of Bacteriology*, **176**, 2044–54.

Ochsner, U. A. & Reiser, J. (1995). Autoinducer-mediated regulation of rhamnolipid biosurfactant synthesis in *Pseudomonas aeruginosa. Proceedings of the National Academy of Sciences, USA*, **92**, 6424–8.

Passador, L., Cook, J. M., Gambello, M. J., Rust, L. & Iglewski, B. H. (1993). Expression of *Pseudomonas aeruginosa* virulence genes requires cell-to-cell communication. *Science*, **260**, 1127–30.

Passador, L., Tucker, K. D., Guertin, K. R., Journet, M. P., Kende, A. S. & Iglewski, B. H. (1996). Functional analysis of the *Pseudomonas aeruginosa* autoinducer PAI. *Journal of Bacteriology*, **178**, 5995–6000.

Pearson, J. P., Gray, K. M., Passador, L., Tucker, K. D., Eberhard, A., Iglewski, B. H. & Greenberg, E. P. (1994). Structure of the autoinducer required for expression of *Pseudomonas aeruginosa* virulence genes. *Proceedings of the National Academy of Sciences, USA*, **91**, 197–201.

Pearson, J. P., Passador, L., Iglewski, B. H. & Greenberg, E. P. (1995). A second *N*-acylhomoserine lactone signal produced by *Pseudomonas aeruginosa. Proceedings of the National Academy of Sciences, USA*, **92**, 1490–4.

Pearson, J. P., Pesci, E. C. & Iglewski, B. H. (1997). Roles of *Pseudomonas aeruginosa las* and *rhl* quorum sensing systems in control of elastase and rhamnolipid biosynthesis genes. *Journal of Bacteriology*, **179**, 5756–67.

Pesci, E. C. & Iglewski, B. H. (1997). The chain of command in *Pseudomonas* quorum sensing. *Trends in Microbiology*, **5**, 132–5.

Pesci, E. C., Pearson, J. P., Seed, P. C. & Iglewski, B. H. (1997). Regulation of *las* and *rhl* quorum sensing in *Pseudomonas aeruginosa. Journal of Bacteriology*, **179**, 3127–32.

Pesci, E. C. & Iglewski, B. H. (1998). Quorum sensing in *Pseudomonas aeruginosa*. In *Cell–Cell Communication in Bacteria*, ed. G. Dunny & S. C. Winans, Washington, DC: American Society for Microbiology Press, in press.

Reimmann, C., Beyeler, M., Latifi, A., Winteler, H., Foglino, M., Lazdunski, A. & Haas, D. (1997). The global activator GacA of *Pseudomonas aeruginosa* PAO positively controls the production of the autoinducer *N*-butyryl-homoserine lactone and the formation of the virulence factors pyocyanin, cyanide, and lipase. *Molecular Microbiology*, **24**, 309–19.

Rust, L., Pesci, E. C. & Iglewski, B. H. (1996). Analysis of the *Pseudomonas aeruginosa* elastase (*lasB*) regulatory region. *Journal of Bacteriology*, **178**, 1134–40.

Seed, P. C., Passador, L. & Iglewski, B. H. (1995). Activation of the *Pseudomonas aeruginosa lasI* gene by LasR and the *Pseudomonas* autoinducer PAI: an autoinduction regulatory hierarchy. *Journal of Bacteriology*, **177**, 654–9.

Storey, D. G., Ujack, E. E., Rabin, H. R. & Mitchell, I. (1998). *Pseudomonas aeruginosa lasR* transcription correlates with the transcription of *lasA*, *lasB*, and *toxA* in chronic lung infections associated with cystic fibrosis. *Infection and Immunity*, **66**, 2521–8.

Tang, H. B., DiMango, E., Bryan, R., Gambello, M., Iglewski, B. H., Goldberg, J. B. & Prince, A.(1996). Contribution of specific *Pseudomonas aeruginosa* virulence factors to pathogenesis of pneumonia in a neonatal mouse model of infection. *Infection and Immunity*, **64**, 37–43.

Telford, G., Wheeler, D., Williams, P., Tomkins, P. T., Appleby, P., Sewell, H., Stewart, G. S. A. B., Bycroft, B. W. & Pritchard, D. I. (1998). The *Pseudomonas aeruginosa* quorum-sensing signal molecule *N*-(3-oxododecanoyl)-L-homoserine lactone has immunomodulatory activity. *Infection and Immunity*, **66**, 36–42.

Toder, D. S., Gambello, M. J. & Iglewski, B. H. (1991). *Pseudomonas aeruginosa* LasA; a second elastase gene under transcriptional control of *lasR*. *Molecular Microbiology*, **5**, 2003–10.

Van Delden, C. & Iglewski, B. H. (1998). Cell-signalling and the pathogenesis of *Pseudomonas aeruginosa* infections. *Emerging Infectious Diseases*, in press.

Vab Delden, C., Pesci, E. C., Pearson, J. P. & Iglewski, B. H. (1998). Starvation selection restores elastase and rhamnolipid production in a *Pseudomonas aeruginosa* quorum-sensing mutant. *Infection and Immunity*, in press.

Winson, M. K., Camara, M., Latifi, A., Foglino, M., Chhabra, S. R., Daykin, M., Bally, M., Chapon, V., Salmond, G. P. C., Bycroft, B. W., Lazdunski, A., Stewart, G. S. A. B. & Williams, P. (1995). Multiple *N*-acyl-L-homoserine lactone signal molecules regulate production of virulence determinants and secondary metabolites in *Pseudomonas aeruginosa*. *Proceedings of the National Academy of Sciences, USA*, **92**, 9427–31.

You, Z., Fukushima, J., Ishiwata, T., Chang, B., Kurata, M., Kawamoto, S., Williams, P. & Okuda, K. (1996). Purifications and characterization of LasR as a DNA-binding protein. *FEMS Microbiology Letters*, **142**, 301–7.

MULTIPLE ROLES FOR ENTEROCOCCAL SEX PHEROMONE PEPTIDES IN CONJUGATION, PLASMID MAINTENANCE AND PATHOGENESIS

GARY M. DUNNY[1], HELMUT HIRT[2] AND STANLEY ERLANDSEN[2]

Department of Microbiology[1] and Cell Biology and Neuroanatomy[2]
University of Minnesota Medical School, Minneapolis, MN 55455,
USA

INTRODUCTION

This chapter reviews some of the important features of a novel conjugative transfer system in the genus *Enterococcus*. In this system the donors of genetic information sense the presence of potential recipients by recognizing peptide signals excreted by the recipient cells during normal growth. These peptide pheromones serve as an indication to the donor cells of the cell density of recipients in a particular ecological niche, such that conjugative transfer functions are only expressed under conditions where the probability that the donor cell will encounter a mating partner is high. In the quarter century in which this type of genetic transfer has been studied, a number of the molecular details of the signalling process have been ascertained, although many important aspects of the system remain mysterious. One important result of the work that has been done during the past few years is the realization that the signal molecules and the gene products they regulate play a much broader role in the biology of these organisms than has been appreciated previously. Several reviews of the enterococcal pheromone systems have appeared in the last few years (Wirth, 1994; Dunny & Leonard, 1997; Clewell, 1998). The aspects of the system that have been discussed extensively in the recent past will not be repeated in great detail. Instead, the general features of these systems will be summarized with some updates, and the rest of the text will focus on some new elements of the system that have emerged from recent work. The review will conclude with an indication of important future directions and potential applications of this research.

A BRIEF HISTORY

For over 20 years following the discovery of conjugation in *Escherichia coli*, there was no evidence for a similar process in the 'low GC' branch of the Gram-positive bacteria. The first hint that a gene transfer process involving direct formation of a mating channel between the transferring bacteria came from the work of Tomura *et al.* (1973). These investigators found that the genetic capacity for bacteriocin production could be transferred between strains of *Enterococcus faecalis* (at the time called *Streptococcus faecalis*) in mixed cultures by a process that was unaffected by the addition of nucleases, and required direct contact between viable donor and recipient cells. At the time this work was published, physical characterization of DNA from Gram-positive bacteria was in its infancy and no molecular analysis of the transferred genes was carried out to support the genetic observations. However, this report was followed in rapid succession by two papers from A. Jacob's group that provided convincing molecular and genetic data for transfer of haemolysin/bacteriocin and antibiotic resistance genes carried by plasmids (Jacob & Hobbs, 1974; Jacob *et al.*, 1975). Dunny and Clewell (1975) then described a similar form of transfer for a haemolysin/bacteriocin plasmid, pAM g1 of *E. faecalis* strain DS5. Further analysis of the kinetics of transfer of pAM g1 and similar plasmids from other strains suggested that efficient transfer only occurred after co-cultivation of donor and recipient cells for about 1–2 h. It was also noted that aggregates of cells, visible to the naked eye, were often present under conditions where high-frequency transfer was occurring. A series of conditioned medium experiments demonstrated that recipient cells were secreting an extracellular factor that could induce expression of aggregation and high frequency transfer phenotypes in donor cells. The low-molecular weight, heat stable, protease sensitive molecule was originally termed clumping inducing agent (CIA), and had the properties of a bacterial sex pheromone (Dunny *et al.*, 1978). Further analysis suggested that plasmid-free recipient cells produced a number of different pheromones, each with activities for a different family of conjugative plasmids (Dunny *et al.*, 1979). The model proposed for pheromone-inducible plasmid transfer in enterococci shown in Fig. 1 has largely remained since its initial formulation (Dunny *et al.*, 1979).

During the next ten years the most important work in this area included the identification of several pheromones and plasmid-determined inhibitor molecules as small hydrophobic peptides (Mori *et al.*, 1984, 1986, 1988; Suzuki *et al.*, 1984; Clewell *et al.*, 1990; Nakayama *et al.*, 1994; and reviewed in Clewell (1998)). (As noted below, the genes required for pheromone production have remained elusive.) In addition, considerable progress was made in identification and characterization of the plasmid genes and gene products whose expression is induced by pheromones. Significant work was also done on the regulatory genetic loci controlling the system. Detailed molecular studies have focused on three plasmids thus far. These include tetracycline-resistance plasmid pCF10 (Dunny *et al.*, 1981) and the haemolysin/bacteriocin plasmid pAD1 (Tomich *et al.*, 1979). Recently considerable analysis has also been done with a second haemolysin plasmid pPD1 (Yagi *et al.*, 1983). The regulatory circuits for each plasmid are very complex and

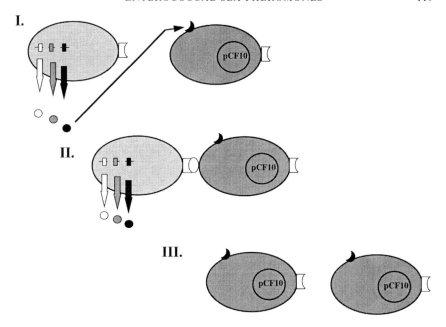

Fig. 1. Model for pheromone inducible conjugation in *Enterococcus faecalis*. I. A plasmid-free recipient cell (left side) excretes several different peptides (filled circles) into its growth medium. II. One of these peptides (cCF10) is recognized by a specific binding protein (half-moon) on a pCF10-containing donor cell (right side), and expression of pCF10-encoded transfer functions is induced. Binding of the induced aggregation substance (PrgB) protein (open circle) to the complementary enterococcal binding substance (EBS) receptor facilitates stable mating pair formation, allowing efficient plasmid transfer. III. The newly created donor cell separates from its mating partner, and behaves subsequently like the original donor, excreting no net cCF10 activity (but continuing to produce the other pheromones), and not expressing PrgB in the absence of exogenously supplied cCF10.

show some common features, as well as several properties unique to each plasmid. The conjugation gene products induced by pheromones that have been characterized so far are quite homologous (Galli & Wirth, 1991; Hirt *et al.*, 1996).

ECOLOGY AND MEDICAL SIGNIFICANCE OF THE ENTEROCOCCI

Enterococci are normal inhabitants of the intestinal tracts of humans and many animals, although they can occasionally be found in other body sites such as the vagina and the oral cavity (Murray, 1990; Devriese *et al.*, 1992). They are also common inhabitants of municipal sewage. The modern sewage treatment facility can serve as an ecological niche in which conjugative plasmid transfer occurs among the resident enterococcal population (Marcinek *et al.*, 1998). The extent to which enterococci may persist as free-living organisms in habitats not influenced by humans or other animals is not completely known, although they have been identified in association with plants (Devriese *et al.*, 1992).

In contrast to many pathogenic streptococcal species, enterococci tend to be much less autolytic, and better able to survive starvation, desiccation, and other forms of stress (Murray, 1990; Devriese *et al.*, 1992). They also have a high level of intrinsic resistance to antimicrobial agents, even in the absence of plasmids and transposons (Murray, 1990). These traits no doubt play a significant role in the ability of the organisms to cause nosocomial infections and in the dissemination of resistance genes. There is also experimental evidence to suggest that environmental conditions can have significant effects on the same transfer systems whose expression is regulated by pheromones (Dunny *et al.*, 1982; Weaver & Clewell, 1991; Marcinek *et al.*, 1998).

There is substantial evidence for persistence of certain epidemic strains for prolonged periods, in particular hospital environments (Murray, 1990; Jett *et al.*, 1994), as well as evidence for the persistence of a particular pheromone plasmid family (pCF10-like) in different enterococcal strains in single hospital for over 15 years (Heaton *et al.*, 1996). It is likely that the ability of certain strains to colonize new individuals is of great importance in enterococcal nosocomial infections. As discussed below, pheromone plasmids may encode gene products (in addition to resistance) that contribute to the ability of their bacterial hosts to colonize humans.

Enterococci have been implicated in urinary tract infections, endocarditis, bacteraemia, surgical wound infections and various polymicrobial infections. There is some controversy about the extent to which these organisms are actual pathogenic agents in some opportunistic infections, but they are clearly capable of causing endocarditis and bladder infections in the absence of other bacteria (see Murray, 1990; Devriese *et al.*, 1992; Jett *et al.*, 1994) for further review of this subject). The best studied example of an enterococcal virulence factor playing a critical role in a specific disease process is in enterococcal endophthalmitis, which occurs in humans and can be reproduced in a rabbit model (Jett *et al.*, 1992). In this model, expression of the plasmid pAD1-encoded cytolysin (haemolysin) greatly increased the progression of ocular damage in infected animals (Jett *et al.*, 1992). Several other potential virulence factors, including protease, hylauronidase, and lipoteichoic acid have been identified, as well as some enterococcal proteins and carbohydrates shown to induce immune responses in infected patients (Jett *et al.*, 1994). However, there is little direct evidence to confirm the involvement of any of these factors in virulence. Roles for pheromone-inducible gene products in virulence will be further discussed in a subsequent section.

PHEROMONES AND THEIR BIOSYNTHESIS

In all the pheromone plasmids analysed to date, it has been found that the expression of transfer functions is induced by a chromosomally encoded peptide (pheromone), whose activity can be antagonized by a plasmid-encoded peptide (inhibitor). To date, about 20 different sex pheromone plasmids have been identified (Wirth *et al.*, 1992). These can be grouped into families of closely related elements that all determine a response to a single peptide. The total number of peptide pheromones encoded by the enterococcal genome is not known, but the minimum number is in the range of 6–10 (Dunny *et al.*, 1979; Wirth *et al.*, 1992). All pheromones and inhibitors are hydrophobic peptides 7–8 amino acids in length; the inhibitors typically share 40–60% identical amino acid residues with their respective pheromone. A nomenclature system has been developed that designates pheromones and inhibitors according to the name of the cognate responding plasmid (Wirth *et al.*, 1991). For example, cCF10 (sequence: LVTLVFV) and iCF10 (sequence: AITLIFI) serve as the pheromone and inhibitor peptides, respectively, in the case of pCF10 (Mori *et al.*, 1988; Nakayama *et al.*, 1994). Comprehensive lists of all the pheromones and inhibitors whose sequences have been determined can be found in recent reviews (Wirth, 1994; Clewell, 1998). In the case of the inhibitors, analysis of the coding sequences has shown that the mature, active peptides each represent the C-terminal portion of a 20–25 amino acid precursor, which itself resembles a signal peptide (Clewell *et al.*, 1990; Nakayama *et al.*, 1994). Apparently, no special processing machinery is required for inhibitor secretion, since expression in an *E. coli* host of the pCF10 Orf (prgQ) encoding iCF10 results in release of active inhibitor into the culture medium (Nakayama *et al.*, 1994). However, it is not clear that pheromone biosynthesis proceeds in the same way, since the biosynthetic genes for the pheromones have not been identified. There is no compelling evidence to indicate that these compounds are synthesized ribosomally, although protein synthesis inhibition studies are consistent with this (B.A.B. Leonard and G.M Dunny, unpublished data). In addition, recent studies (M. Antiporta, B.A.B. Leonard and G.M. Dunny, unpublished data) have shown that both membrane and wall fractions of pheromone-producing *E. faecalis* cells contain substantial pheromone activity equal to, or greater than, the amounts released into the medium by the same cells. It remains to be determined whether this cell-associated activity is reflective of a post-translational processing pathway at the cell membrane similar to those involved in synthesis of peptide pheromones and bacteriocins in other Gram-positive bacteria (Nes *et al.*, 1996). At the time of writing, no successful attempts to clone pheromone biosynthetic genes have been reported. The final identification will depend on the availability of the complete genome sequence for *E. faecalis*. A search of the available sequence, as of June, 1998, from the Institute for Genomic Research (ftp:ftp.tigr.org/pub/data/e_faecalis/) revealed two potential genes whose proteolytic products could give rise to cCF10; it remains to be determined whether either represents the structural gene. A remarkable aspect of this form of signalling is the high selectivity of each of these peptides for a specific plasmid (or family of plasmids), in spite of the fact that all these peptides are very hydrophobic and similar in size. As will be discussed below, there are probably at least three distinct specific cellular targets for the cCF10 peptide, implying an impressive range of function for such a small molecule.

AN OVERVIEW OF THE MATING RESPONSE TO PHEROMONES

Figure 1 depicts the current understanding of the cycle of pheromone-inducible conjugation between two *E. faecalis* cells. For illustrative purposes, pCF10-encoded gene products will be described. However, similar sets of proteins with corresponding functions are encoded by the other pheromone plasmids, such as pPD1 and pAD1, that have been examined in detail (Clewell, 1998). The plasmid-free recipient cell secretes several peptides, one of which (cCF10) is recognized as a pheromone by the donor cell carrying the plasmid, pCF10. Initially, the specificity of the response is determined by the plasmid-encoded pheromone binding protein, PrgZ. This protein is a homologue of the chromosomal oligopeptide permease (Opp) binding protein OppA, but is highly specific for the cCF10 peptide (Ruhfel *et al.*, 1993; Leonard *et al.*, 1996). Pheromone signalling involves recruitment of the chromosomal Opp system by PrgZ to import the peptide (Leonard *et al.*, 1996). Once inside the cell, the peptide interacts with one or more cytoplasmic effector molecules to initiate the mating response (Leonard *et al.*, 1996; Bensing *et al.*, 1997).

Up-regulation of transfer functions includes increased cell surface expression of an aggregation substance protein encoded by the *prgB* gene (Kao *et al.*, 1991; Olmsted *et al.*, 1991). (The protein product of the *prgB* gene has been previously denoted Asc10, following a suggested convention for designation of pheromone plasmid gene products (Wirth *et al.*, 1991). However, to simplify the present discussion, the transfer proteins will be named according to standard bacterial genetic nomenclature; thus PrgB will refer to the protein product of the *prgB* gene.) Formation of a mating pair (or mating aggregates in dense cultures) occurs by interaction of PrgB with the cognate chromosomally encoded surface receptor, enterococcal binding substance (EBS), on the recipient cell. Both the genetics and biochemistry of EBS are complex; however, several lines of evidence suggest that lipoteichoic acid is a component of EBS (Ehrenfeld *et al.*, 1986; Trotter & Dunny, 1990). The initial attachment of the cells is probably followed by formation of a mating channel that facilitates transfer of a copy of the plasmid via a mechanism believed to be similar to that of the *E. coli* sex factor F and similar plasmids (Clewell, 1991). The available evidence suggests that AS-EBS binding can be separated from the subsequent steps that appear to involve other transfer gene products (Olmsted *et al.*, 1991). Little information about the steps that follow mating pair formation is available at present.

Following transfer, the new donor cell created by the conjugation event has the genetic capacity to both produce cCF10 and respond to cCF10. However this cell generally behaves like the original donor, only responding to exogenous cCF10, but not undergoing self-induction by endogenously produced pheromone. The two regulatory circuits controlling these two phenotypes are each controlled by separate sets of pCF10 genes, as described in the next section. In addition to avoiding self-induction, the new donor strain is not capable of inducing other strains carrying pCF10-like plasmids. However, culture supernatants of the pCF10-containing strain continue to produce other pheromones, such as cAD1 and cPD1, maintaining the ability of the strain to acquire unrelated pheromone plasmids via conjugation. Clinical isolates often carry several different conjugative plasmids, each determining a response to a unique pheromone (Dunny *et al.*, 1979; Wirth *et al.*, 1992; Hirt *et al.*, 1996).

REGULATORY CONTROL CIRCUITS

As described in the previous section, negative control in the pheromone systems is manifested in the ability of the donor cell to avoid self-induction by endogenous pheromone. In the pCF10 system, three negative regulatory genes have been identified, and two of these appear to act at the level of self-induction. Figure 2 illustrates the current working model for control of

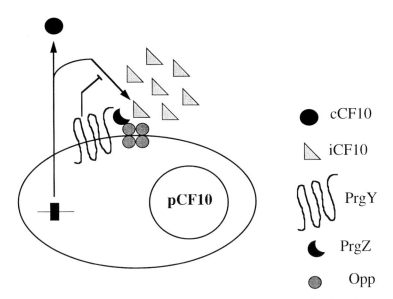

cCF10

iCF10

PrgY

PrgZ

Opp

pCF10

Fig. 2. Model for control of self-induction in pCF10-containing donor cells. The cell excretes the same amount of cCF10 into the medium as an isogenic plasmid-free cell, but this activity is neutralized by plasmid-encoded iCF10. PrgY blocks self-induction resulting from re-internalization of cCF10 as it is secreted across the membrane, but prior to its release from the cell, at the membrane.

endogenous pheromone activity. When unfractionated culture supernatants of pCF10-containing strains are analysed, no cCF10 activity is found (Nakayama *et al.*, 1994). However, when the same material is fractionated by high-pressure liquid chromatography, two distinct peptides with opposing biological activities can be isolated. One of these corresponds to cCF10 and is produced in approximately the same amount as is found for plasmid-free strains. A second peak of pheromone inhibitor activity can also be identified, mediated by a plasmid-encoded peptide, iCF10, which is the product of a short open reading frame encoded by the *prgQ* gene (Nakayama *et al.*, 1994). (As described below, the RNA products of the *prgQ* region are much larger than the polypeptide-coding segment, and these RNAs play a major role in positive control.) Biochemical and physiological analysis of the two peptides indicates that their production is co-regulated, such that a pure culture of donor cells produces them in a ratio (about 40-fold molar excess of inhibitor) that is equivalent in biological activity. This balance is extremely delicate, since a very small amount ($\leqslant 5$ molecules/cell) of exogenously added cCF10 peptide is sufficient to induce a mating response (Mori *et al.*, 1988). Taken together, these data suggest that the primary function of iCF10 is to control self-induction in the extracellular culture medium of donor cells.

In view of the previously described evidence for cell-associated pheromone activity, an autocrine induction circuit at the cell membrane mediated by endogenously synthesized pCF10 could also be envisioned. It is conceivable that cCF10 could be recognized by PrgZ as it is secreted across the membrane, and be immediately re-internalized by the Opp system, causing self-induction. Several lines of evidence suggest that PrgY, a putative membrane protein encoded in the same operon as *prgZ* (Ruhfel *et al.*, 1993), acts to prevent such a process. PrgY may function by direct interaction with cCF10, or by interaction with PrgZ, or both. More detailed models for PrgY activity will need to account for the fact that the signalling from endogenous peptide is blocked without interfering with the ability of the cell to sense the same peptide added exogenously. This multicomponent model for negative control predicts self-induction could occur either: (i) extracellularly by altering production, activity or stability of the secreted peptides; or (ii) intracellularly, by alteration of the expression or activity of the cell-associated pheromone, or of PrgY.

In the case of the pAD1 and pPD1 systems, *prgY* homologues (*traB*) have been identified (An & Clewell, 1994; Nakayama *et al.*, 1995). It has been observed that pAD1-containing (TraB$^+$) strains seem to have only iAD1 but not cAD1 in the medium. Thus the *traB* gene has been termed a 'shutdown' gene for endogenous pheromone synthesis (An & Clewell, 1994). While this might imply different mechanisms of action for TraB versus PrgY, it seems more likely that the observed differences are quantitative rather than qualitative. They could also reflect differential abilities of the respective pheromones to bind to cell envelope components prior to release into the growth medium.

A 1.5–2.0 kb positive control segment of pCF10 encompassing the *prgQ-PrgS* loci is required for expression of *prgB* and other pheromone-inducible transfer functions (Chung & Dunny, 1992; Chung *et al.*, 1995). The current model for positive regulation in this system is illustrated in Fig. 3. The published data upon which this model is based has been discussed in detail in two recent reviews (Dunny & Leonard, 1997; Clewell, 1998). In discussing the regulatory system, the *prgB* gene will represent the target for regulation. The most important feature of the model is that all relevant transcription of *prgB* is initiated from the very strong, constitutive *prgQ* promoter located about 5 kb upstream of *prgB*. In the uninduced cells, transcription is believed to proceed through the *prgS* gene, with termination frequently occurring past the 3′ end of *prgS* (Bensing *et al.*, 1996, 1997), well upstream of *prgB*. Although the positive control region mRNA is produced in abundance in the absence of pheromone induction, there is little or no translation of the message, with the exception of the inhibitor precursor peptide at the extreme 5′ end of the transcript. Since the amount of inhibitor produced is quite low

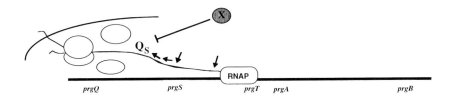

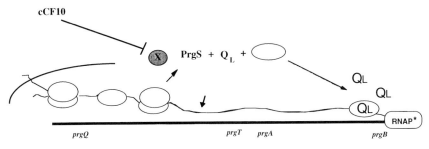

Fig. 3. Positive regulation of *prgB* expression during the intracellular phase of pheromone induction. In the top panel, a high level of transcription occurs from the *prgQ* promoter, generally terminating near the 3′ end of *prgS*. However, there is a barrier (possibly mediated by PrgX) to efficient translation of this message, and it is rapidly degraded (arrows) to stable processing products such as QS. Pheromone induction, possibly via direct cCF10–PrgX interaction, overcomes the translation block, and allows PrgS to be made. This results in stable association of QL with ribosomes, increasing their ability to translate *prgB* mRNA. RNA polymerase is also termination-resistant (denoted by the asterisk), such that more *prgB* message is made.

in comparison to the level of its mRNA, inhibitor translation may also be relatively inefficient. This could provide a means to keep inhibitor synthesis coordinated with the level of pheromone in the culture medium. The gene product responsible for translational inhibition is not known. However, the best candidate at the present time is the product of *prgX*, known to be a negative regulator (Hedberg *et al.*, 1996), and probably acting in the cytoplasm. In the absence of translation, there is rapid degradation of the message. Some relatively stable processing products observed include a 430 nt RNA, QS, from the *prgQ* region, some smaller segments of the same region, and a transcript extending from the central portion of *prgS*, into a 3′ downstream region (Bensing *et al.*, 1997).

Once exogenous pheromone is added and imported into the responder cell, it appears to interact with one or more intracellular effector molecules, with the result being the relief of the translational block, resulting in the synthesis of the PrgS protein. The expression of PrgS is correlated with a stable association of a 530 nt RNA, QL with ribosomes; the QL-modified ribosomes apparently have increased ability to translate *prgB* mRNA (Bensing & Dunny, 1997). The association of QL with ribosomes is believed to stabilize this RNA, inhibiting its processing to the QS form (Bensing & Dunny, 1997). Induced cells also display termination resistant extension of the transcripts initiating from the *prgQ* promoter such that they proceed beyond the 3′ end of *prgS*, into the downstream transfer genes such as *prgB* (Bensing *et al.*, 1997). Whether the transcription and translational effects are two independent responses to pheromone induction or different manifestations of a single response is not clear at present. One important feature of the system is that cCF10 has a direct, active role in the induction, as opposed to simply displacing iCF10 from an intracellular target. It is likely that the primary site of competition between the two peptides is at the external face of the cytoplasmic membrane with PrgZ being the target receptor (Ruhfel *et al.*, 1993; Leonard *et al.*, 1996). Although iCF10 may have a minor intracellular role in negative control, the demonstration of inducibility of a mutant strain unable to produce iCF10 provided compelling genetic evidence for a direct, active role of pheromone in the induction process (Bensing *et al.*, 1997). A key question that remains unanswered at present is the molecular identity of the intracellular target(s) for cCF10 binding. Previous biochemical data suggested ribosomal components (Leonard *et al.*, 1996; Bensing & Dunny, 1997) and QL as potential targets, while recent data (T. Bae and G.M. Dunny, unpublished results) implicates PrgX as a possible target. It is quite conceivable that some, or all, of these candidates form a ribonucleoprotein complex in the cell which mediates the mating response (Leonard *et al.*, 1996), with pheromone directly interacting with only one component.

In the case of pAD1, there is production of mRNAs with almost identical size and sequence as those encoded by the Q region of pCF10 (Galli *et al.*, 1992; Tanimoto & Clewell, 1993). However, the 3′ portions of the positive control regions of the two plasmids are completely different, with no *prgS* homologue being present in pAD1. Instead, pAD1 encodes a protein called TraE1, which is a transcriptional activator (Tanimoto & Clewell, 1993; Muscholl *et al.*, 1993). In addition, *asa1*, the *prgB* homologue in pAD1, is preceded by a small Orf whose 5' end likely represents the transcription initiation site for *asa1* (Muscholl *et al.*, 1993). The model currently favoured by Clewell and coworkers (Clewell, 1998) postulates a cascade mechanism where induction of pheromone results in antitermination in a region (IRS1) that would correspond to the 3' end of QS. Antitermination allows transcription to extend into *traE1*, whose protein product then activates both its own transcription, as well as that of *asa1*. It has been suggested that the key intracellular mediator is TraA, which is a homologue of PrgX in the pCF10 system (Pontius & Clewell, 1992). An *in vitro* TraA binding site in the DNA near the upstream promoter in the positive control region has been identified, and TraA has also been shown to bind pheromone (Fujimoto & Clewell, 1998). It is suggested that the DNA binding of TraA in the uninduced cell enhances downstream termination of the positive control transcripts before they reach the *traE1* gene. Pheromone binding is believed to release TraA from the DNA and relieve antitermination. One potential inconsistency with the model is that iAD1 binding to TraA seems to have a larger inhibitory effect on the DNA binding activity of TraA than that observed for cAD1 (Fujimoto & Clewell, 1998). If this occurred *in vivo*, iCF10 would be predicted to induce, rather than inhibit expression. An alternative possibility is that TraA could also have either an *in vivo* DNA binding activity that is different from that observed *in vitro*, or an *in vivo* RNA binding activity. Alteration of this hypothetical activity would then be predicted to affected by pheromone binding. Since the TraA negative regulator binds pheromone in both the pAD1 (Fujimoto & Clewell, 1998) and pPD1 (Nakayama *et al.*, 1998) systems; it is quite likely that the same thing is true for PrgX. It is also likely that pheromone binding somehow interferes with the negative regulatory function of the proteins in each system. An additional negative regulator has been identified in the pAD1 system. A genetic locus called *traD*, encoding an antisense RNA for the positive control region, has been identified as an element that can modulate the level of transfer gene expression in the pAD1 system. This RNA is decreased in induced cells (Freire Bastos *et al.*, 1997).

HOST RANGE OF PHEROMONE PLASMIDS: PHEROMONE PRODUCTION BY THE HOST STRAIN MAY BE REQUIRED FOR PLASMID MAINTENANCE

To date, pheromone plasmids have only been identified in *Enterococcus* spp. (*E. faecalis* and *E. faecium*) (Handwerger *et al.*, 1990; Heaton & Handwerger, 1995; Clewell, 1998). Strains from a few other Gram-positive bacterial species have been found to produce a peptide that can induce a specific pheromone plasmid (Clewell, 1998), but only enterococci produce the full range of peptide pheromones. Thus, host range is correlated with pheromone production. The minimal replicons of pAD1 and pCF10 are in regions of the plasmids that also determine functions involved in pheromone sensing and in controlling self-induction by endogenous pheromone (Weaver et al., 1993; Hedberg et al., 1996). These results suggest the possibility of some sort of involvement of endogenous pheromone in plasmid maintenance. If this is the case, the pheromone-dependent regulatory circuits involved must be distinct from those involved in induction of conjugation, since the latter only occurs in the presence of exogenously supplied pheromone.

Two recent observations provide more evidence for involvement of endogenous pheromone in plasmid maintenance. The first of these is that PrgW can bind cCF10, as demonstrated by chromatography experiments utilizing a cCF10-coupled affinity column (B.A.B. Leonard and G.M. Dunny, unpublished data). The second line of evidence is that the host range of the pCF10 replicon can be extended to *Lactococcus lactis* by insertion into this species a mutant derivative of *prgQ* that encodes production of cCF10 rather than iCF10 (B.A.B. Leonard, A. Colwell and G.M. Dunny, unpublished data). This indicates that plasmid replication may be regulated by an autocrine circuit involving endogenous cCF10. This regulatory loop must be spatially or temporally separated from the circuits controlling the induction of conjugation by exogenous pheromone. In any case, the ability of these plasmids to be maintained in enterococcal populations may be related to the role of pheromone peptides in replication.

BEHAVIOUR OF PHEROMONE PLASMIDS IN ENTEROCOCCI INHABITING MAMMALIAN HOSTS

An important question regarding these plasmids is the efficiency at which conjugative transfer occurs in a mammalian host. As indicated in Table 1, plasmid transfer in the rodent GI tract, in blood, or in tissues is remarkably high. Transfer in these environments compares with frequencies observed in laboratory matings where the mating bacteria were co-cultured for sufficient lengths of time to allow the donors to respond to the pheromone produced by the recipients. These observations suggest that induction of transfer functions occurs *in vivo*. This, in turn, implies that donor cells are occupying an ecological niche where the pheromone produced by recipients can accumulate to a level sufficient for induction, or that induction can occur by other mechanisms.

In the course of analysing pheromone-inducible conjugation, numerous strains have been constructed in several laboratories with well-defined genotypes relating to cell surface properties. Since differences in such properties could affect the interaction of the organism with a mammalian host, there has been considerable interest in examining whether plasmid and chromo-

Table 1. *Transfer of pheromone plasmids* in vivo *and* in vitro

	In vitro (4 h)	Hamster GI tract (24–72 h)	Rabbit subdermal transplant (48 h)	Rabbit heart vegetation (72 h)
Transconjugants/donor (pAD1)	10^{-1}–10^{-3}	10^{-1}–10^{-2}	–	–
Transconjugants/donor (pCF10)	10^{-1}–10^{-3}	–	10^{-2}	$3 \cdot 10^{-2}$–$9 \cdot 10^{-2}$

Transfer frequencies expressed as transconjugants/donor of the plasmids pAD1 and pCF10 in different host environments.
The transfer frequencies in the hamster gastro intestinal tract were determined by Huycke *et al.* (1992). For the subdermal transplant a hollow plastic ball, approximately 3 cm in diameter was implanted in rabbits. A donor and a recipient strain were co-inoculated and the number of donor, recipient and transconjugant cells determined after two days. For determination of plasmid transfer in heart vegetations, experimental endocarditis was established as described previously (Schlievert *et al.*, 1998). The rabbits were co-challenged by a donor and recipient strain and the formed vegetations were assessed after 3 days for cell numbers.

somal gene products involved in pheromone-inducible conjugation might affect the pathogenic properties of enterococcal strains. Figure 4 illustrates some of the cell surface components that have been identified and examined in the pCF10 system. At least two of these components affect the virulence of the host bacteria in one or more model systems, as described below.

Several published reports have suggested that pheromone plasmid genes can increase the pathogenicity of enterococcal strains in experimental systems. Kreft *et al.* (1992) demonstrated the potential enhancement of urinary tract infections by plasmid pAD1, by showing that adherence of enterococci to a canine kidney cell line was increased in plasmid-containing strains. The aggregation substance protein Asa1 was implicated as a critical adhesin, and preliminary evidence for upregulation of its expression in the presence of serum was presented. Chow *et al.* (1993), using an endocarditis model carried out another study involving pAD1. The results from these studies suggested that both Asa1 and the plasmid-encoded cytolysin (hae-molysin) contributed to virulence, with increased lethality associated with cytolysin and increased mass of the induced cardiac vegetations associated with Asa1. Olmsted *et al.* (1994) examined the effects of surface expression of the pCF10-encoded, pheromone-inducible aggregation substance and surface exclusion proteins (encoded by the *prgB* and *prgA* genes, respectively; Kao *et al.*, 1991), on the internalization of enterococci by HT29 intestinal epithelial cells. While the latter protein had no effect on internalization, PrgB expression caused a significant increase, suggesting that this protein could contribute to invasiveness of these organisms.

This same pCF10-encoded aggregation protein was recently shown by Schlievert *et al.* (1998) to enhance *E. faecalis* virulence in the rabbit endocarditis model in the absence of any cytolysin. Expression of wild-type EBS was also required for full virulence. In addition to increased vegetation

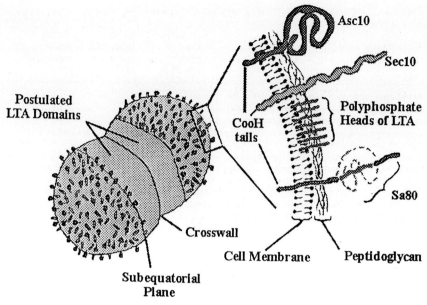

Fig. 4. Cell surface components of *E. faecalis* identified during analysis of pheromone-inducible transfer of pCF10. The distribution of the aggregation substance (Asc10) and surface exclusion (Sec10) proteins (encoded by *prgB* and *prgA*, respectively; Kao *et al.*, 1991)) was inferred from field-emission scanning electron microscopic analysis (Olmsted *et al.*, 1993), and the predicted structures derived by computer analysis of the amino acid sequences (Kao *et al.*, 1991). The polyphosphate 'tails' of lipoteichoic acid (LTA) are depicted extending from the cell surface. As noted in the text, LTA is likely to be a major component of EBS (Trotter & Dunny, 1990). The use of recombinant strains carrying cloned fragments of pCF10, as well as strains with insertion mutations in EBS genes, enabled the examination of the role of these components in virulence (Olmsted *et al.*, 1994; Schlievert *et al.*, 1998).

sizes, the rabbits infected with the PrgB⁺, EBS⁺ strains showed other signs of severe illness, including fever, swollen spleens and acute lethargy. Several of these animals actually died in the course of the experiments with a toxic shock-like illness characterized by extensive tissue destruction of both the heart and lungs, and a complete lack of an inflammatory response. This invasive infection was never observed with strains deficient in expression of either EBS or PrgB; EBS⁻, PrgB⁻ strains were completely avirulent in the rabbits (Schlievert *et al.*, 1998). A spleen cell mitogenic factor could be extracted from the cell wall of the PrgB⁺, EBS⁺ strains. The systemic effects on the animals could be related to the activity of this factor. These studies were carried out with recombinant strains carrying cloned fragments of pCF10, constructed such that the PrgB⁺ strains expressed the protein constitutively. Given the apparent importance of the *prgB* gene product in enterococcal virulence in this system, determination of the extent to which this gene would be expressed in the animal host by bacteria carrying wild-type pCF10 became very important.

To assess the extent of *in vivo prgB* expression in the context of wild-type pCF10, isogenic strains differing only in plasmid content were again employed in the rabbit model (H. Hirt, P.M. Schlievert and G.M. Dunny, unpublished data). A strain carrying pCF10 produced large vegetations comparable to those seen previously (Schlievert *et al.*, 1998) in rabbits infected with the recombinant PrgB[+] strains, while the virulence of a *prgB*::Tn917 derivative of pCF10 was identical to that of a plasmid-free strain. This result not only confirmed the importance of the Aggregation Substance encoded by *prgB* in virulence, but also suggested that its expression was induced in the animal host following the introduction of the organisms into the rabbit bloodstream.

In subsequent experiments it was found that there is a factor in rabbit and human plasma that can induce both expression of *prgB*, and conjugative transfer of pCF10. Further analysis suggests that a complex of serum albumin with fatty acids can interact with both cCF10 and iCF10. The current hypothesis is that this binding alters the normal ratio of inhibitor versus pheromone activity (Fig. 2) in the medium of a donor cell culture. Because of this alteration, the pheromone is not completely neutralized by the inhibitor. This would cause self-induction. The process may actually occur *in vivo* in a microenvironment created when the bacterial cells are coated with albumin, while colonizing a surface such as a heart valve, or in the bloodstream. The current model for *in vivo* induction is shown in Fig. 5. Although further work is needed to confirm the molecular details of the model, the presently available data are highly suggestive of a mechanism involving disruption of the bacterial mechanism for control of self-induction as opposed to induction by an animal host-encoded pheromone-like factor.

The evidence for effects of pheromone inducible gene products on virulence-related phenotypes, and for efficient plasmid transfer in vivo, is very consistent an important role of these plasmids in the ecology of enterococci in the hospital environment. It is clear that this role extends beyond the fact that the pheromone plasmids often carry resistance genes, which are selected for by antibiotic use.

PRIORITIES FOR FUTURE INVESTIGATION

Experimental analysis of the enterococcal pheromone plasmids has reached a stage where detailed understanding of many of the important aspects of the system in the near future is a realistic expectation. In particular, the identification of all the direct molecular targets for the peptides, and of the other regulatory components with which these targets interact should be achieved within the next few years. This information should allow for the confirmation, refinement, or modification of the models for positive and negative regulation of the system described above. In addition, more emphasis on the mechanism of conjugative plasmid transfer (as opposed to

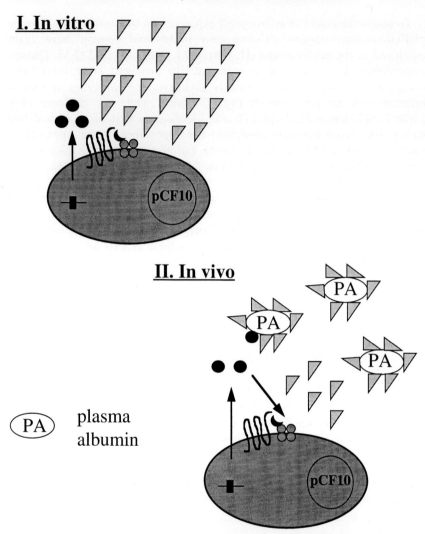

Fig. 5. Model for induction of *prgB* expression *in vivo* in *E. faecalis* cells carrying pCF10. In the bloodstream, or on the surface of colonized heart valves plasma albumin (PA) complexed to lipid is believed to bind to the bacterial surface, and to also bind some of the inhibitor and pheromone molecules released by the bacterial cell. This changes the ratio of the two peptides available to interact with the bacteria resulting in self-induction. Expression of *prgB* increases the virulence of these bacteria in experimental endocarditis in rabbits (Schlievert *et al.*, 1998).

regulation of transfer functions), including cell–cell interactions and DNA processing steps, will clearly be an important and interesting priority for future work. Perhaps of most significance will be future investigations into several novel roles of pheromones and their cognate plasmids in aspects of enterococcal biology not directly related to conjugation. Results of such studies could have significant implications for research areas ranging from microbial physiology, to ecology, to pathogenesis. Finally, it is possible that there could be important direct practical applications of studies of the pheromone plasmids. For example, the positive control model discussed previously (Fig. 3) postulates changes in ribosome specificity resulting from binding of ribosomes to pCF10-encoded QL RNA. It might be predicted that such modified ribosomes could be impaired for translation of all the cellular mRNAs not induced by the pheromome. Preliminary results (G. Dunny, A. Colwell and L. Lynch, unpublished data) indicate that this is the case. The binding site on the ribosome for the active component of QL RNA may thus represent a novel drug target. Pathogenesis studies could ultimately lead to the identification of novel vaccine candidates or to improved chemotherapeutic strategies. Whether or not these specific practical benefits are realized, the pheromone system will continue to be a challenging and important paradigm of intercellular signalling in Gram-positive bacteria.

ACKNOWLEDGEMENTS

We thank Pat Schlievert and all the members of the Dunny laboratory for their contributions to the research described in this review. Our pheromone research is supported by NIH grants GM49530 and HL51987.

REFERENCES

An, F. Y. & Clewell, D. B. (1994). Characterization of the determinant (*traB*) encoding sex pheromone shutdown by the hemolysin/bacteriocin plasmid, pAD1 in *Enterococcus faecalis*. *Plasmid*, **31**, 215–21.

Bensing, B. A. & Dunny, G. M. (1997). Pheromone-inducible expression of an aggregation protein in *Enterococcus faecalis* requires interaction of a plasmid-encoded RNA with components of the ribosome. *Molecular Microbiology*, **24**, 285–94.

Bensing, B. A., Manias, D. A. & Dunny, G. M. (1997). Pheromone cCF10 and plasmid pCF10-encoded regulatory molecules act post-transcriptionally to activate expression of downstream conjugation functions. *Molecular Microbiology*, **24**, 295–308.

Bensing, B. A., Meyer, B. J. & Dunny, G. M. (1996). Sensitive detection of bacterial transcription initiation sites and differentiation from RNA processing sites in the pheromone-induced plasmid transfer system of *Enterococcus faecalis*. *Proceedings of the National Academy of Sciences, USA*, **93**, 7794–9.

Chow, J. W., Thal, L. A., Perri, M. B., Vazquez, J. A., Donabedian, S. M., Clewell, D. B. & Zervos, M. J. (1993). Plasmid-associated hemolysin and aggregation substance production contribute to virulence in experimental enterococcal endocarditis. *Antimicrobial Agents and Chemotherapy*, **37**, 2474–7.

Chung, J. W. & Dunny, G. M. (1992). Cis-acting, orientation-dependent, positive control system activates pheromone-inducible conjugation functions at distances greater than 10 kilobases upstream from its target in *Enterococcus faecalis*. *Proceedings of the National Academy of Sciences, USA*, **89**, 9020–4.

Chung, J. W., Bensing, B. A. & Dunny, G. M. (1995). Genetic analysis of a region of the *Enterococcus faecalis* plasmid pCF10 involved in positive regulation of conjugative transfer functions. *Journal of Bacteriology*, **177**, 2107–17.

Clewell, D. B. (1991). *Bacterial Conjugation*. New York: Plenum.

Clewell, D. B. (1998). Sex pheromone systems in enterococci. In *Cell–Cell Signaling in Bacteria*, ed. G. M. Dunny and S. C. Winans. Washington, DC: ASM Press, in press.

Clewell, D. B., Pontius, L. T., An, F. Y., Ike, Y., Suzuki, A. & Nakayama, J. (1990). Nucleotide sequence of the sex pheromone inhibitor (iAD1) determinant of *Enterococcus faecalis* conjugative plasmid pAD1. *Plasmid*, **24**, 156–61.

Devriese, L. A., Collins, M. D. & Wirth, R. (1992). The genus *Enterococcus*. In *The Procaryotes*, ed. A. Balows, H. G. Truper, M. Dworkin, K. Harder & K-H. Schliefer, pp. 1465–81. New York: Springer-Verlag.

Dunny, G., Funk, C. & Adsit, J. (1981). Direct stimulation of the transfer of antibiotic resistance by sex pheromones in *Streptococcus faecalis*. *Plasmid*, **6**, 270–8.

Dunny, G., Yuhasz, M. & Ehrenfeld, E. (1982). Genetic and physiological analysis of conjugation in *Streptococcus faecalis*. *Journal of Bacteriology*, **151**, 855–9.

Dunny, G. M. & Clewell, D. B. (1975). Transmissible toxin (hemolysin) plasmid in *Streptococcus faecalis* and its mobilization of a noninfectious drug resistance plasmid. *Journal of Bacteriology*, **124**, 784–90.

Dunny, G. M. & Leonard, B. A. B. (1997). Cell–cell communication in Gram-positive bacteria. *Annual Reviews of Microbiology*, **51**, 527–64.

Dunny, G. M., Brown, B. L. & Clewell, D. B. (1978). Induced cell aggregation and mating in *Streptococcus faecalis*: evidence for a bacterial sex pheromone. *Proceedings of the National Academy of Sciences, USA*, **75**, 3479–83.

Dunny, G. M., Craig, R. A., Carron, R. L. & Clewell, D. B. (1979). Plasmid transfer in *Streptococcus faecalis*: production of multiple pheromones by recipients. *Plasmid*, **2**, 454–65.

Ehrenfeld, E. E., Kessler, R. E. & Clewell, D. B. (1986). Identification of pheromone-induced surface proteins in *Streptococcus faecalis* and evidence of a role for lipoteichoic acid in the formation of mating aggregates. *Journal of Bacteriology*, **168**, 6–12.

Freire Bastos, M. C., Tanimoto, K. & Clewell, D. B. (1997). Regulation of transfer of the *Enterococcus faecalis* pheromone responding plasmid pAD1: temperature-sensitive transfer mutants and identification of a new regulatory determinant, traD. *Journal of Bacteriology*, **97**, 3250–9.

Fujimoto, S. & Clewell, D. B. (1998). Regulation of the pAD1 sex pheromone response of *Enterococcus faecalis* by direct interaction between the cAD1 peptide mating signal and the negatively regulating, DNA-binding TraA protein. *Proceedings of the National Academy of Sciences, USA*, **95**, 6430–5.

Galli, D. & Wirth, R. (1991). Comparative analysis of *Enterococcus faecalis* sex pheromone plasmids identifies a single homologous DNA region which codes for aggregation substance. *Journal of Bacteriology*, **173**, 3029–33.

Galli, D., Friesnegger, A. & Wirth, R. (1992). Transcriptional control of sex-pheromone-inducible genes on plasmid pAD1 of *Enterococcus faecalis* and sequence of a third structural gene for (pPD1–encoded) aggregation substance. *Molecular Microbiology*, **6**, 1297–308.

Handwerger, S., Pucci, M. J. & Kolokathis, A. (1990). Vancomycin resistance is encoded on a pheromone response plasmid in *Enterococcus faecium* 228. *Antimicrobial Agents and Chemotherapy*, **34**, 358–60.

Heaton, M. P. & Handwerger, S. (1995). Conjugative mobilization of a vancomycin resistance plasmid by a putative *Enterococcus faecium* sex pheromone response plasmid. *Microbial Drug Resistance*, **1**, 177–83.

Heaton, M. P., Discotto, L. F., Pucci, M. J. & Handwerger, S. (1996). Mobilization of vancomycin resistance by transposon-mediated fusion of a VanA plasmid with an *Enterococcus faecium* sex pheromone-response plasmid. *Gene*, **171**, 9–17.

Hedberg, P. J., Leonard, B. A. B., Ruhfel, R. E. & Dunny, G. M. (1996). Identification and characterization of the genes of *Enterococcus faecalis* plasmid pCF10 involved in replication and in negative control of pheromone-inducible conjugation. *Plasmid*, **35**, 46–57.

Hirt, H., Wirth, R. & Muscholl, A. (1996). Comparative analysis of 18 sex pheromone plasmids from *Enterococcus faecalis*: detection of a new insertion element on pPD1 and hypotheses on the evolution of this plasmid family. *Molecular and General Genetics*, **252**, 640–7.

Huycke, M. M., Gilmore, M. S., Jett, B. D. & Booth, J. L. (1992). Transfer of pheromone-inducible plasmids between *Enterococcus faecalis* in the Syrian hamster gastrointestinal tract. *Journal of Infectious Diseases*, **166**, 1188–91.

Jacob, A. E. & Hobbs, S. J. (1974). Conjugal transfer of plasmid-borne multiple antibiotic resistance in *Streptococcus faecalis* var. *zymogenes*. *Journal of Bacteriology*, **117**, 360–72.

Jacob, A. E., Douglas, G. J. & Hobbs, S. J. (1975). Self-transferable plasmids determining the hemolysin and bacteriocin of *Streptococcus faecalis* var. *zymogenes*. *Journal of Bacteriology*, **121**, 863–72.

Jett, B. D., Huycke, M. M. & Gilmore, M. S. (1994). Virulence of Enterococci. *Clinical Microbiology Reviews*, **7**, 462–78.

Jett, B. D., Jensen, H. G., Nordquist, R. E. & Gilmore, M. S. (1992). Contribution of the pAD1-encoded cytolysin to the severity of experimental *Enterococcus faecalis* endophthalmitis. *Infection and Immunity*, **60**, 2445–52.

Kao, S-M., Olmsted, S. B., Viksnins, A. S., Gallo, J. C. & Dunny, G. M. (1991). Molecular and genetic analysis of a region of plasmid pCF10 containing positive control genes and structural genes encoding surface proteins involved in pheromone-inducible conjugation in *Enterococcus faecalis*. *Journal of Bacteriology*, **173**, 7650–64.

Kreft, B., Marre, R., Schramm, U. & Wirth, R. (1992). Aggregation substance of *Enterococcus faecalis* mediates adhesion to cultured renal tubular cells. *Infection and Immunity*, **60**, 25–30.

Leonard, B. A. B., Podbielski, A., Hedberg, P. J. & Dunny, G. M. (1996). *Enterococcus faecalis* pheromone binding protein, PrgZ, recruits a chromosomal oligopeptide permease system to import sex pheromone cCF10 for induction of conjugation. *Proceedings of the National Academy of Sciences, USA*, **93**, 260–4.

Marcinek, H., Wirth, R., Muscholl-Silberhorn, A. & Gauer, M. (1998). *Enterococcus faecalis* gene transfer under natural conditions in municipal sewage water treatment plants. *Applied and Environmental Microbiology*, **64**, 626–32.

Mori, M., Isogai, A., Sakagami, Y., Fujino, M., Kitada, C., Clewell, D. B. & Suzuki, A. (1986). Isolation and structure of the *Streptococcus faecalis* sex pheromone inhibitor, iAD1, that is excreted by the donor strain harboring plasmid pAD1. *Agricultural and Biological Chemistry*, **50**, 539–41.

Mori, M., Sakagami, Y., Ishii, Y., Isogai, A., Kitada, C., Fujino, M., Adsit, J. C., Dunny, G. M. & Suzuki, A. (1988). Structure of cCF10, a peptide sex pheromone which induces conjugative transfer of the *Streptococcus faecalis* tetracycline resistance plasmid, pCF10. *Journal of Biological Chemistry*, **263**, 14 574–8.

Mori, M., Sakagami, Y., Narita, M., Isogai, A., Fujino, M., Kitada, C., Craig, R. A., Clewell, D. B. & Suzuki, A. (1984). Isolation and structure of the bacterial sex pheromone, cAD1, that induces plasmid transfer in *Streptococcus faecalis*. *FEBS Letters*, **178**, 97–100.

Murray, B. E. (1990). The life and times of the Enterococcus. *Clinical Microbiology Reviews*, **3**, 46–65.

Muscholl, A., Galli, D., Wanner, G. & Wirth, R. (1993). Sex pheromone plasmid pAD1-encoded aggregation substance of *Enterococcus faecalis* is positively regulated in trans by *traE1*. *European Journal of Biochemistry*, **214**, 333–8.

Nakayama, J., Ruhfel, R. E., Dunny, G. M., Isogai, A. & Suzuki, A. (1994). The *prgQ* gene of the *Enterococcus faecalis* tetracycline resistance plasmid pCF10 encodes a peptide inhibitor, iCF10. *Journal of Bacteriology*, **176**, 7405–8.

Nakayama, J., Takanami, Y., Horii, T., Sakuda, S. & Suzuki, A. (1998). Molecular mechanism of peptide-specific pheromone signaling in *Enterococcus faecalis*: functions of pheromone receptor TraA and pheromone-binding protein TraC encoded by plasmid pPD1. *Journal of Bacteriology*, **180**, 449–56.

Nakayama, J., Yoshida, K., Kobayashi, H., Isogai, A., Clewell, D. B. & Suzuki, A. (1995). Cloning and characterization of a region of *Enterococcus faecalis* plasmid pPD1 encoding pheromone inhibitor (ipd), pheromone sensitivity (*traC*), and pheromone shutdown (*traB*) genes. *Journal of Bacteriology*, **177**, 5567–73.

Nes, I. F., Diep, D. B., Havarstein, L. S., Brurberg, M. B., Eijsink, V. & Holo, H. (1996). Biosynthesis of bacteriocins in lactic acid bacteria. *Antonie van Leeuwenhoek Journal of Microbiology and Serology*, **70**, 113–28.

Olmsted, S. B., Dunny, G. M., Erlandsen, S. & Wells, C. L. (1994). A plasmid-encoded surface protein on *Enterococcus faecalis* augments its internalization by cultured intestinal epithelial cells. *Journal of Infectious Diseases*, **170**, 1549–56.

Olmsted, S. B., Erlandsen, S. L., Dunny, G. M. & Wells, C. L. (1993). High-resolution visualization by field emission scanning electron microscopy of *Enterococcus faecalis* surface proteins encoded by the pheromone-inducible conjugative plasmid pCF10. *Journal of Bacteriology*, **175**, 6229–37.

Olmsted, S. B., Kao, S-M., van Putte, L. J., Gallo, J. C. & Dunny, G. M. (1991). Role of the pheromone-inducible surface protein Asc10 in mating aggregate formation and conjugal transfer of the *Enterococcus faecalis* plasmid pCF10. *Journal of Bacteriology*, **173**, 7665–72.

Pontius, L. T. & Clewell, D. B. (1992). Regulation of the pAD1-encoded sex pheromone response in *Enterococcus faecalis*: nucleotide sequence analysis of *traA*. *Journal of Bacteriology*, **174**, 1821–7.

Ruhfel, R. E., Manias, D. A. & Dunny, G. M. (1993). Cloning and characterization of a region of the *Enterococcus faecalis* conjugative plasmid, pCF10, encoding a sex pheromone binding function. *Journal of Bacteriology*, **175**, 5253–9.

Schlievert, P. M., Gahr, P. J., Assimacopoulos, A. P., Dinges, M. M., Stoehr, J. A., Harmala, J. W., Hirt, H. & Dunny, G. M. (1998). Aggregation and binding substances enhance pathogenicity in rabbit models of *Enterococcus faecalis* endocarditis. *Infection And Immunity*, **66**, 218–23.

Suzuki, A., Mori, M., Sakagami, Y., Isogai, A., Fujino, M., Kitada, C., Craig, R. A. & Clewell, D. B. (1984). Isolation and structure of bacterial sex pheromone, cPD1. *Science*, **226**, 849–50.

Tanimoto, K. & Clewell, D. B. (1993). Regulation of the pAD1-encoded sex pheromone response in *Enterococcus faecalis*: expression of the positive regulator TraE1. *Journal of Bacteriology*, **175**, 1008–18.

Tomich, P. K., An, F. Y., Damle, S. P. & Clewell, D. B. (1979). Plasmid related transmissibility and multiple drug resistance in *Streptococcus faecalis* subspecies *zymogenes* strain DS16. *Antimicrobial Agents and Chemotherapy*, **15**, 828–30.

Tomura, T., Hirano, T., Ito, T. & Yoshioka, M. (1973). Transmission of bacteriocinogenicity by conjugation in Group D streptococci. *Japanese Journal of Microbiology*, **17**, 445–52.

Trotter, K. M. & Dunny, G. M. (1990). Mutants of *Enterococcus faecalis* deficient as recipients in mating with donors carrying pheromone-inducible plasmids. *Plasmid*, **24**, 57–67.

Weaver, K. E. & Clewell, D. B. (1991). Control of *Enterococcus faecalis* sex pheromone cAD1 elaboration: effects of culture aeration and pAD1 plasmid-encoded determinants. *Plasmid*, **25**, 177–89.

Weaver, K. E., Clewell, D. B. & An, F. (1993). Identification, characterization, and nucleotide sequence of a region of *Enterococcus faecalis* pheromone-responsive plasmid pAD1 capable of autonomous replication. *Journal of Bacteriology*, **175**, 1900–9.

Wirth, R. (1994). The sex pheromone system of *Enterococcus faecalis*: more than just a plasmid-collection mechanism? *European Journal of Biochemistry*, **222**, 235–46.

Wirth, R., Friesenegger, A. & Horaud, T. (1992). Identification of new sex pheromone plasmids in *Enterococcus faecalis*. *Molecular and General Genetics*, **233**, 157–60.

Wirth, R., Olmsted, S. B., Galli, D. & Dunny, G. M. (1991). Comparative analysis of cAD1 and cCF10 induced aggregation substances of *Enterococcus faecalis*. In *Genetics and Molecular Biology of Streptococci, Lactococci, and Enterococci*, ed. G. M. Dunny, P. P. Cleary & L. L. McKay, pp. 34–8. Washington, DC: American Society for Microbiology.

Yagi, Y., Kessler, R. E., Shaw, J. H., Lopatin, D. E., An, F. & Clewell, D. B. (1983). Plasmid content of *Streptococcus faecalis* strain 39–5 and identification of a pheromone (cPD1)-induced surface antigen. *Journal of General Microbiology*, **129**, 1207–15.

INTERCELLULAR SIGNALLING FOR MULTICELLULAR MORPHOGENESIS

DALE KAISER

Department of Biochemistry and Developmental Biology, Stanford University School of Medicine, Stanford, CA 94305, USA

INTRODUCTION

Myxobacteria have adopted multicellularity as their strategy for survival. When they begin to exhaust their available food supply, they construct fruiting bodies (Reichenbach, 1993). Each fruiting body contains about 100 000 cells differentiated as asexual spores. Multicellular sporulation is thought to improve their long-term survival by enhancing the dispersion of those spores and by providing a high cell density for cooperative feeding when the spores germinate (Reichenbach, 1984). In their vegetative phase they also feed co-operatively. Figure 1 shows *Myxococcus xanthus*, the 'golden' myxobacterium, and the favourite for research, due to its rapid and reliable development. There are many species of myxobacteria, but all are phylogenetically related. Myxococcus and Stigmatella, which has a stalk and multiple sporangioles, have the same size circular genome of 9.5 Mb, and common initial steps of their morphological development (Reichenbach & Dworkin, 1981; Shimkets & Woese, 1992).

Fruiting body morphogenesis calls to mind a moving picture run backward. In this reverse film, shards of a tea cup and saucer pick themselves up from the floor and assemble as they rise to the table top. Watching fruiting body development take place begs us to identify the sources of its structural and temporal order, which appears to arise from a total disorder. Physical chemists tell us that thermodynamics poses no difficulty for the creation of order, that it can always be purchased by expending enough ATP. But this formal answer avoids an important issue. In the living world, order arises from prior order. This chapter is an attempt to identify the relevant prior order and to trace its transformations.

One possible source of order is the asymmetry of myxobacterial cells. They are rods with the same diameter as *E. coli* but are roughly five times longer, giving them a 10:1 aspect ratio. Myxobacteria also differ from most other bacteria in their means of locomotion. Having no flagella, they don't swim. Rather, they 'glide', a form of translocation which requires a surface (Burchard, 1984). The surface may be a particle of clay in the soil where they live and recycle particulate organic matter. Myxobacteria can also glide

FRUITING BODY DEVELOPMENT OF *Myxococcus xanthus*

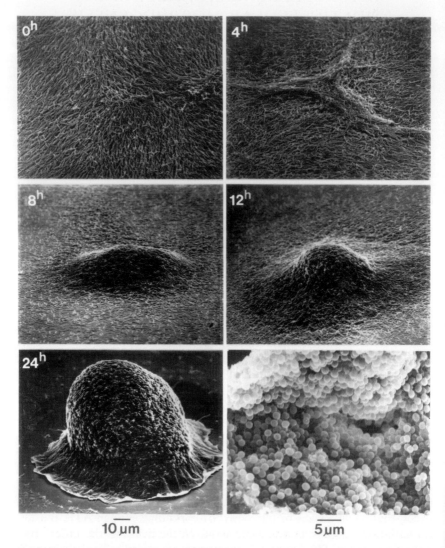

Fig. 1. Fruiting body development in *M. xanthus*. Development was initiated at 0 hours by replacing nutrient medium with a buffer devoid of a usable carbon or nitrogen source. The lower right panel shows a fruiting body which has split open, revealing spores inside. This frame is three times the magnification of the others. Scanning electron microscopy by J. Kuner. Reproduced from Kaiser *et al.*, 1985. Copyright 1985 Cold Spring Harbor Laboratory.

on the surface of an agar plate. Importantly, they can glide over the surfaces of other cells. The coordinated gliding of many cells, called swarming, is regulated by more than 50 genes (Hartzell & Youderian, 1995). This amounts to 1% of their genome. Its magnitude is an indication of the importance myxobacteria attach to swarming. The activity of almost all those genes is used in fruiting body morphogenesis.

Using its capacity to glide on the surface of other cells, *M. xanthus* builds a fruiting body as shown in the multiple views of Fig. 1. The culture shown in Fig. 1 was started by allowing cells from a uniform, high density suspension to settle onto the floor of a culture dish (Kuner & Kaiser, 1982). There, the asymmetry of individual cells driven by the high cell density brings their long axes into alignment. The favoured orientation changes from one locale to another, so that the layer of cells is a patchwork quilt of intersecting facets (Fig. 1, panel 1). Within a few hours, the facet pattern gives way to a punctate distribution of small asymmetric aggregates (panel 2).

More and more cells enter some of these early aggregates, and after about 10^5 cells (panel 5), have entered, a mound becomes a steep-sided hemisphere. Then the cells differentiate from long rods into spherical spores. Spores have thick walls, are metabolically dormant, are resistant to radiation and desiccation, and are long-lived (White, 1993). A mature fruiting body is entirely filled with close-packed spores (Fig. 1, panel 6).

To carry out their programme of morphological development, the cells communicate with each other by emitting and responding to extracellular chemical signals. Three signals have been chemically identified: A, E, and C. This review will focus on the signals, what is known of their production and signal transduction pathways and their roles in morphogenesis.

Mutants that are unable to produce either the A, C or E signals prematurely arrest the assembly of fruiting bodies (Hagen *et al.*, 1978; Toal *et al.*, 1995). Each mutant stops at a different stage, which corresponds to the developmental checkpoint at which the signal is needed, Fig. 2. These morphological stages are linked with the developmental expression of sets of genes. Transcriptional fusions of developmentally regulated promoters to *lacZ*, the structural gene for β-galactosidase, report the activity of each promoter as expression of the enzyme. More than nine-tenths of the transcriptional *lacZ* fusion mutants modify development in a limited way, but are not necessary for its completion (Kroos *et al.*, 1986). Hence the fusions serve as reporters for the various regulatory stages of development without interrupting the overall flow of that programme.

A-SIGNALLING

According to reporter expression, mutants defective in producing extracellular signal A arrest at 1 to 2 hours of development – their terminal morphology is a flat film of cells with no sign of focal aggregation. A-signal

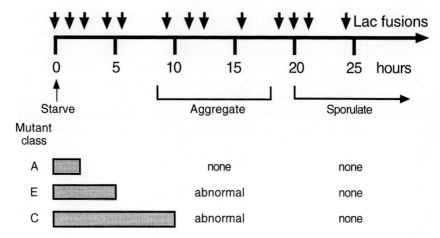

Fig. 2. Three classes of signal-defective developmental mutants of *M. xanthus*. Vertical arrow heads point to the time at which one of the *lacZ* fusions to a developmentally regulated promoter begins to be expressed. These fusions are reporters of normal development. The horizontal bars indicate the period of normal expression of reporters for each of the indicated mutants. Rightward from the end of the bar development is defective (either none or greatly reduced reporter expression). The columns to the right indicate the morphological phenotype with respect to aggregation and sporulation of the indicated mutants.

mutants are capable of sensing starvation, however (Singer & Kaiser, 1995). Mutants defective in producing extracellular signal E are blocked at 3 to 5 hours, and C mutants arrest after about 6 hours, partially aggregated and having expressed more genes. This chapter includes all three signals but will emphasize A- and C-signalling.

A-signal-production-defective mutants have been found in three genes. One, *asgA*, encodes a protein with a two-component receiver domain followed by a histidine protein kinase domain (Davis *et al.*, 1995; Plamann *et al.*, 1994, 1995). This protein has autokinase activity (Li & Plamann, 1996). The second gene, *asgB*, encodes an HTH protein that appears to be a transcription factor for the −35 region of a promoter; and the third, *asgC*, encodes an *rpo*D homologue (Davis *et al.*, 1995). These three proteins are thought to function together in a signal transduction pathway that, in response to starvation, is required for generation of extracellular A-factor.

An A-signal-dependent transcriptional *lacZ* reporter supplied the bioassay that was used to identify A-factor. Isolated and purified from medium conditioned by developing cells, A-factor proved to be a set of six amino acids: trp, pro, phe, tyr, leu and ile, peptides containing these six amino acids, or proteases capable of releasing these amino acids from *M. xanthus* cells (Kuspa *et al.*, 1992a; Plamann & Kaplan, 1998; Plamann *et al.*, 1992).

Myxococcus releases small quantities of these amino acids about 2 hours into development, then proceeds to take them back (Fig 3). This seemingly futile release and uptake helps Myxococcus choose between two alternative

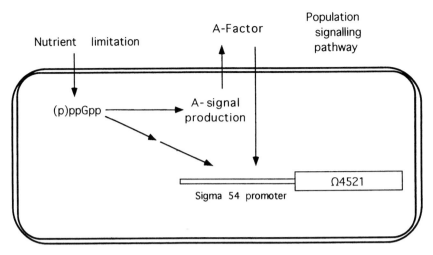

Fig. 3. Dual control of gene expression. As nutrient levels decrease and the protein synthetic capacity of the cell falls, the (p)ppGpp levels increase. Production of A-factor is induced by that increase and it is released into the medium. Each starved cell releases a fixed amount of the set of A-factor amino acids, which pool in the fluid surrounding the cells. In responding to A-factor, each cell perceives the pooled concentration of these amino acids in its vicinity. Promoters of genes that are A-factor and starvation dependent, like Tn5 lac Ω4521, receive one input from the cell's (p)ppGpp level and a second input from A-factor; both inputs are necessary for Ω4521 expression.

responses to nutrient limitation. The choice is between entering stationary phase with very slow growth, on the one hand, and fruiting body development with differentiation of spores, on the other. Neither response to starvation is biologically perfect, for each leads to the death of most of the cells. If the nutrient is on its way to exhaustion, then slowing growth to match the level of residual nutrient will lead to slower and slower growth, and ultimately to death from starvation. Fruiting body development, on the other hand, kills the majority of cells. Only about 1–10 % of cells will survive as differentiated myxospores, as development is usually carried out in the laboratory (Kim & Kaiser, 1990c). At least 30 new proteins are made during fruiting body development (Inouye et al., 1979), so a capacity to synthesize protein must be retained well into the sporulation phase. Accordingly, the cells must make their choice before any nutrient essential for protein synthesis has been totally depleted. The 'wiser' choice would thus seem to depend on cells predicting whether nutrients are on their way to total exhaustion or whether the nutrient shortage is more likely to be temporary.

To make this calculation, which could be a matter of life or death, myxobacteria use their ribosomes. Starvation for any amino acid, or starvation for carbon, energy, or phosphorous induces fruiting body development (Manoil & Kaiser, 1980). However, neither purine nor pyrimidine starvation induces fruiting body development (Kimsey & Kaiser, 1991). The set of effective inducing conditions implicate the availability of a complete set of amino-acylated tRNA. Lack of carbon, energy, or phosphorous would limit their availability. In Myxococcus as in other bacteria, the absence or shortage of any one of the charged tRNAs leads a ribosome, sensing with a codon 'hungry' for its cognate amino acylated tRNA, to synthesize guanosine tetra (and penta)-phosphate [(p)ppGpp] by condensing ATP and GTP. A rise in this highly phosphorylated nucleotide sets off a stringent response that stops the synthesis of new ribosomes, and of the other major polymers of the cell, including DNA, phospholipids, and peptidoglycan (Cashel et al., 1996). Stringent conditions do allow for expression of certain genes as well as for selective protein synthesis, in so far as activated amino acids are available. Thus the genes for new fruiting body proteins can be expressed and translated.

Accumulation of (p)ppGpp is both necessary and sufficient for fruiting body development. On the one hand, ectopic production of (p)ppGpp in M. xanthus initiates early developmentally specific gene expression (Singer & Kaiser, 1995). The E. coli relA gene was introduced into M. xanthus for this purpose; its introduction was followed by production of the E. coli relA protein and (p)ppGpp accumulation without any prior starvation. Moreover, the rise in (p)ppGpp also induces production of A-factor (Singer & Kaiser, 1995). On the other hand, M. xanthus has its own relA gene, and either a point mutation or a deletion mutation in that gene blocks starvation-initiated development at the flat biofilm stage seen in Fig. 1, panel 1 (Harris et al., 1998). In fact, these relA mutants arrest before expression of any of the developmentally regulated reporters.

A-signalling and dual control

Each starved cell, which has accumulated ppGpp, releases a fixed amount of the set of A-factor amino acids. Being soluble, these amino acids then pool in the fluid surrounding the cells. In responding to A-factor, each cell perceives the pooled concentration of these amino acids in its vicinity (Kuspa et al., 1992b). Promoters of genes that are A-factor and starvation dependent, like Tn5 lac Ω4521, receive one input from the cell's (p)ppGpp level and a second input from A-factor. Both inputs are necessary for Ω4521 expression (Kuspa et al., 1986). The promoter for Ω4521 recognizes sigma 54 rather than the more common sigma 70 (Keseler & Kaiser, 1995). All known sigma 54 promoters require an upstream activator protein to initiate open-complex formation and then transcription (Wedel & Kustu, 1995). The activator protein and the sigma-54 holo enzyme are two spatially distinct input sites for controlling transcription of Ω4521 (Fig 3). There is experimental evidence for a specific activator protein of the sigma-54 type for this promoter (H. Kaplan, unpublished data; L. Gorski, 1998, personal communication). By comparing these two distinct inputs, a cell could extrapolate to probable future nutrient levels. Specifically, the (p)ppGpp level indicates the level of nutrient currently available, since this nucleotide is produced by any ribosome with a hungry codon, and its level tracks nutrient changes. A-factor recalls the level of starvation 2 h previously. Two hours earlier, a cell having assessed serious starvation produced the A-factor. The expression of $\Omega4521$ requires both inputs; very low expression occurs when either input is absent. In this way, the A-factor pool summarizes the votes of all the cells. Thus A-factor belongs to the class of extracellular signals called quorum sensors (Kaiser, 1996). Given the concentration window for an A-factor response, it is expected that the reliability of the judgement whether to enter stationary phase or to initiate fruiting body development is increased when it has been made by the whole population of cells.

C-SIGNALLING

Intracellular (p)ppGpp and extracellular A-factor advance morphological development from a uniform, flat, film-like distribution of cells (Fig. 1, panel 1) through the A-signalling checkpoint, to asymmetric aggregates (Fig. 1, panel 2). E-signalling follows (Toal et al., 1995). Then two very different morphological processes come into view: (i) recruiting 10^5 cells to a select set of asymmetric foci until each of the chosen ones have mounded into a hemisphere; (ii) inside a mound, initiating and completing sporulation. Sporulation includes the transformation of a motile rod cell into a non-motile spherical spore cell, as well as the assembly of spore coats. Despite the fundamental differences between these processes, both are initiated by and require the same extracellular signal molecule, called C-factor. Null mutations in the csgA gene, which encodes C-factor protein, have defective aggregation. They arrest at the second check point of irregular aggregates (Fig. 1, panel 2). They also do not sporulate; the sporulation frequency is $<10^{-5}$ of csg$^+$. Sporulation requires C-factor for its initiation and for expression of genes involved in the cell shape change.

C-factor was purified and identified from detergent extracts of whole cells, using a bioassay similar to that used to identify A-factor, except that a C-signal deficient mutant was employed as the reporter strain. For this bioassay, a mutant defective in the structural gene for C-factor, the csgA gene was employed (Hagen & Shimkets, 1990; Lee et al., 1995). After column chromatography, a 17 kDa protein was obtained (Kim & Kaiser, 1990d). Antibodies to C-factor have been used to localize it to the cell surface (Shimkets & Rafiee, 1990). The chemical properties of A- and C-factors match the cell density at which they signal (Fig. 4).

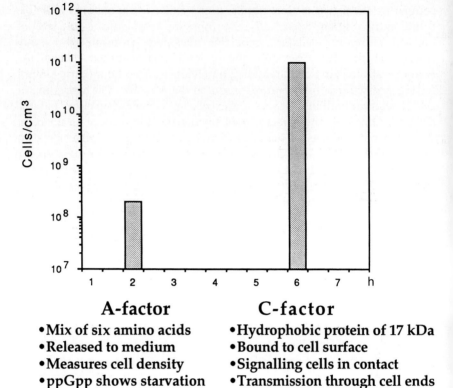

A-factor   C-factor

- Mix of six amino acids   - Hydrophobic protein of 17 kDa
- Released to medium   - Bound to cell surface
- Measures cell density   - Signalling cells in contact
- ppGpp shows starvation   - Transmission through cell ends

Fig. 4. Signals and cell density. As development proceeds, the cell density increases due to aggregation. A-factor is released when the density is 2×10^8 cells/cm^3, and A-factor diffuses between cells. As aggregation proceeds, the density rises to 10^{11} cells/cm^3. Cells are in close-packed contact with each other at the time of C-signalling.

A-factor

The cells are separate, so the signal must diffuse between them in a generally aqueous environment; A-factor amino acids are water soluble.

C-factor

The cells are touching each other; a cell-bound signal molecule can be used.

The C-factor check point is defined by the arrested morphology of a *csgA* mutant and, at the level of gene expression, by the set of reporters expressed in a *csgA*⁻ culture. As noted above, the C-factor phase of aggregation consists of enlargement and rounding of the early asymmetrical focal aggregates. Neither an aggregation reporter like the Tn5lac insertion Ω4499, nor a sporulation reporter, like Ω4435, are expressed in *csgA*⁻ cells. However, both are expressed if purified C-factor is added to the *csgA*⁻ cells

(Kim & Kaiser, 1991). Expression of C-factor-dependent genes is very sensitive to the concentration of C-factor. Both the Ω4499 and the Ω4435 reporter were expressed when 1 unit of partially purified factor were added; neither was expressed when 0.6 units were added to the same volume. The aggregation reporter (Ω4499) was expressed when 0.8 units were added (again to the same volume), but the sporulation reporter was not. When the amount of C-factor produced *in vivo* was limited by deleting segments of the region upstream of the *csg*A transcription start, a similar separation and ordering of aggregation and sporulation was observed (Li *et al.*, 1992). These two types of experiments demonstrate clearly that higher levels of C-factor are needed for sporulation than for aggregation.

Little, if any, C-factor is found in extracts of growing cells. Around 4 h of development, low levels can be recovered. The amount of C-factor in cell extracts rises steeply between 8 and 18 h (Kim & Kaiser, 1990*a*). Once a sporulating cell loses its rod shape, it also loses its capacity to glide and to build a fruiting body. The observed rising concentration of C-factor allows aggregation to go forward with its lower required concentration, before sporulation is induced.

The steep rise in C-factor concentration is a consequence of a positive feedback in the C-signalling circuit. One of the genes whose expression is controlled by C-factor is the *csg*A gene itself. As *csg*A expression rises and more C-factor is displayed on the cell surface, the amount of C-signalling increases. This positive feedback circuit can be demonstrated both *in vitro* and *in vivo*. A *csg*A transcriptional fusion to Lac Z in a *csg*A⁻ mutant increases β-galactosidase expression when either purified C-factor or wild type cells are added (Kim & Kaiser, 1991). CsgA gene expression rises even further as a consequence of an augmentation of positive feedback, as the local density of cells increases within a nascent fruiting body. Increased density increases the number of C-signalling contacts between cells.

C-signal transduction

When first approaching C-signal transduction, it is useful to ask how one pathway can control such physiologically and biochemically diverse processes as cell aggregation, expression of a variety of genes, including *csg*A, and sporulation. Sporulation brings about a change in the shape of the cell, transforming it from a long cylinder into a sphere. Several spore coat proteins are synthesized; and the spore cell loses water. These profound changes in cell structure are induced by C-factor. Aggregation, on the other hand, follows from a change in a cell's movement behaviour. Gene expression typically involves transcription factors, often factors that are specific for a particular target gene. The answer to this 'how' question is that the transduction circuit branches, one branch for each of the three distinct

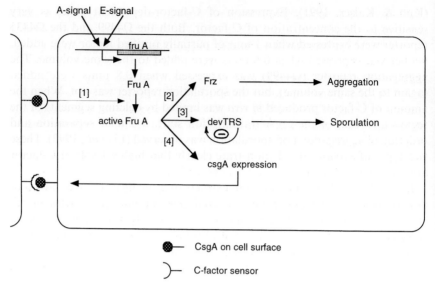

Fig. 5. C-signal transduction pathway. *fru*A is the structural gene for a basic helix–turn–helix transcription factor. Steps [1], [2], [3], [4] are described in the text. Frz is a set of proteins that make up a phosphorelay pathway. *dev*TRS is an autoregulatory operon. The large, rounded rectangle represents one member of a pair of rod cells that are C-signalling to each other through their ends.

processes. The branch point and organization of these diverse processes is evident in the circuit for C-signal transduction, shown in Fig. 5.

C-factor on the surface of one cell interacts with a receptive cell. Little is currently known about the mechanism by which C-signal is transmitted between cells except that C-signalling requires direct contact between motile cells (Kim & Kaiser, 1990*b,c*; Kroos *et al.*, 1988). It is thought that cell movement aligns cells, bringing them into end-to-end contact with each other (Kim & Kaiser, 1990*b*; Sager & Kaiser, 1994; Wall & Kaiser, 1998), and this evidence will be sketched out at the end of this chapter. C-factor bears sequence homology to a family of dehydrogenase enzymes (Baker, 1994; Lee & Shimkets, 1994). It has been suggested that signalling involves a de-hydrogenation, though its nature and consequences are unknown (Lee *et al.*, 1995).

The branch point in the signal transduction pathway immediately follows *fru*A. Fru A protein is a DNA binding response regulator with a HTH motif (Ellehauge *et al.*, 1998; Ogawa *et al.*, 1996). The first event following C-signal transmission, Fig. 5 [1], is the activation of FruA protein by a post-translational modification (Ellehauge *et al.*, 1998). Activated FruA is necessary for all three outputs of C-signalling. The functions assigned to FruA are based on sequence homologies to the FixJ subfamily of proteins

(Parkinson & Kofoid, 1992), and on molecular modelling of the FruA polypeptide (Ellehauge *et al.*, 1998). Most particularly, there is conservation of aspartate 59, which in this subfamily of proteins becomes phosphorylated. Expression of *fru*A depends on A-factor and E-factor, but not on C-factor (Fig. 5). However, C-factor is needed for a post-translational modification, most likely a phosphorylation of FruA on aspartate 59. Changing this residue to alanine destroyed FruA activity, while mutating aspartate 59 to glutamate generated a functional allele, thought to mimic aspartyl-phosphate (Ellehauge *et al.*, 1998).

FruA, once activated, has at least three targets: *frz*, *dev* TRS, and *csg*A. The *frz* target is a phosphorelay (Fig. 5, [2]), which controls the frequency of reversal of gliding direction (Blackhart & Zusman, 1985). The frizzy proteins constitute a two-component signal transduction cascade, related by sequence to those of the *che* genes in *E. coli* and *Salmonella* (McBride *et al.*, 1993). Amino acid residues important for a phosphorelay are conserved in *frz* proteins. Even though this phosphorelay resembles a chemotaxis cascade, the protein domains are joined in different combinations, and its methyl accepting protein, frzCD, unlike those of chemotaxis, is not a membrane receptor; it is found in the cytoplasm; and it has no transmembrane or extracellular domain in its amino acid sequence. This opens the possibility of an input different from classical chemotaxis. Nevertheless, FrzCD protein has a carboxy-domain that resembles a methyl-accepting chemotaxis protein (McBride *et al.*, 1989). Also surprising is the fact that FruA, a response regulator, controls neither the transcription nor translation of *frz* (Søgaard-Anderson & Kaiser, 1996). Instead, activated FruA sends a signal along the *frz* phosphorelay. Although the molecular nature of the link between activated FruA and *frz* is not known, that link is clearly manifest in the methylation of the *frz*CD protein (Søgaard-Anderson & Kaiser, 1996), and in a striking modulation of the movement behaviour of cells, which leads them to aggregate into the hemispherical mounds which mature into fruiting bodies.

At the end of growth, about half the *frz*CD protein is found methylated, half non-methylated (Søgaard-Anderson & Kaiser, 1996). Then, during fruiting body development there are two sequential shifts in methylation. First, as cells recognize starvation in preparation for development, their *frz*CD protein shifts to a fully non-methylated state. Secondly, as cells aggregate, the *frz*CD protein gradually shifts to methylated states, so that, by 9 h, the time of symmetrical mound building, all the *frz*CD protein is found methylated. The former change in FrzCD methylation is found in a mutant which produces no C-factor (*csg*A$^-$), but the latter change is not. The *csg*A mutant arrests with almost equal amounts of methylated and non-methylated *frz*CD as it stops at the C-factor check point. Extracellular addition of purified C-factor to the C-factor-less mutant cells directly and specifically induces the full methylation of their *frz*CD protein, completing the second change in a way that parallels wild-type development (Søgaard-Anderson & Kaiser, 1996). Thus, only the latter shift depends on C-signalling. The first shift to fully non-methylated *frz*CD protein implies another signal input channel to *frz*CD protein that comes from the sense of starvation. Perhaps FruA signals that putative input channel.

As mentioned above, the response of the *frz* phosphorelay to C-signalling is aggregation. However, the cellular mechanism of aggregation has not yet been established, and two models have been proposed. One, emphasizing the sequence homologies of *frz* proteins to *che* proteins, postulates chemotaxis toward an attractant thought to be produced by nascent aggregates (Lev, 1954; McVittie & Zahler, 1962; Shi *et al.*, 1993; Ward & Zusman, 1997). The putative attractant has not yet been isolated or identified despite multiple attempts. A second model is based on the observation that aggregating myxobacterial cells form chain-like streams, which flow into nascent aggregates (Kuhlwein & Reichenbach, 1968). Considering those streaming cells, it has been pointed out that random movement would occasionally bring a free cell to the end of such a stream (Søgaard-Anderson & Kaiser, 1996). Obviously, random movement would also allow cells to leave the stream. However, once an *M. xanthus* cell had made end-to-end contact with the cell at the tip of a stream, that cell would exchange C-signal with the former tip cell. C-signalling would trigger the positive feedback loop between these cells thus intensifying their C-signal exchange. The new cell would thus be trapped at the end of the chain and trapped in the state of gliding toward the aggregate as a consequence of C-signal induced decrease in gliding reversal frequency (Søgaard-Anderson & Kaiser, 1996). FrzCD methylation is reported to be correlated with a decrease in reversal frequency (McBride *et al.*, 1992; Shi *et al.*, 1993).

Transduction pathway mutants, an aggregation screen

Both *fru*A and *frz* mutants were found in a screen of Tn5 insertion strains that arrested development in a state of partial aggregation. Mutants that arrested at the C-factor check point were visually selected from that screen, These mutants arrest aggregation at the same stage as *csg*A mutants (Søgaard-Anderson *et al.*, 1996). Unlike *csg*A mutants, however, the targeted mutants would be cell-autonomous; they would not be rescued by addition of wild-type cells, or by addition of purified C-factor. Transduction pathway mutants give an abnormal (possibly partial) response to C-factor. To ensure that the new mutants would be null and to facilitate subsequent genetic analysis, they were produced by insertion of transposon Tn5 Lac (Søgaard-Anderson *et al.*, 1996).

A collection of such aggregation-defective mutants was obtained and eight of the first nine had their Tn5 insertions in one of two regions of the myxococcus genome: *fru*A or *frz*. Comparison of the properties of the *fru*A and *frz* mutants immediately indicated that the signal transduction pathway is branched. The FruA function is needed for aggregation, sporulation, and late gene expression including *csg*A. The *frz* function is only needed for those responses involving cell movement, specifically aggregation and fruiting body morphogenesis. *Frz*CD, *frz*E and *frz*F genes were all hit by transposon insertion in this screen for *csg*A-like aggregation defective mutants. Since three different *frz* genes were hit in this limited sample, C-factor-induced aggregation evidently requires a functioning phosphorelay.

The second target of activated FruA is the *dev*TRS operon (Fig. 5 [3]). *Dev*TRS expression as measured by the extent and time course of β-galactosidase expression from a Tn5 Lac transcriptional fusion to *dev*R depends on FruA in the same manner as it depends on C-factor (Ellehauge *et al.*, 1998). *Dev*TRS, in turn, is necessary for fruiting body sporulation; *dev*RS null mutants sporulate at 0.1% the efficiency of wild-type cells (Thony-Meyer & Kaiser, 1993). As mentioned above, the morphological differentiation of spores, or more properly myxospores to distinguish them from the endo-spores of bacilli, occurs after aggregation is complete. In *Sigmatella auran-tiaca*, whose fruiting bodies have a more complex morphology with a stalk, branches and multiple cysts containing the spores, and aggregation has many stages, again myxospores form as the fruiting bodies are assuming their final shape (Qualls *et al.*, 1978).

Wild-type cells can bypass the multicellular steps of fruiting body con-struction and proceed directly to sporulation if certain substances that interfere with peptidoglycan turnover are added to a growing culture of vegetative cells (O'Connor, 1997; Jacobs, 1997). Glycerol (Dworkin & Gibson, 1964) or dimethylsulphoxide (Komano *et al.*, 1980) induces more than 90% of cells to become spores. This conversion of a rod-shaped cell to a spherical spore requires about 2 h, and thus is faster than sporulation during fruiting body development; it is also more synchronous. *Asg* and *csg* mutants, which fail to sporulate in fruiting bodies, nevertheless can be induced to sporulate by the addition of glycerol or dimethylsulphoxide (Hagen *et al.*, 1978; LaRossa *et al.*, 1983). These experiments show that all wild-type cells have a cell-intrinsic capacity for sporulation which awaits activation, by *dev*TRS. A Tn5 Lac insertion (Ω7536), which depends on C-signalling and *dev*RS, blocks sporulation but not aggregation (E. Licking, unpublished data). This insertion therefore lies downstream of *dev*TRS in the branch labelled 'sporulation'. Fruiting body development induces β-galacto-sidase production from the Ω7536 reporter, and glycerol does so as well, suggesting that glycerol induction enters the sporulation branch after *dev*TRS.

The ordered, branching circuit shown in Fig. 5 accounts for the pheno-types of double mutants constructed from *csg*A, *fru*A, *frz*, and *dev*RS (Ellehauge *et al.*, 1998). In these double mutants, a *csg*A mutant, or a *fru*A mutant is found to be epistatic to *dev*RS with respect to aggregation and sporulation. A *frz* mutant is epistatic to *dev*RS with respect to aggregation, whereas *dev*RS is epistatic to *frz* with respect to sporulation.

The third target of activated FruA is expression of *csg*A (Fig. 5 [4]). The upstream regulatory region of *csg*A is about one kilobase-pairs long, and sequential deletion from its 5' end progressively decreases *csg*A expression, as if there were several regulatory sites within that region (Li *et al.*, 1992). FruA, as a transcription factor, may interact with one or more of those sites.

DevTRS as a switch

As mentioned above, activated FruA starts transcription of *dev*TRS. The *dev* operon has a switch-like quality, evident in the bimodality of operon expression in populations of developing cells (Russo-Marie *et al.*, 1993). The two states are consequences of the fact that, once the expression of T, R, and S has risen to a certain point, these proteins extinguish further synthesis of *dev*TRS mRNA (Thony-Meyer & Kaiser, 1993). The rate of *dev*TRS mRNA synthesis is observed to fall eight-fold when T, R, and S proteins are made (B. Julien & D. Kaiser, unpublished data). The extent of *dev* expression is thus limited by this negative feedback to an initial burst. However, the negative feedback loop does not limit sporulation, since wild-type strains, in which the loop operates, sporulate with high efficiency (Thony-Meyer & Kaiser, 1993). The *dev*TRS negative feedback loop lies in tandem with the positive feedback loop of C-signalling. As the *dev* operon is transcribed, translated, and its proteins accumulated, the *dev*TRS operon should turn its own expression down. As a consequence of this shut-off of *dev*, sporulation would be initiated by a pulse of limited duration of the *dev*T, R, and S proteins.

If this is the case, then changing the balance between the positive and negative loops would be expected to change either aggregation or sporulation or both. In-frame deletions of *dev*T, *dev*R, and *dev*S have been constructed, and their aggregation and sporulation properties examined. In brief, all three null mutants changed both aggregation and sporulation. But each mutant changed them in a different way, suggesting that, indeed, the two tandem loops normally hold each other in a delicate balance, which is upset in different ways by the loss of T, R, or S function.

Cell alignment and C-signalling

The mechanism of C-signalling is closely related to the multicellular organization of a nascent fruiting body. Previous observations, described above, had implied that C-signalling cells must be aligned. That C-signalling requires a special kind of aligned contact between cells, was first suggested by the observation that non-swarming mutants (which are effectively non-motile) arrested development precisely at the C-signalling check point

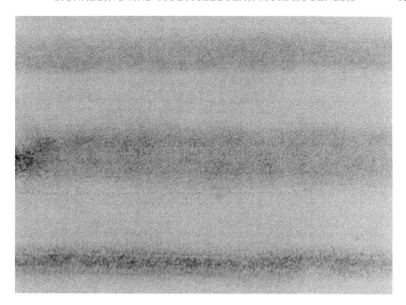

Fig. 6. C-factor-dependent gene expression and cell differentiation in non-motile cells depends on cell position in *M. xanthus*. Mutant non-motile cells containing a C-factor-dependent *lacZ* fusion were allowed to settle from suspension onto an agar surface containing microscopic grooves made by scratching an agar surface with emery paper. The darker areas indicate C-factor-dependent *β*-galactosidase activity from aligned cells in the grooves. Photograph was taken 3 days after starvation had initiated development.

(Kroos *et al.*, 1988). This unexpected observation led to the hypothesis that the necessary correctly aligned contacts are brought about by the normal movements of cells within a nascent fruiting body. Non-motile cells, unable to align, would fail to transmit the C-signal, even though they could make C-factor and could respond to added C-factor (Kim & Kaiser, 1990*b*). To test this hypothesis, non-swarming cells were forced into a proper end-to-end alignment by mechanical means. The asymmetry of the long, rod-shaped *Myxococcus* cells was used to orientate them lengthwise as they fell into the narrow grooves produced by scoring agar with a fine-grained aluminum oxide abrasive paper. The aluminium oxide crystals on this paper were 5 to 10 micrometres in diameter, a size comparable to the length of *M. xanthus* cells. Interference contrast microscopy revealed that cells which had settled into the grooves were indeed orientated with their long axes parallel to the axis of the groove.

The aligned cells, which are non-motile but csg^+, turn on a C-factor dependent *lacZ* fusion, causing the cells to turn blue (Fig. 6). An equal number of non-aligned cells, outside a groove failed to activate this C-factor-dependent promoter and thus remain yellow, the natural colour of these cells which are rich in yellow pigments, including carotenoids. If the medium is

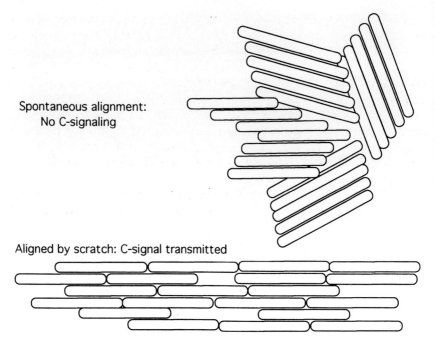

Spontaneous alignment:
No C-signaling

Aligned by scratch: C-signal transmitted

Fig. 7. Sketch interpreting the effects of grooves in the surface of the agar on the alignment of cells. The same surface density of cells is represented in the upper and lower parts of the sketch. The depth of a groove is 10 microns, while the droplet of cell suspension applied to it is more than 1000 microns high.

nutrient deficient and the plates are incubated longer, spores can be seen, but again, only in the grooves (Kim & Kaiser, 1990*b*). Although the blue stripes indicate that the C-dependent, sporulation-related reporter $\Omega4401$ has been activated as a consequence of alignment, spore formation implies that the entire set of C-dependent genes necessary for spore morphogenesis have been activated. Since sporulation requires the highest levels of C-signalling, the efficiency of C-signal transmission in the grooves must have been high.

Cellular organization within the grooves is represented schematically in Fig. 7. Cells that settled outside grooves formed very small rafts of side by side cells but the rafts orientated randomly, as judged by interference contrast microscopy. The same number of cells per unit area are present in both parts of this diagram; the difference is not in the surface density of cells, but in their arrangement. Cells falling from suspension in a fluid droplet whose diameter is 100-times greater than the depth of a groove would be expected to deposit at the same density on the flat surface between grooves as within a groove. In particular, alignment increases the number of end-to-end contacts. The restriction of C-signalling to the grooves strongly suggests that the C-signal passes through the ends of aligned cells, as represented in Fig. 5.

A second, independent line of experiments leads to the same conclusion about C-signal transmission. As fruiting bodies form and enlarge on the surface of the initial mat of cells (Fig. 1, panel 1), cells between the aggregates migrate as travelling waves. The waves are dependent on C-signalling (Shimkets & Kaiser, 1982). Each wave crest is a linear heap of orientated cells. Cultures of traveling waves can be prepared which have many collisions between crests. The collisions occur at all possible angles between colliding crests. Surprisingly, when two wave crests collide, they fail to interfere with each other. Instead, they appear to pass through each other – a physical impossibility for two dense heaps of cells. An experimental analysis of this phenomenon led to the conclusion that colliding waves reflect from each other (Sager & Kaiser, 1994). In order to reflect with the accuracy required, the C-signal must be transmitted through the ends of cells in contact (Sager & Kaiser, 1994).

Mechanical alignment in grooves and the travelling waves approximate to the conditions for cell–cell interactions that are found within nascent fruiting bodies. A surprisingly high degree of cell organization is found in the outer domain of a fruiting body, where C-signalling is believed to take place. Note the regular striations in that area (Fig. 8) (Sager & Kaiser, 1993).

PERSPECTIVES

Morphogenesis

Myxobacteria grow and divide as separate cells, yet they constitute a primitive multicellular organism. Despite the relative simplicity of their fruiting bodies, the progressive creation of structure and order from initial disorder, which is the hallmark of biological development, is clearly evident and exposed for experimental study. The size of the cells and their growth rate render them convenient for culture as well as for biochemical investigation. The size of their fruiting bodies is convenient for fluorescence-based light microscopy of the individual cells within them. This paper has attempted to illustrate how multicellular order can arise from the combination of cell arrangement and cell–cell signalling contingent on that arrangement.

Genetics

The process of fruiting body development is gratuitous: it is completely dispensable for the growth of a cell or of a culture. Null mutants that fail to form multicellular aggregates can be propagated indefinitely by maintaining cells in their growth phase. Three extracellular signals that control *M. xanthus* fruiting body development were discovered via such mutants. Much remains to be learned about their signal transduction pathways. *M. xanthus*

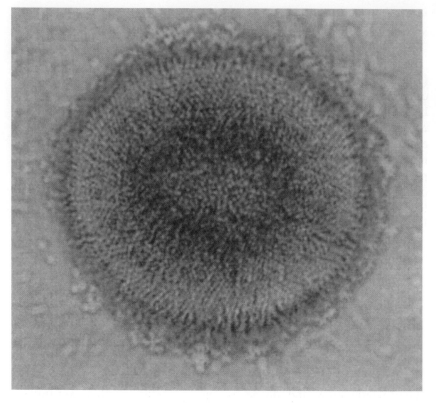

Fig. 8. Fruiting body viewed from below in bright field optics. The focal plane passed through the very bottom layer of cells. Fruiting bodies of strain DK4299 exposed to X-gal (a chromogenic substrate for β-galactosidase) reveal an inner and outer domain. The inner domain of low cell-density contains spores and is expressing β-galactosidase at high level. The outer domain has a very high cell density indicated by the regularly striated pattern that results from optical interference with the higher cell layers, producing a Moiré effect (Sager & Kaiser, 1993).

benefits from a wide range of molecular genetic tools, whose variety and power approach those for *E. coli* (Gill & Shimkets, 1993).

REFERENCES

Baker, M. (1994). *Myxococcus xanthus* C-factor, a morphogenetic paracrine signal, is homologous to *E. coli* 3-ketoacyl-acyl carrier protein reductase and human 17β-hydroxysteroid dehydrogenase. *The Biochemistry Journal*, **301**, 311–12.

Blackhart, B. D. & Zusman, D. (1985). The frizzy genes of *Myxococcus xanthus* control directional movement of gliding motility. *Proceedings of the National Academy of Sciences, USA*, **82**, 8767–70.

Burchard, R. P. (1984). Gliding motility and taxes. In *Myxobacteria*, ed. E. Rosenberg, pp. 139–161. New York: Springer.

Cashel, M., Gentry, D. R., Hernandez, V. J. & Vinella, D. (1996). The stringent response. In *Escherichia coli and Salmonella*, ed. F. Neidhardt, 2nd edn. pp. 1458–96. Washington, DC: ASM Press.

Davis, J. M., Mayor, J. & Plamann, L. (1995). A missense mutation in *rpoD* results in an A-signaling defect in *Myxococcus xanthus*. *Molecular Microbiology*, **18**, 943–52.

Dworkin, M. & Gibson, S. (1964). Rapid formation of microcysts in *Myxococcus xanthus*. *Science*, **146**, 243–4.

Ellehauge, E., Norregaard-Madsen, M. & Sogaard-Anderson, L. (1998). The FruA signal transduction protein provides a checkpoint for the temporal coordination of intercellular signals in *M. xanthus* development. *Molecular Microbiology*, **30**, 807–13.

Gill, R. E. & Shimkets, L. J. (1993). Genetic approaches for analysis of myxobacterial behavior. In *Myxobacteria II*, ed. M. Dworkin & D. Kaiser, pp. 129–55. Washington, DC: American Society of Microbiology.

Hagen, T. J. & Shimkets, L. J. (1990). Nucleotide sequence and transcriptional products of the *csg* locus of *Myxococcus xanthus*. *Journal of Bacteriology*, **172**, 15–23.

Hagen, D. C., Bretscher, A. P. & Kaiser, D. (1978). Synergism between morphogenetic mutants of *Myxococcus xanthus*. *Developmental Biology*, **64**, 284–96.

Harris, B. Z., Kaiser, D. & Singer, M. (1998). The guanosine nucleotide (p)ppGpp initiates development and A-factor production in *Myxococcus xanthus*. *Genes and Development*, **12**, 1022–35.

Hartzell, P. H. & Youderian, P. (1995). Genetics of gliding motility and development in *Myxococcus xanthus*. *Archives of Microbiology*, **164**, 309–23.

Inouye, M., Inouye, S. & Zusman, D. (1979). Gene expression during development of *Myxococcus xanthus*: pattern of protein synthesis. *Developmental Biology*, **68**, 579–91.

Jacobs, C., Frere, J. M. & Normark, S. (1997). Cytosolic intermediates for cell wall biosynthesis and degradation control inducible beta-lactam resistance in gram-negative bacteria. *Cell*, **88**, 823–32.

Kaiser, D. (1996). Bacteria also vote. *Science*, **272**, 1598–9.

Kaiser, D., Kroos, L. & Kuspa, A. (1985). Cell interactions govern the temporal pattern of Myxococcus development. *Cold Spring Harbor Symposia on Quantitative Biology*, **50**, 823–30.

Keseler, I. M. & Kaiser, D. (1995). An early A-signal-dependent gene in *Myxococcus xanthus* has a sigma-54-like promoter. *Journal of Bacteriology*, **177**, 4638–44.

Kim, S. K. & Kaiser, D. (1990*a*). C-factor: a cell–cell signalling protein required for fruiting body morphogenesis of *M. xanthus*. *Cell*, **61**, 19–26.

Kim, S. K. & Kaiser, D. (1990*b*). Cell alignment required in differentiation of *Myxococcus xanthus*. *Science*, **249**, 926–8.

Kim, S. K. & Kaiser, D. (1990*c*). Cell motility is required for the transmission of C-factor, an intercellular signal that coordinates fruiting body morphogenesis of *Myxococcus xanthus*. *Genes & Development*, **4**, 896–905.

Kim, S. K. & Kaiser, D. (1990*d*). Purification and properties of *Myxococcus xanthus* C-factor, an intercellular signaling protein. *Proceedings of the Academy of Sciences, USA*, **87**, 3635–9.

Kim, S. K. & Kaiser, D. (1991). C-factor has distinct aggregation and sporulation thresholds during *Myxococcus development*. *Journal of Bacteriology*, **173**, 1722–8.

Kimsey, H. H. & Kaiser, D. (1991). Targeted disruption of the *Myxococcus xanthus* orotidine 5′-monophosphate decarboxylase gene: effects on growth and fruiting-body development. *Journal of Bacteriology*, **173**, 6790–7.

Komano, T., Inouye, S. & Inouye, M. (1980). Patterns of protein production in *Myxoccocus xanthus* during spore formation induced by glycerol, dimethylsulf-oxide and phenethyl alcohol. *Journal of Bacteriology*, **144**, 1076–82.

Kroos, L., Hartzell, P., Stephens, K. & Kaiser, D. (1988). A link between cell movement and gene expression argues that motility is required for cell–cell signalling during fruiting body development. *Genes and Development*, **2**, 1677–85.

Kroos, L., Kuspa, A. & Kaiser, D. (1986). A global analysis of developmentally regulated genes in *Myxococcus xanthus*. *Developmental Biology*, **117**, 252–66.

Kuhlwein, H. & Reichenbach, H. (1968). Swarming and morphogenesis in *Myxobacteria*. Inst. Wiss Film, Film C893/1965.

Kuner, J. & Kaiser, D. (1982). Fruiting body morphogenesis in submerged cultures of *Myxococcus xanthus*. *Journal of Bacteriology*, **151**, 458–61.

Kuspa, A., Kroos, L. & Kaiser, D. (1986). Intercellular signaling is required for developmental gene expression in *Myxococcus xanthus*. *Developmental Biology*, **117**, 267–76.

Kuspa, A., Plamann, L. & Kaiser, D. (1992*a*). Identification of heat-stable A-factor from *Myxococcus xanthus*. *Journal of Bacteriology*, **174**, 3319–26.

Kuspa, A., Plamann, L. & Kaiser, D. (1992*b*). A-signaling and the cell density requirement for *Myxococcus xanthus* development. *Journal of Bacteriology*, **174**, 7360–9.

LaRossa, R., Kuner, J., Hagen, D., Manoil, C. & Kaiser, D. (1983). Developmental cell interactions in *Myxococcus* analysis of mutants. *Journal of Bacteriology*, **153**, 1394–404.

Lee, K. & Shimkets, L. J. (1994). Cloning and characterization of the *socA* locus which restores development to *Myxococcus xanthus* C-signaling mutants. *Journal of Bacteriology*, **176**, 2200–9.

Lee, B-U., Lee, K., Mendez, J. & Shimkets, L. J. (1995). A tactile sensory system of *Myxococcus xanthus* involves an extracellular NAD(P)$^+$-containing protein. *Genes and Development*, **9**, 2964–73.

Lev, M. (1954). Demonstration of a diffusible fruiting factor in Myxobacteria. *Nature*, **173**, 501.

Li, Y. & Plamann, L. (1996). Purification and phosphorylation of *Myxococcus xanthus* AsgA protein. *Journal of Bacteriology*, **178**, 289–92.

Li, S., Lee, B. U. & Shimkets, L. (1992). *csgA* expression entrains *Myxococcus xanthus* development. *Genes and Development*, **6**, 401–10.

McBride, M. J., Hartzell, P., & Zusman, D. R. (1993). Motility and tactic behavior of *Myxococcus xanthus*. In *Myxobacteria II*, ed. M. Dworkin & D. Kaiser, pp. 285–305. Washington, DC: American Society of Microbiology.

McBride, M., Kohler, T. & Zusman, D. (1992). Methylation of FrzCD, a methyl-accepting taxis protein of *Myxococcus xanthus* is correlated with factors affecting cell behavior. *Journal of Bacteriology*, **174**, 4246–57.

McBride, M. J., Weinberg, R. A. & Zusman, D. R. (1989). Frizzy aggregation genes of the gliding bacterium *Myxococcus xanthus* show sequence similarities to the chemotaxis genes of enteric bacteria. *Proceedings of the National Academy of Sciences, USA*, **86**, 424–8.

McVittie, A. & Zahler, S. A. (1962). Chemotaxis in Myxococcus. *Nature*, **194**, 1299–300.

Manoil, C. & Kaiser, D. (1980). Guanosine pentaphosphate and guanosine tetraphosphate accumulation and induction of *Myxococcus xanthus* fruiting body development. *Journal of Bacteriology*, **141**, 305–15.

O'Connor, K. and Zusman, D. R. (1997). Starvation-independent sporulation in *Myxoccus xanthus* involves the pathway for b-lactamase induction and provides a mechanism for competitive cell survival. *MolecularMicrobiology*, **24**, 839–50.

Ogawa, M., Fujitani, S., Mao, X., Inouye, S. & Komano, T. (1996). FruA, a putative transcription factor essential for the development of *Myxococcus xanthus*. *Molecular Microbiology*, **22**(4), 757–67.

Parkinson, J. S. & Kofoid, E. C. (1992). Communication modules in bacterial signaling proteins. *Annual Review of Genetics*, **26**, 71–112.

Plamann, L. & Kaplan, H. B. (1999). Cell-density sensing during early development in *Myxococcus xanthus*. In press.

Plamann, L., Davis, J. M., Cantwell, B. & Mayor, J. (1994). Evidence that *asgB* encodes a DNA-binding protein essential for growth and development of *Myxococcus xanthus*. *Journal of Bacteriology*, **176**, 2013–20.

Plamann, L., Kuspa, A. & Kaiser, D. (1992). Proteins that rescue A-signal-defective mutants of *Myxococcus xanthus*. *Journal of Bacteriology*, **174**, 3311–18.

Plamann, L., Li, Y., Cantwell, B. & Mayor, J. (1995). The *Myxococcus xanthus asg*A gene encodes a novel signal transduction protein required for multicellular development. *Journal of Bacteriology*, **177**(8), 2014–20.

Qualls, G., Stephens, K. & White, D. (1978). Morphogenetic movements and multicellular development in the fruiting myxobacterium, *Stigmatella aurantiaca*. *Developmental Biology*, **66**, 270–4.

Reichenbach, H. (1984). Myxobacteria: a most peculiar group of social prokaryotes. In *Myxobacteria*, ed. E. Rosenberg, pp. 1–50. New York: Springer-Verlag.

Reichenbach, H. (1993). Biology of the Myxobacteria: ecology and taxonomy. In *Myxobacteria II*, ed. M. Dworkin & D. Kaiser, pp. 13–62. Washington, DC: American Society of Microbiology.

Reichenbach, H. & Dworkin, M. (1981). The Order *Myxobacterales*. In *The Prokaryotes*, ed. M. P. Starr, H. Stolp, H. G. Truper, A. Balows & H. G. Schlegel, Chapter 20, pp. 328–55. Berlin, Heidelberg: Springer.

Russo-Marie, F., Roederer, M., Sager, B., Herzenberg, L. A. & Kaiser, D. (1993). β-Galactosidase activity in single differentiating bacterial cells. *Proceedings of the National Academy of Sciences, USA*, **90**, 8194–8.

Sager, B. & Kaiser, D. (1993). Spatial restriction of cellular differentiation. *Genes and Development*, **7**, 1645–53.

Sager, B. & Kaiser, D. (1994). Intercellular C-signaling and the traveling waves of *Myxococcus*. *Genes and Development*, **8**, 2793–804.

Shi, W., Köhler, T. & Zusman, D. R. (1993). Chemotaxis plays a role in the social behaviour of *Myxococcus xanthus*. *Molecular Microbiology*, **9**, 601–11.

Shimkets, L. & Kaiser, D. (1982). Induction of coordinated movement of *Myxococcus xanthus* cells. *Journal of Bacteriology*, **152**, 451–61.

Shimkets, L. J. & Rafiee, H. (1990). CsgA, an extracellular protein essential for *Myxococcus xanthus* development. *Journal of Bacteriology*, **172**, 5299–306.

Shimkets, L. & Woese, C. R. (1992). A phylogenetic analysis of the myxobacteria: basis for their classification. *Proceedings of the National Academy of Sciences, USA*, **89**, 9459–63.

Singer, M. & Kaiser, D. (1995). Ectopic production of guanosine penta-and tetraphosphate can initiate early developmental gene expression in *Myxococcus xanthus*. *Genes and Development*, **9**, 1633–44.

Søgaard-Anderson, L. & Kaiser, D. (1996). C-factor, a cell-surface-associated intercellular signaling protein, stimulates the cytoplasmic Frz signal transduction system in *Myxococcus xanthus*. *Proceedings of the National Academy of Sciences, USA*, **93**, 2675–9.

Søgaard-Anderson, L., Slack, F., Kimsey, H. & Kaiser, D. (1996). Intercellular C-signaling in *Myxococcus xanthus* involves a branched signal transduction pathway. *Genes and Development*, **10**, 740–54.

Thony-Meyer, L. & Kaiser, D. (1993). *dev*RS, an autoregulated and essential genetic locus for fruiting body development in *Myxococcus xanthus*. *Journal of Bacteriology*, **175**, 7450–62.

Toal, D. R., Clifton, S. W., Roe, B. A. & Downard, J. (1995). The *esg* locus of *Myxococcus xanthus* encodes the E1α and E1β subunits of a branched-chain keto acid dehydrogenase. *Molecular Microbiology*, **16**, 177–89.

Wall, D. & Kaiser, D. (1998). Alignment enhances the cell-to-cell transfer of pilus phenotype. *Proceedings of the National Academy of Sciences, USA*, **95**, 3054–8.

Ward, M. J. & Zusman, D. R. (1997). Regulation of directed motility in *Myxococcus xanthus*. *Molecular Microbiology*, **24**, 885–93.

Wedel, A. & Kustu, S. (1995). The bacterial enhancer-binding protein NTRC is a molecular machine: ATP hydrolysis is coupled to transcriptional activation. *Genes and Development*, **9**, 2042–52.

White, D. (1993). Myxospore and fruiting body morphogenesis. In *Myxobacteria II*, ed. M. Dworkin & D. Kaiser, pp. 307–32. Washington, DC: American Society of Microbiology.

REGULATION OF CARBAPENEM ANTIBIOTIC AND EXOENZYME SYNTHESIS BY CHEMICAL SIGNALLING

SIMON J. McGOWAN AND GEORGE P. C. SALMOND

Department of Biochemistry, University of Cambridge, Tennis Court Road, Cambridge CB2 1QW, UK

INTRODUCTION

Erwinia carotovora is a Gram-negative enterobacterial plant pathogen and the aetiological agent of soft rot diseases. During the invasion of plant tissue, *E. carotovora* produces a variety of enzymes including cellulases, proteases and a wide range of pectinases, intended to enzymatically digest the plant cell wall. The expression of this arsenal of enzymes is tightly regulated and reflects the nature of the environment, the status of the individual cell and, importantly in the context of this chapter, the status of the population of infecting *Erwinia* cells. In order to maintain communication with its neighbours, *E. carotovora* uses the signalling molecule, *N*-(3-oxohexanoyl)-L-homoserine lactone (OHHL; Fig. 1). OHHL is freely diffusible and is both produced and sensed by each member of the colony or culture to provide an estimation of the size of the overall population. This dynamic process, known as 'quorum sensing' (Fuqua *et al.*, 1994), is employed by other, often pathogenic, bacterial species, where it controls expression of a diverse range of phenotypes (for review, see Fuqua *et al.*, 1996; Robson *et al.*, 1997).

In common with some other strains from the *Erwinia* genus, *E. carotovora* is also an antibiotic producer (Parker *et al.*, 1982; Axelrood *et al.*, 1988; Ishimaru *et al.*, 1988). In response to the same OHHL signalling molecule used to induce the cell wall-degrading enzymes, it produces a β-lactam antibiotic, 1-carbapen-2-em-3-carboxylic acid (Fig. 2). It is thought that the carbapenem antibiotic might be targeted at competing pathogenic and saprophytic bacteria to reduce competition for the newly acquired nutritional resources liberated from plant cell wall degradation during an *Erwinia*

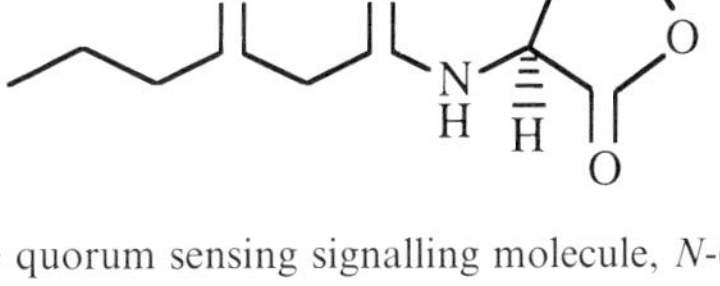

Fig. 1. Structure of the quorum sensing signalling molecule, *N*-(3-oxohexanoyl)-L-homoserine lactone (OHHL).

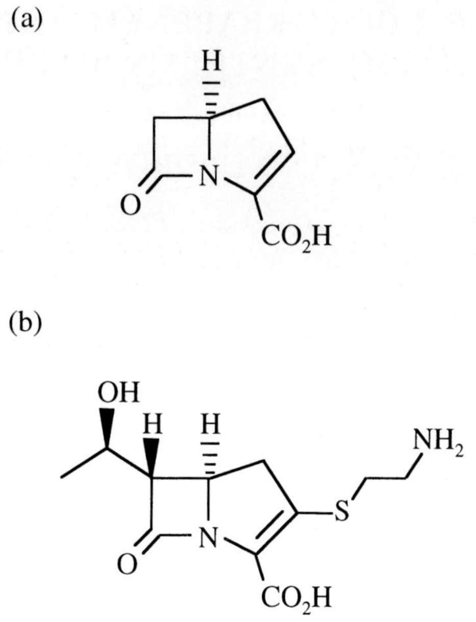

(a)

(b)

Fig. 2. Carbapenem antibiotics. (a) 1-carbapen-2-em-3-carboxylic acid produced by *Erwinia carotovora* and *Serratia marcescens* (Parker *et al.*, 1982) and (b) thienamycin produced by *Streptomyces cattleya* and *Streptomyces penemifaciens* (Williamson *et al.*, 1985).

infection *in planta*. The antibiotic-mediated suppression of its competitors by the closely related *E. herbicola* is so successful that its use in the biological control of apple fire blight disease, caused by *Erwinia amylovora*, has been advocated (Vanneste *et al.*, 1992; Kearns & Manhanty, 1998).

The antibiotic made by *E. carotovora* is similar to the structurally more complex carbapenem, thienamycin, produced by several species of Streptomycetes (Williamson *et al.*, 1985; Fig. 2). Thienamycin, like the majority of this class of β-lactams, is a broad spectrum antibiotic and has attracted a great deal of interest from the pharmaceutical industry. When appropriate modifications are made to improve its chemical stability and resistance to degradation within the kidneys, the resulting compounds, imipenem and meropenem, respectively, have proved to be highly successful drugs. Both are prescribed in the treatment of intra-abdominal infections, skin infections, septicaemia and, in the case of meropenem, meningitis (ABPI).

The demands of modern medicine long ago outstripped the ability of Biology to provide large quantities of novel carbapenem antibiotics, and total chemical synthesis now fulfils this role (for review, see Coulton & Hunt, 1996). Nevertheless, the desire to study the apparently novel carbapenem biosynthetic pathway, as well as its regulation via OHHL, in *E. carotovora* remains important as it may generate information enabling production of new carbapenems with therapeutic potential. The genetics, enzymes and biosynthetic pathway leading to the formation of the penicillins – the more famous β-lactam cousins of the carbapenems – are now well understood (for review, see Cohen & Aharonowitz, 1995). After work to elucidate the precursors of carbapenem biosynthesis in a second Car producer, *Serratia marcescens*, it was clear that the two biosynthetic routes leading to penicillin and carbapenem were likely to be different (Bycroft *et al.*, 1988). Because of the relative structural simplicity of the carbapenem antibiotic produced by the Gram-negative bacteria *E. carotovora* and *S. marcescens* and because of the genetic tractability of both species, these organisms have become model systems for understanding the biosynthesis and regulation of carbapenems.

QUORUM SENSING

carI and carR – the fundamental carbapenem control regulon

In the early 1990s, a number of reports began to appear in the literature describing quorum sensing regulation in an ever-increasing number of Gram-negative bacteria. Before this period, gene regulation studies involving acylated homoserine lactones had been largely restricted to one genus. The transcriptional regulation by OHHL of the *lux* gene operon, responsible for bioluminescence in *Vibrio fischeri*, had remained an elegant, if curious, mechanism since its discovery in the 1970s (Eberhard, 1972). Following the explosion of interest in the phenomenon of quorum sensing, however, the large body of work that now exists in this area has made *lux* gene regulation the point of reference for the entire field.

The control of carbapenem antibiotic biosynthesis in *Erwinia carotovora* subspecies *carotovora* was one of the first reports to define a regulatory system involving acylated homoserine lactones outside the *Vibrio* genus (Bainton *et al.*, 1992*a*). Subsequent work has shown that *E. carotovora* subspecies *carotovora*, has two genes, *carI* and *carR*, that are essential for quorum sensing (Swift *et al.*, 1993; McGowan *et al.*, 1995; Fig. 3). Homo-logues of *carI* and *carR* have now been found within the majority of bacteria employing a quorum sensing regulatory mechanism (including the related *E. carotovora* subspecies *betavasculorum*; Costa & Loper, 1997). These genes are members of the *V. fischeri luxI* and *luxR* gene families, respectively.

luxI encodes an enzyme responsible for catalysing the formation of the *N*-acyl homoserine lactone (AHL) signalling molecule (Schaefer *et al.*, 1996; More *et al.*, 1996) and *luxR* encodes an OHHL-responsive, DNA-binding transcriptional activator. Once the intracellular concentration of OHHL is high enough, it is believed to bind LuxR, causing a conformational change in the activator, thereby allowing it to transcriptionally activate the *lux* operon (Sitnikov *et al.*, 1995; Stevens & Greenberg, 1997). Although coincidentally, both *E. carotovora* and *V. fischeri* produce and sense OHHL, it is clear that

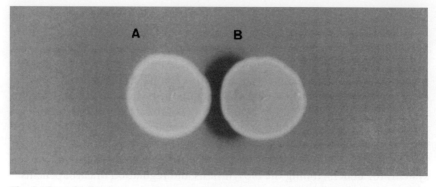

Fig. 3. Cross-feeding among carbapenem-non-producing strains of *Erwinia carotovora*. The *carR* gene in strain A and the *carI* gene in strain B have each been insertionally inactivated, resulting in their inability to respond to, or to produce, *N*-(3-oxohexanoyl)-L-homoserine lactone, respectively. However, when strain A, which still produces the diffusible OHHL, is grown alongside strain B, carbapenem production is restored in the latter strain. The presence of the antibiotic is indicated by a zone of clearing in the lawn of a β-lactam supersensitive strain of *Escherichia coli*.

an *N*-acyl homoserine lactone with a six-carbon side chain is not a prerequisite for quorum sensing. A spectrum of molecules has now been described, involved in bacterial signalling in, for example, *Serratia liquefaciens* and *Rhizobium leguminosarum* where acylated homoserine lactones with side chains of four and fourteen carbon atoms, respectively, act as pheromones (Eberl *et al.*, 1996; Schripsema *et al.*, 1996; Williams *et al.*, this volume).

Most bacteria are reported as producing predominantly a single species of signalling molecule, although there are some notable exceptions. *Pseudomonas aeruginosa*, for example, has two LuxI homologues, LasI and RhlI, responsible for production of two different homoserine lactones (Pesci & Iglewski, this volume). With carbon side chains of four and twelve carbons respectively, BHL (*N*-butanoyl-L-homoserine lactone) and OdDHL (*N*-(3-oxododecanoyl)-L-homoserine lactone) are 'sensed' by two distinct LuxR homologues (Latifi *et al.*, 1996; Pearson *et al.*, 1997). In contrast, *Yersinia enterolitica* has only a single LuxI homologue, but is capable of producing two signalling molecules (*N*-hexanoyl-L-homoserine lactone (HHL) and OHHL), which vary in the number of carbonyl groups in the common six-carbon side chain (Throup *et al.*, 1995; Swift *et al.*, this volume).

Work carried out on the single LuxI homologue of *E. carotovora*, CarI, suggests that, in common with other such proteins (Shaw *et al.*, 1997), it is capable of varying the length of the carbon side chain. Extensive analysis of the spent culture supernatants using a variety of sensitive biosensors has revealed that, in addition to OHHL, *E. carotovora* also produces at least three other AHL molecules (Holden, personal communication). When the spent culture supernatant of a *carI* mutant strain of *E. carotovora* was analysed, none of these molecules was detected. One of the additional molecules appears to be HHL and the identities of the others are currently been investigated. The physiological significance of these observations is not yet understood.

When *carI* and *carR* (the gene encoding the transcriptional activator of the carbapenem genes) were first cloned from *E. carotovora*, their genetic organization was compared with that of the *luxIR* genes. In *V. fischeri*, *luxR* and *luxI* are transcribed divergently, with *luxI* transcribed as the first gene of the *lux* operon. This genetic organization leads to a genetic feedback loop resulting in autoinduction of *luxI*. When *luxI* is transcribed with the rest of the *lux* operon, it leads to the production of more OHHL and ultimately therefore, to further LuxR-activated transcription of the operon. This positive feedback involving *luxI* and OHHL has led to the term 'autoinducer' being broadly applied, perhaps misleadingly, to the signalling molecules of the subsequently discovered quorum sensing systems. The genetic arrangement of the two homologues in *E. carotovora* and in most (but by no means all) subsequently discovered quorum sensing systems is different. Crucially, *carI* is not part of the *car* gene operon under the transcriptional control of CarR and does not therefore cause its own autoinduction in a manner analogous to that of *luxI*.

The expression of OHHL in *E. carotovora* has been analysed using a number of different assays in an attempt to determine the kinetics of its production throughout the growth curve. A bioluminescent sensor plasmid has been used to assess the levels of OHHL both *in vivo* (Swift *et al.*, 1993) as well as in spent culture supernatants (Harris *et al.*, 1998; Holden *et al.*, 1998), and in addition, the insertion of a promoterless *lacZ* reporter gene in *carI* was used to measure transcription (Rivet, M., personal communication). All of the resulting data support the conclusion, in direct contrast to *V. fischeri*, that OHHL is produced constitutively, as has been found for homologues of *luxI* elsewhere (for example, *Yersinia enterolitica*, Throup *et al.*, 1995; *Erwinia stewartii*, Beck von Bodman *et al.*, 1998). On the face of it therefore, the concentration of OHHL produced by a population of cells builds up gradually throughout the growth curve (rather than by a steep induction step) until an intracellular threshold concentration is attained, allowing activation of the *car* genes by CarR and the exoenzyme biosynthetic genes presumably by a hypothetical (and as yet unidentified) homologue of LuxR.

In the context of its pathogenic lifestyle, the advantage to the cell of this regulatory mechanism in the control of both exoenzyme and carbapenem production is clear. In response to the various oligosaccharide elicitors released from the plant cell wall during attack by pathogenic bacteria, plants have evolved a large array of defences (for reviews, see Baron & Zambryski (1995) and Benhamou (1996)). Only a large body of bacterial cells is capable of producing the local, high concentrations of cell wall-degrading enzymes thought necessary to overwhelm these defences. It is an obvious evolutionary advantage therefore, to limit expression of the relevant virulence genes to a population above a certain size, using quorum sensing regulation. A single cell or small number of cells that are unable to produce enough of the freely diffusible OHHL to attain the intracellular threshold required for expression, would also be unlikely to produce enough enzymes to result in a successful infection.

Fine tuning the quorum sensing signal

The simple quorum sensing regulatory system outlined so far is not the whole story, however, and bacteria have added a variety of extra features to enhance its effectiveness. Data already exist to show that the products of other *E. carotovora* genes are also involved in the regulation of carbapenem and exoenzymes. The product of the *rsmA* gene (regulator of *s*econdary *m*etabolites) for example, was first discovered following the isolation of a mutant strain in which expression of the cell wall-degrading enzymes was independent of normal OHHL-dependent control (Chatterjee *et al.*, 1995). This protein, consisting of only 61 amino acids, apparently exerts its effect by modifying the overall levels of intracellular OHHL. Homologues of *rsmA* are found in a number of the Enterobacteriaceae as well as elsewhere among the eubacteria, and although not necessarily acting via OHHL, are likely to act as global regulators of a range of phenotypes (Cui *et al.*, 1995; White *et al.*, 1996). In particular, RsmA has homology to the product of the *csrA* gene of *Escherichia coli*, with 95% identity at the protein level. CsrA (*c*arbon *s*torage *r*egulator) has been shown to bind to, and facilitate the decay of, certain species of mRNA (Cui *et al.*, 1995; Liu & Romeo, 1997), and it is highly likely that RsmA acts in a similar manner.

Following the isolation of *rsmA*, a second regulator that possibly interacts with the quorum sensing machinery was discovered. Originally isolated in *S. marcescens*, where it was named *rap* (regulator of *a*ntibiotic and *p*igment), it was subsequently shown to have homologues in *E. carotovora* as well as in other members of the Enterobacteriaceae (Thomson *et al.*, 1997). In both *S. marcescens* and *E. carotovora* (where the gene was named *hor* – homologue of *rap*), the Rap/Hor proteins are required for carbapenem antibiotic production. Using *lacZ* reporter fusions in target genes, the Rap/Hor-dependent regulation was shown to act at the level of transcription of the *car* biosynthetic genes and did not affect transcription of either *carI* or *carR* (Fig. 4). An analysis of the primary amino acid sequence of the Hor protein does not indicate the presence of any known DNA-binding motif, and although an interaction with either CarR or OHHL is possible (certainly, each of the genera identified as encoding a homologue of *rap* also carry a *luxR* homologue), it is unclear at the time of writing, how Rap/Hor exerts its influence.

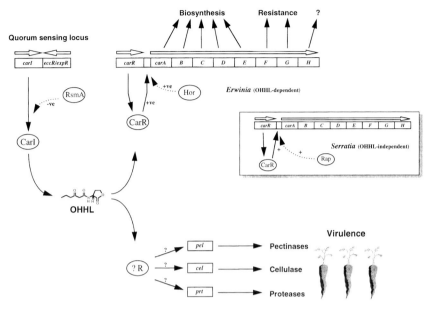

Fig. 4. The quorum sensing regulon in *Erwinia carotovora*. The quorum sensing genetic locus is shown (*carI*, *eccR/expR*). The function of *eccR* is not known. The *carI* gene encodes the enzyme responsible for synthesis of OHHL. OHHL is thought to bind to the CarR regulator protein, enabling it to transcriptionally activate the carbapenem biosynthetic cluster. The carbapenem gene cluster is unlinked to the quorum sensing locus. The genes for biosynthesis (*carA-E*) and the resistance genes (*carFG*) are operonic. The function of *carH* is not known. However, a homologue of *carH* is also present in the *Serratia* cluster (inset) which encodes homologues of all of the corresponding *Erwinia car* gene cluster proteins, including CarR. The *Serratia* CarR does not appear to require OHHL for function, but both the *Erwinia* and the *Serratia* carbapenem gene clusters require Hor/Rap for expression. In *Erwinia*, OHHL is also required for full expression of the exoenzyme virulence factor genes – encoding pectinases (*pel*), cellulase (*cel*) and proteases (*prt*) (Jones *et al.*, 1993). Thus, in *Erwinia*, both carbapenem antibiotic production and the synthesis of the exoenzyme virulence factors are under quorum sensing control, mediated by OHHL.

As more work is carried out in this area, further genes are identified that influence the level of control mediated by the quorum sensing regulatory system. One such is the recently identified *hexA* gene (Harris *et al.*, 1998). The product of this gene, a member of the LysR family of DNA-binding, transcriptional regulators, modulates expression of the virulence genes in *E. carotovora*. This effect is apparently achieved by direct binding of HexA upstream of the virulence genes themselves, as well as by a reduction in the amount of OHHL produced throughout the growth curve (Harris *et al.*, 1998).

An unexpected means of regulation has also recently been uncovered in *E. carotovora*, layered on top of the quorum sensing system. Very few natural carbapenem producing strains of *E. carotovora* have been isolated. However, many *E. carotovora* strains have been shown to retain the genes required for biosynthesis of the antibiotic. These genes are cryptic because the CarR protein, encoded by *carR* in single copy on the chromosome, is inactive (Holden *et al.*, 1998). However, if complemented by the *carR* gene from a normally antibiotic producing strain, carbapenem production in these strains can be restored. It was presumed that a spontaneous mutation to inactivate the *carR* gene product had occurred in the cryptic strains, and a hypothesis was proposed suggesting that this allowed the regulation of carbapenem expression at the level of the population (Holden *et al.*, 1998). This means of regulation has both advantages and disadvantages for the bacterial population. Each individual bacterium retains all the genetic information required to manufacture the antibiotic, but is not constrained to do so. However, as there is currently no evidence for a precise genetic switch to reverse the lesion within *carR*, the bacteria are apparently at the mercy of random mutations in order to reconstitute the production of carbapenem.

Before the discovery of the widespread occurrence of quorum sensing regulation in bacteria, it had been realized that the regulation of carbapenem production in *S. marcescens* was different from that of *E. carotovora* (Bycroft *et al.*, 1988). Whereas production of the antibiotic is growth phase dependent in the plant pathogen, it was reported to be produced constitutively in *S. marcescens*. Recently, a study was undertaken to determine the nature of the *S. marcescens* carbapenem biosynthetic and regulatory genes. A library of DNA fragments derived from the *S. marcescens* chromosome was used to transform carbapenem non-producing strains of *E. carotovora* in which the *car* biosynthetic genes had been insertionally inactivated (Cox *et al.*, 1998). Surprisingly, the two cosmids that were isolated were also able to complement *E. carotovora carR* mutations. Although this suggested the presence of a homologue of *carR* on the *S. marcescens* chromosome, none had been anticipated because attempts to find an AHL using the *lux* plasmid sensor (Bainton *et al.*, 1992*b*) had been unsuccessful.

Sequence analysis of part of the cosmid insert identified an open reading frame, the predicted protein product of which had 72% similarity at the amino acid level to CarR. Later work showed that *S. marcescens* possesses homologues of all of the *car* genes in addition to *carR* (Cox, unpublished observations; see below). An explanation of the different carbapenem expression patterns observed in the *Erwinia* and *Serratia* genera, was thought to be due to the OHHL-independent nature of the activation of the *car* genes in *S. marcescens* by CarR (Cox *et al.*, 1998). It is important to remember however, that the inability to detect a freely diffusible signalling molecule in the carbapenem-producing strain of *S. marcescens* via the lux-sensor assay is only circumstantial evidence that there is no AHL produced. An alternative signal(s) could still be elaborated by this strain. Nevertheless, it is possible that this *S. marcescens* CarR protein has evolved to activate expression outside the influence of any quorum sensing mechanism, following its transfer into a non-AHL genetic background by horizontal transfer from another species.

The *carR* and *carI* genes of *E. carotovora* were originally cloned on separate cosmids and are unlinked (Swift *et al.*, 1993; McGowan *et al.*, 1995). CarR is a dedicated activator of the *car* genes, and is not responsible for the OHHL-induced expression of the cell wall-degrading enzymes. The hypothetical gene encoding the LuxR homologue proposed to play a role in regulating the latter has yet to be discovered. An initial candidate, *expR*, was found convergently transcribed with and overlapping the *carI* gene (Fig. 4; Pirhonen *et al.*, 1993). However, in at least two strains of *E. carotovora*, this gene has no obvious function; *expR* mutants made apparently wild-type levels of exoenzymes and antibiotic (Rivet, M., personal communication). In contrast, *carR* is located immediately upstream of the genes under its control – a cluster of eight genes, the products of which are variously responsible for biosynthesis of carbapenem and for intrinsic resistance to carbapenem – in both *E. carotovora* and *S. marcescens* (Fig. 4; McGowan *et al.*, 1996; Cox, unpublished observations). As predicted from results showing differences in the biosynthetic precursors for carbapenems and the related class of β-lactams, the penicillins, none of the predicted protein products were homo-logues of enzymes involved in penicillin biosynthesis (McGowan *et al.*, 1996). Surprisingly, however, the first gene of the cluster, *carA*, was predicted to encode a homologue of a protein involved in the biosynthesis of clavulanic acid.

Clavulanic acid is a metabolite of *Streptomyces clavuligerus* and is a β-lactam. Unlike the members of the other β-lactam classes, clavulanic acid possesses little intrinsic antibacterial activity. A great deal of interest has been shown in this compound however, by virtue of its β-lactamase inhibitory properties. Many formerly highly successful β-lactam antibiotics have been rendered ineffective against some bacterial species by alterations in their target sites, the penicillin binding proteins, or by a reduction in the permeability of the outer membrane. However, by far the most important form of resistance is the physical rupture of the β-lactam ring of the antibiotic by β-lactamase enzymes. Rather than drop these antibiotics from clinical use, it was found that combining them with clavulanic acid to combat any β-lactamase activity could extend their functional lifetime. This clinical efficacy of clavulanic acid stimulated interest in its biosynthesis and now a consider-able amount is known concerning the biochemistry of the pathway. A number of genes involved in clavulanic acid production have also been identified.

The *S. clavuligerus* homologue of CarA has thus far received little attention, however, and has not yet been named. Mutant strains of *S.*

clavuligerus have been isolated, in which a transposon has insertionally inactivated this gene (strains dclH65 and dclH50; Hodgson *et al.*, 1995), and some preliminary work has been carried out on the metabolites accumulated in these strains. Two novel, opine-type derivatives of one of the precursors, arginine, were isolated from the culture broth of strain dclH65 (Elson *et al.*, 1993*a*). The significance of this finding becomes clear following further work by Elson *et al.* (1993*b*), showing that the opine derivatives are the immediate precursors of the first compound in the clavulanic acid pathway containing a β-lactam ring. It was concluded that the function of the enzyme encoded by the gene in question, is to construct the β-lactam ring of the final product, by a novel amide bond-forming mechanism. Therefore, the involvement of a homologue of this protein in the biosynthetic pathway of carbapenem – another β-lactam ring containing compound – is potentially of great significance. The third gene in the cluster, *carC*, also encodes a homologue of an enzyme of the clavulanic acid biosynthetic pathway – clavaminic acid synthase (McGowan *et al.*, 1996). This oxygenase plays a key role in the construction of all the clavam class of β-lactams, including clavulanic acid (Baggaley *et al.*, 1997; Egan *et al.*, 1997).

Both CarA and CarC have been shown to be essential in the biosynthesis of carbapenem (McGowan *et al.*, 1997). *E. carotovora* utilizes different precursors (glutamate and acetate; Bycroft *et al.*, 1988) in the manufacture of carbapenem from those used by *S. clavuligerus* in the biosynthesis of clavulanic acid (arginine and pyruvate; Valentine *et al.*, 1993; Thirkettle *et al.*, 1997). A simple interpretation of the two pairs of homologues, however, is that *E. carotovora* uses a similar mechanism to form the β-lactam ring of carbapenem. In contrast to earlier reports (Townsend, 1993), a possible evolutionary relationship between the two biosynthetic pathways has been advanced (McGowan *et al.*, 1998).

CONCLUDING REMARKS

During this decade, research into the biosynthesis of carbapenem antibiotics has thrown up some genuine surprises. The early discovery of the quorum sensing regulation of carbapenem contributed to the recognition that such regulatory systems were widespread among Gram-negative bacteria. In addition, the discovery of evidence for a previously unknown antibiotic resistance mechanism (McGowan *et al.*, 1997) and the uncovering of homologies with enzymes involved in the biosynthesis of clavulanic acid, had not been anticipated, and have now stimulated research in novel β-lactam synthesis and resistance mechanisms.

This research is continuing. Apart from advances in the understanding of the biosynthetic pathway itself, recent work in this laboratory has started to focus directly upon the signalling process. Using both gel filtration and sucrose density centrifugation, it is apparent that CarR naturally forms a dimer and that it does so in either the presence or absence of OHHL (Welch, unpublished observations). In addition, by monitoring the quenching of the intrinsic tryptophan fluorescence of CarR, it has been possible to monitor the binding of different homoserine lactones to CarR. Although the natural *E. carotovora* signalling molecule is OHHL, OOHL (*N*-(3-oxooctanoyl)-L-homoserine lactone) with an eight- rather than six-carbon side chain, appeared to bind CarR with the greatest efficiency. The apparent K_d of the binding of all the AHLs, in the range of $1–10\,\mu M$, was surprisingly high, given the lower levels required to activate carbapenem *in vivo* (Welch, unpublished observations).

Both the nature of the carbon source and temperature of growth, also seem to play a role in the quorum sensing signalling process in *E. carotovora*. Transcription of the *car* genes is reduced by 75% when the temperature of culture is raised to 36 °C compared with transcription at 35 °C (Bozgelmez, unpublished observations). A rise of a further one degree Celsius to 37 °C, totally abolishes transcription, and these results indicate an acutely sensitive response to temperature in the quorum sensing regulon. Further work suggests that this effect may be mediated through the amount of OHHL produced (Barnard, unpublished observations).

It is important to appreciate that OHHL is likely to be involved in multiple physiological processes. In addition to the regulation of expression of the antibiotic and virulence factors, OHHL has now been shown to regulate a variety of other genes. The phenotypes involved, and the mechanism of OHHL-mediated regulation – both by repression as well as activation – are currently being investigated (Whitehead, personal communication). These observations clearly imply that quorum sensing signalling molecules play a far wider role in bacterial physiology than has been realized to date.

This chapter has examined quorum sensing and in particular, its influence on just one phenotype – the production of carbapenem antibiotic. A brief description of its mechanism might lead one to assume that this regulatory mechanism is a simple on/off switch in response to cell density. However, many further inputs have apparently been added by evolution to increase the flexibility of the system. It is quite unlikely that *Erwinia carotovora* is alone in having such complexity layered on top of the basic quorum sensing system, and as other examples are examined in similar detail, equally intricate responses will almost certainly be discovered.

ACKNOWLEDGEMENTS

This work was supported by the BBSRC, UK. We would like to thank A. Barnard, G. Bosgelmez, M. Holden, M. Rivet, N. Robson, M. Sebaihia, M. Welch and N. Whitehead for generous access to unpublished results during the writing of this chapter.

REFERENCES

ABPI – The Association of the British Pharmaceutical Industry compendium of data sheets and summaries of product characteristics, 1998–1999, Datapharm Publications Limited, London.

Axelrood, P. E., Rella, M. & Schroth, M. N. (1988). Role of antibiosis in competition of *Erwinia* strains in potato infection courts. *Applied and Environmental Microbiology*, **54**, 1222–9.

Baggaley, K. H., Brown, A. G. & Schofield, C. J. (1997). Chemistry and biosynthesis of clavulanic acid and other clavams. *Natural Product Reports*, **14**, 309–33.

Bainton, N. J., Stead, P., Chhabra, S. R., Bycroft, B. W., Salmond, G. P. C., Stewart, G. S. A. B. & Williams, P. (1992*a*). *N*-(3-oxohexanoyl)-L-homoserine lactone regulates carbapenem antibiotic production in *Erwinia carotovora*. *Biochemical Journal*, **288**, 997–1004.

Bainton, N. J., Bycroft, B. W., Chhabra, S. R., Stead, P., Gledhill, L., Hill, P. J., Rees, C. E. D., Winson, M. K., Salmond, G. P. C., Stewart, G. S. A. B. & Williams, P. (1992*b*). A general role for the *lux* autoinducer in bacterial cell signalling: control of antibiotic biosynthesis in *Erwinia*. *Gene*, **116**, 87–91.

Baron, C. & Zambryski, P. C. (1995). The plant-response in pathogenesis, symbiosis, and wounding – variations on a common theme. *Annual Review of Genetics*, **29**, 107–29.

Beck von Bodman, S., Majerczak, D. R. & Coplin, D. L. (1998). A negative regulator mediates quorum-sensing control of exopolysaccharide production in *Pantoea stewartii* subsp. *stewartii*. *Proceedings of the National Academy of Science, USA*, **95**, 7687–92.

Benhamou, N. (1996). Elicitor-induced plant defense pathways. *Trends in Plant Science*, **1**, 233–40.

Bycroft, B. W., Maslen, C., Box, S. J., Brown, A. & Tyler, J. W. (1988). The biosynthetic implications of acetate and glutamate incorporation into (3*R*,5*R*)-carbapenem-3-carboxylic acid and (5*R*)-carbapen-2-em-3-carboxylic acid by *Serratia* sp. *Journal of Antibiotics*, **41**, 1231–42.

Chatterjee, A., Cui, Y. Y., Liu, Y., Dumenyo, C. K. & Chatterjee, A. K. (1995). Inactivation of *rsmA* leads to overproduction of extracellular pectinases, cellulases, and proteases in *Erwinia carotovora* subsp. *carotovora* in the absence of the starvation cell density-sensing signal, *N*-(3-oxohexanoyl)-L-homoserine lactone. *Applied and Environmental Microbiology*, **61**, 1959–67.

Cohen, G. & Aharonowitz, Y. (1995). Molecular genetics of antimicrobials: a case study of β-lactam antibiotics. In *Fifty Years of Antimicrobials: Past Perspectives and Future Trends*, ed. P. A. Hunter, G. K. Darby & N. J. Russell, *Society for General Microbiology Symposium Vol. 53*, pp. 139–63. Cambridge: Cambridge University Press.

Costa, J. M. & Loper, J. E. (1997). EcbI and EcbR: homologs of LuxI and LuxR affecting antibiotic and exoenzyme production by *Erwinia carotovora* subsp. *betavasculorum*. *Canadian Journal of Microbiology*, **43**, 1164–71.

Coulton, S. & Hunt, E. (1996). Recent advances in the chemistry and biology of carbapenem antibiotics. *Progress in Medicinal Chemistry*, **33**, 99–145.

Cox, A. R. J., Thomson, N. R., Bycroft, B., Stewart, G. S. A. B., Williams, P. & Salmond, G. P. C. (1998). A pheromone-independent CarR protein controls carbapenem antibiotic synthesis in the opportunistic human pathogen *Serratia marcescens*. *Microbiology*, **144**, 201–9.

Cui, Y., Chatterjee, A., Liu, Y., Dumenyo, C. K. & Chatterjee, A. K. (1995). Identification of a global repressor gene, *rsmA*, of *Erwinia carotovora* subsp. *carotovora* that controls extracellular enzymes, *N*-(3-oxohexanoyl)-L-homoserine lactone, and pathogenicity in soft-rotting *Erwinia* spp. *Journal of Bacteriology*, **177**, 5108–15.

Eberhard, A. (1972). Inhibition and activation of bacterial luciferase synthesis. *Journal of Bacteriology*, **109**, 1101–5.

Eberl, L., Winson, M. K., Sternberg, C., Stewart, G. S. A. B., Christiansen, G., Chhabra, S. R., Bycroft, B., Williams, P., Molin, S. & Givskov, M. (1996). Involvement of *N*-acyl-L-homoserine lactone autoinducers in controlling the multicellular behavior of *Serratia liquefaciens*. *Molecular Microbiology*, **20**, 127–36.

Egan, L. A., Busby, R. W., Iwata-Reuyl, D. & Townsend, C. A. (1997). Probable role of clavaminic acid as the terminal intermediate in the common pathway to clavulanic acid and the antipodal clavam metabolites. *Journal of the American Chemical Society*, **119**, 2348–55.

Elson, S. W., Baggaley, K. H., Fulston, M., Nicholson, N. H., Tyler, J. W., Edwards, J., Holms, H., Hamilton, I. & Mousdale, D. M. (1993a). Two novel arginine derivatives from a mutant of *Streptomyces clavuligerus*. *Journal of the Chemical Society – Chemical Communications*, **15**, 1211–12.

Elson, S. W., Baggaley, K. H., Davison, M., Fulston, M., Nicholson, N. H., Risbridger, G. D. & Tyler, J. W. (1993b). The identification of three new biosynthetic intermediates and one further biosynthetic enzyme in the clavulanic acid pathway. *Journal of the Chemical Society – Chemical Communications*, **15**, 1212–14.

Fuqua, W. C., Winans, S. C. & Greenberg, E. P. (1994). Quorum sensing in bacteria – the LuxR-LuxI family of cell density-responsive transcriptional regulators. *Journal of Bacteriology*, **176**, 269–75.

Fuqua, W. C., Winans, S. C. & Greenberg, E. P. (1996). Census and consensus in bacterial ecosystems – the LuxR-LuxI family of quorum-sensing transcriptional regulators. *Annual Review of Microbiology*, **50**, 727–51.

Harris, S. J., Shih, Y., Bentley, S. D. & Salmond, G. P. C. (1998). The *hexA* gene of *Erwinia carotovora* encodes a LysR homologue and regulates motility and the expression of multiple determinants. *Molecular Microbiology*, **28**, 705–17.

Hodgson, J. E., Fosberry, A. P., Rawlinson, N. S., Ross, H. N. M., Nea, R. J., Arnell, J. C., Earl, A. J. & Lawlor, E. J. (1995). Clavulanic acid biosynthesis in *Streptomyces clavuligerus* – gene cloning and characterisation. *Gene*, **166**, 49–55.

Holden, M. T. G., McGowan, S. J., Bycroft, B. W., Stewart, G. S. A. B., Williams, P. & Salmond, G. P. C. (1998). Cryptic carbapenem antibiotic production genes are widespread in *Erwinia carotovora*: facile *trans* activation by the *carR* transcriptional activator. *Microbiology*, **144**, 1495–508.

Ishimaru, C. A., Klos, E. J. & Brubaker, R. R. (1988). Multiple antibiotic production by *Erwinia herbicola*. *Phytopathology*, **78**, 746–50.

Jones, S., Yu, B., Bainton, N. J., Birdsall, M., Bycroft, B. W., Chhabra, S. R., Cox, A. J. R., Golby, P., Reeves, P. J., Stephens, S., Winson, M. K., Salmond, G. P. C., Stewart, G. S. A. B. & Williams, P. (1993). The *lux* autoinducer regulates the production of exoenzyme virulence determinants in *Erwinia carotovora* and *Pseudomonas aeruginosa*. *EMBO Journal*, **12**, 2477–82.

Kearns, L. P. & Manhanty, H. K. (1998). Antibiotic production by *Erwinia herbicola* Eh1087: its role in inhibition of *Erwinia amylovora* and partial characterisation of antibiotic biosynthesis genes. *Applied and Environmental Biology*, **64**, 1837–44.

Latifi, A., Foglino, M., Tanaka, K., Williams, P. & Lazdunski, A. (1996). A hierarchical quorum-sensing cascade in *Pseudomonas aeruginosa* links the transcriptional activators LasR and RhiR (VsmR) to expression of the stationary-phase sigma-factor RpoS. *Molecular Microbiology*, **21**, 1137–46.

Liu, M. Y. & Romeo, T. (1997). The global regulator CsrA of *Escherichia coli* is a specific mRNA-binding protein. *Journal of Bacteriology*, **179**, 4639–42.

McGowan, S. J., Bycroft, B. W. & Salmond, G. P. C. (1998). Bacterial production of carbapenems and clavams: evolution of β-lactam antibiotic pathways. *Trends in Microbiology*, **6**, 203–8.

McGowan, S., Sebaihia, M., Jones, S., Yu, B., Bainton, N., Chan, P. F., Bycroft, B., Stewart, G. S. A. B., Williams, P. & Salmond, G. P. C. (1995). Carbapenem antibiotic production in *Erwinia carotovora* is regulated by CarR, a homologue of the LuxR transcriptional regulator. *Microbiology*, **141**, 541–50.

McGowan, S. J., Sebaihia, M., O'Leary, S., Hardie, K. R., Williams, P., Stewart, G. S. A. B., Bycroft, B. W. & Salmond, G. P. C. (1997). Analysis of the carbapenem gene cluster of *Erwinia carotovora*: definition of the antibiotic biosynthetic genes and evidence for a novel β-lactam resistance mechanism. *Molecular Microbiology*, **26**, 545–56.

McGowan, S. J., Sebaihia, M., Porter, L. E., Stewart, G. S. A. B., Williams, P., Bycroft, B. W. & Salmond, G. P. C. (1996). Analysis of bacterial carbapenem antibiotic production genes reveals a novel β-lactam biosynthesis pathway. *Molecular Microbiology*, **22**, 415–26.

More, M. I., Finger, L. D., Stryker, J. L., Fuqua, C., Eberhard, A. & Winans, S. C. (1996). Enzymatic synthesis of a quorum-sensing autoinducer through use of defined substrates. *Science*, **272**, 1655–8.

Parker, W. L., Rathnum, M. L., Wells, J. S., Trejo, W. H., Principe, P. A. & Sykes, R. B. (1982). SQ27860, a simple carbapenem produced by species of *Serratia* and *Erwinia*. *Journal of Antibiotics*, **35**, 653–60.

Pearson, J. P., Pesci, E. C., Iglewski, B. H. (1997). Roles of *Pseudomonas aeruginosa las* and *rhl* quorum-sensing systems in control of elastase and rhamnolipid biosynthesis genes. *Journal of Bacteriology*, **179**, 5756–67.

Pirhonen, M., Flego, D., Heikinheimo, R. & Palva, E. T. (1993). A small diffusible signal molecule is responsible for the global control of virulence and exoenzyme production in the plant pathogen *Erwinia carotovora*. *EMBO Journal*, **12**, 2467–76.

Robson, N. D., Cox, A. R. J., McGowan, S. J., Bycroft, B. W. & Salmond, G. P. C. (1997). Bacterial *N*-acyl-homoserine-lactone-dependent signalling and its potential biotechnological applications. *Trends in Biotechnology*, **15**, 458–64.

Schaefer, A. L., Val, D. L., Hanzelka, B. L., Cronan, J. E. & Greenberg, E. P. (1996). Generation of cell-to-cell signals in quorum sensing: acyl homoserine lactone synthase activity of a purified *Vibrio fischeri* LuxI protein. *Proceedings of the National Academy of Science, USA*, **93**, 9505–9.

Schripsema, J., Derudder, K. E. E., Vanvliet, T. B., Lankhorst, P. P., Devroom, E., Kijne, J. W. & Vanbrussel, A. A. N. (1996). Bacteriocin small of *Rhizobium leguminosarium* belongs to the class of *N*-acyl-L-homoserine lactone molecules, known as autoinducers and as quorum sensing co-transcription factors. *Journal of Bacteriology*, **178**, 366–71.

Shaw, P. D., Ping, G., Daly, S. L., Cha, C., Cronan, J. E., Rinehart, K. L. & Farrand, S. K. (1997). Detecting and characterizing *N*-acyl-homoserine lactone signal molecules by thin-layer chromatography. *Proceedings of the National Academy of Science, USA*, **94**, 6036–41.

Sitnikov, D. M., Schineller, J. B. & Baldwin, T. O. (1995). Transcriptional regulation of bioluminescence genes from *Vibrio fischeri*. *Molecular Microbiology*, **17**, 801–12.

Stevens, A. M. & Greenberg, E. P. (1997). Quorum sensing in *Vibrio fischeri*: Essential elements for activation of the luminescence genes. *Journal of Bacteriology*, **179**, 557–62.

Swift, S., Winson, M. K., Chan, P. F., Bainton, N. J., Birdsall, M., Reeves, P. J., Rees, C. E. D., Chhabra, S. R., Hill, P. J., Throup, J. P., Bycroft, B. W., Salmond, G. P. C., Williams, P. & Stewart, G. S. A. B. (1993). A novel strategy for the isolation of *luxI* homologues: evidence for the widespread distribution of a *luxR:luxI* super-family in enteric bacteria. *Molecular Microbiology*, **10**, 511–20.

Thirkettle, J. E., Baldwin, J. E., Edwards, J., Griffin, J. P. & Schofield, C. J. (1997). The origin of the beta-lactam carbons of clavulanic acid. *Chemical Communications*, **11**, 1025–6.

Thomson, N. R., Cox, A., Bycroft, B. W., Stewart, G. S. A. B., Williams, P. & Salmond, G. P. C. (1997). The Rap and Hor proteins of *Erwinia, Serratia* and *Yersinia*: a novel subgroup in a growing superfamily of proteins regulating diverse physiological processes in bacterial pathogens. *Molecular Microbiology*, **26**, 531–44.

Throup, J. P., Camara, M., Briggs, G. S., Winson, M. K., Chhabra, S. R., Bycroft, B. W., Williams, P. & Stewart, G. S. A. B. (1995). Characterization of the *yenI/yenR* locus from *Yersinia enterocolitica* mediating the synthesis of 2 *N*-acylhomoserine lactone signal molecules. *Molecular Microbiology*, **17**, 345–56.

Townsend, C. A. (1993). Oxidative amino acid processing in β-lactam antibiotic biosynthesis. *Biochemical Society Transactions*, **21**, 208–13.

Valentine, B. P., Bailey, C. R., Doherty, A., Morris, J., Elson, S. W., Baggaley, K. H. & Nicholson, N. H. (1993). Evidence that arginine is a later metabolic intermediate than ornithine in the biosynthesis of clavulanic acid by *Streptomyces clavuligerus*. *Journal of the Chemical Society – Chemical Communications*, **15**, 1210–11.

Vanneste, J. L., Yu, J. & Beer, S. V. (1992). Role of antibiotic production by *Erwinia herbicola* Eh252 in biological control of *Erwinia amylovora*. *Journal of Bacteriology*, **174**, 2785–96.

White, D., Hart, M. E. & Romeo, T. (1996). Phylogenetic distribution of the global regulatory gene *csrA* among eubacteria. *Gene*, **182**, 221–3.

Williamson, J. M., Inamine, E., Wilson, K. E., Douglas, A. W., Liesch, J. M. & Albers-Schonberg, G. (1985). Biosynthesis of the β-lactam antibiotic, thienamycin, by *Streptomyces cattleya*. *Journal of Biological Chemistry*, **260**, 4637–47.

AUTOREGULATORY FACTORS AND REGULATION OF ANTIBIOTIC PRODUCTION IN *STREPTOMYCES*

YASUHIRO YAMADA

Department of Biotechnology, Graduate School of Engineering, Osaka University, 2-1 Yamadaoka, Suitashi, Osaka 565, Japan

INTRODUCTION

The streptomycetes comprise one of the major groups of actinomycetes. Although actinomycetes belong to the group of prokaryotic Gram-positive bacteria, they have a remarkable life cycle accompanied by morphological differentiation; starting from substrate mycelium, aerial mycelia give rise to arthrospore formation and germination in a manner similar to eukaryotic fungi. Additionally, actinomycetes are effective producers of diverse secondary metabolites including many antibiotics. In total about 11 900 antibiotics were discovered in the years up to 1994 and 6600 (55%) of these are products from *Streptomyces* spp. and 1300 (11%) from other actinomycetes. In comparison, 2600 (22%) are from fungi and 1300 (11%) have been isolated from other microorganisms (Strohl, 1997).

Since the discovery of streptomycin from cultures of *Streptomyces griseus* by Waksman in 1940s, during this half a century, actinomycetes have afforded a number of physiologically active secondary metabolites, including antibiotics, anticancer agents, antimetabolites and immunosuppressors as mentioned above. These secondary metabolites are derived from a wide variety of biosynthetic origins, including aminoglycosides, polyketides, terpenes, peptides, β-lactam and nucleoside-type compounds. When the diversity and complexity of the structures of these metabolites are considered, it appears that actinomycetes have evolved wide-ranging biosynthetic pathways for secondary metabolites, even in comparison with fungi which are supposed to be more highly evolved microorganisms in other respects. As for the unusual range of actinomycete products, there are many examples, such as mitomycin with a fused three-membered heterocyclic ring, aziridine and bialaphos with a C–P bond and esperamycin with a enediyne moiety. Additionally, streptomycetes are producers of useful enzymes such as glucose isomerase and P450-type oxygenases. Thus the diversity of their products has made actinomycetes crucially important microorganisms from the viewpoint of fermentation and the pharmaceutical industries.

The importance of their biotechnology applications is mirrored by their vital role in the microbial ecology of the soil where they are regarded as a second major family next to the eubacteria, and as major consumers of organic compounds and producers of soil particles. As a consequence of their productivity, these industrially important microorganisms have been subjected to extensive studies of the structures, chemical synthesis and biosynthetic pathways of their products, development of fermentation technology to increase yields, strain development, and in-depth taxonomic investigation. On the other hand, problems include the high GC content of these organisms and the barrier of endogenous restriction enzymes has hindered the progress of biochemical, molecular biological studies on the mechanisms and regulatory systems underlying secondary metabolite biosyntheses, antibiotic resistance and morphological differentiation. However, D. Hopwood and the John Innes group in the UK have pioneered work in the molecular biology of *Streptomyces coelicolor* and *S. lividans* which has led to the recent progression of this field in other *Streptomyces* species.

The involvement in the regulation of morphological differentiation and secondary metabolite production concurrently with physiological differentiation in *Streptomyces*, of low molecular weight signalling molecules has been known for nearly three decades. These signalling substances appear in the culture broth mainly during exponential growth phase, at around 10 nM concentration, where they induce morphogenesis and secondary metabolite production. They have a common 2,3-disubstituted butyrolactone skeleton and are effective at very low concentration, typically a few nM. In this chapter, these signal substances are designated autoregulators in common with the first one discovered, A-factor (autoregulating factor), isolated and identified by Khokhlov and co-workers in Russia (Khokhlov, 1980). In this review, the three known types of butyrolactone autoregulators of *Streptomyces* are described in detail.

STRUCTURES AND FUNCTIONS OF BUTYROLACTONE AUTOREGULATORS IN *STREPTOMYCES*

A-Factor of Streptomyces griseus

A-factor was the first butyrolactone autoregulator identified; found in the culture broth of *Streptomyces griseus*, a streptomycin producer, by Khokhlov *et al.* at the Shemyakin Institute of Bioorganic Chemistry in Russia. The history of the A-factor discovery has been described in a review with ten Russian references written by Khokhlov in English (Khokhlov, 1980). In the late 1960s, Khokhlov *et al.* prepared many streptomycin non-producing mutants of *S. griseus* and they found one set of mutants, Nos. 1439 and 751, produced streptomycin as effectively as the wild type when they were cultured together. They discovered that an extremely small amount of a low molecular weight compound in the culture broth of No. 751 restored the production of streptomycin in No. 1439. This compound also restored spore formation of

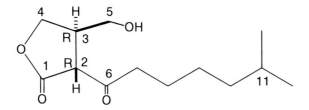

A-factor

Fig. 1. Structure of A-factor.

these non-sporulating mutants. Thus, Khokhlov and co-workers formed the idea of the autoregulator cascade which induces secondary metabolite production and morphological differentiation. This signalling molecule was named A-factor, after its autoregulatory role. As with other signalling substances such as hormones or pheromones, the level of A-factor in the culture broth required is very low, just a few nM. It proved, therefore, very difficult to isolate in order to determine its structure and it took nearly 10 years to propose the structure of A-factor as 2-isocaproyl-3-hydroxymethyl-γ-butyrolactone. However, they proposed the wrong absolute configuration of 2S, 3S on the basis of the CD (circular dichroism) spectrum. Later, Mori and Yamane revised it to 2R, 3R as shown in Fig. 1, using chemical synthesis (Mori & Yamane, 1982; Mori, 1983). In the early 1980s, Beppu and co-workers at the University of Tokyo started studies of A-factor using molecular biological methods and they confirmed the pioneering results of the Russian group (Hara & Beppu, 1982*a*). They additionally found that A-factor induces streptomycin resistance in its producer through the expression of the Sm-6-phosphotransferase gene (Hara & Beppu, 1982*b*).

Virginiae butanolides of Streptomyces virginiae

At the end of the 1960s, Yanagimoto and Terui at Osaka University found a low molecular weight inducer of virginiamycin in the culture broth of *Streptomyces virginiae* (Yanagimoto & Terui, 1971). They named this signal substance IM (*i*nducing *m*aterial). IM was produced at 12 h of cultivation and was able to induce almost full production of virginiamycin 4 hours later, after 16 h of cultivation. They tried to isolate IM and partially purified it. IM was extracted from the culture broth with ethyl acetate and the existence of a lactone ring and hydroxyl group in this small molecule was suggested (Yanagimoto *et al.*, 1979). In the 1980s, Yamada and coworkers purified IM (Yamada *et al.*, 1987) and revealed that IM is a mixture of five related 2,3-disubstituted butyrolactone derivatives. These five homologous butyrolactone autoegulators were named Virginiae Butanolide A, B, C, D, E (VB A,

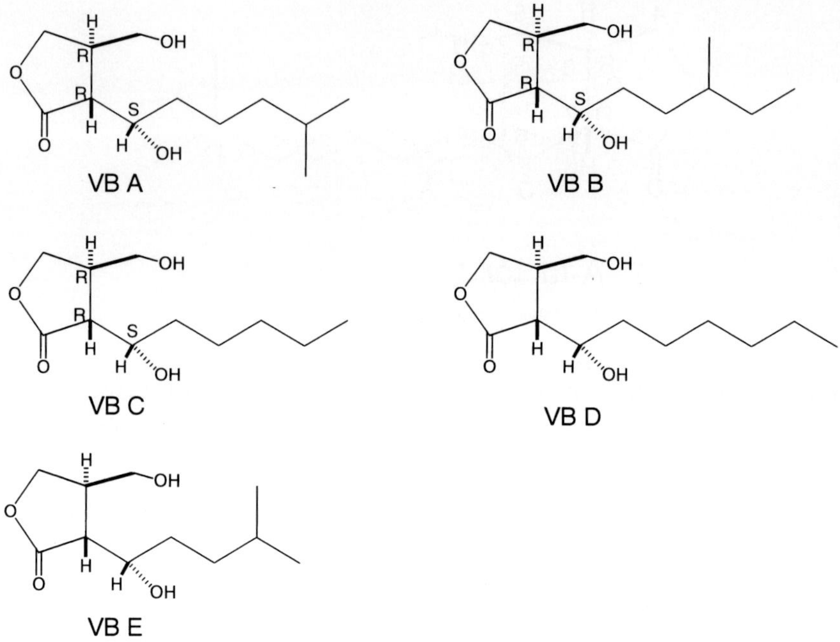

Fig. 2. Structures of virginiae butanolides.

B, C, D, E) (Kondo *et al.*, 1989). The yield of each autoregulator from a 1000 L-scale culture broth was only about 1 mg or less. The structure of the VBs were determined on the basis of their NMR and confirmed by chemical synthesis of VB C. At that time the stereochemistry of two substituents on the butyrolactone ring was wrongly assigned as *cis* and the configuration of hydroxyl group on C6 was not determined until later. Subsequently, the configuration of the two substituents was revised to *trans* and that of C6 of the hydroxyl group was determined to be α (Sakuda & Yamada, 1991). The absolute configuration of VB A, B, C was determined by Mori and co-workers as 2R, 3R, 6S by chemical synthesis (Mori & Chiba, 1990). The structures of VB A, B, C, D, E are shown in Fig. 2.

VBs are produced in late exponential growth phase and induce the production of both virginiamycin M and S, but there is no evidence that they are involved in morphological differentiation of *S. virginiae*. The addition of VB at an early phase of culture, for example, at the starting point of the culture, dramatically reduces virginiamycin production. These facts suggest that VBs trigger pleiotropic regulatory systems which control both up and down regulation of secondary metabolite production, depending upon the internal clocks controlling cellular gene expression.

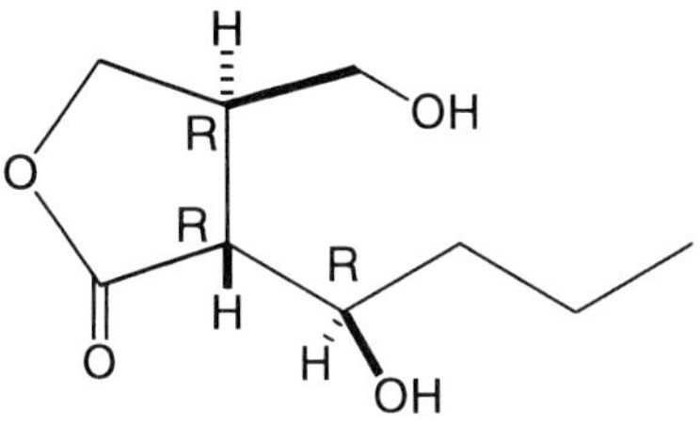

IM-2

Fig. 3. Structure of IM-2.

IM-2 of Streptomyces sp. FRI-5

Streptomyces sp. FRI is a producer of D-cycloserine. It also produces blue pigments when cultivated in a rather nutrient-poor medium. Yanagimoto and Enatu found an endogenous signal substance involved in controlling this phenomenon (Yanagimoto & Enatu, 1983). Later, Sato *et al.* isolated this signal substance and determined its structure (Sato *et al.*, 1989). From more than 1 ton of culture broth, only about 700 μg of the pure autoregulator was isolated and it was named IM-2. The structure of IM-2 was assigned as 2-(1′-hydroxylbutyl)-3-hydroxymethyl butyrolactone and the stereochemistry of the two substituents was deduced to be *trans*. The configuration of the hydroxyl group on C6 was not determined at this stage. Later it was assigned as β (Sakuda & Yamada, 1991) and the absolute configuration of IM-2 was determined as 2-R, 3-R, 6-R (Mizuno *et al.*, 1994) as shown in Fig. 3. As the chain length of the 1′-hydroxyalkyl group is shorter than those of VBs, IM-2 has more hydrophilic properties.

IM-2 appears in mid-exponential growth and induces the production of blue pigment. It also switches the production pattern of antibiotics from D-cycloserine to nucleoside antibiotics such as showdomycin, minimycin and their congeners as shown in Fig. 4 (Hashimoto *et al.*, 1992*b*).

Gräfe's factors

Gräfe and co-workers searched for autoregulators among streptomycetes using a morphological mutant of *Streptomyces griseus*: ZIMET 43682 which also lacks anthracycline production. They isolated one autoregulator from *Streptomyces viridochromogenes* named Factor I which restored the production of antibiotic and mycelium in ZIMET 43682 (Gräfe *et al.*, 1982). They also isolated other autoregulators from *S. bikiniensis* and *S. cyaneofuscatus* with the same activities as Factor I (Gräfe *et al.*, 1983). Gräfe *et al.* assigned

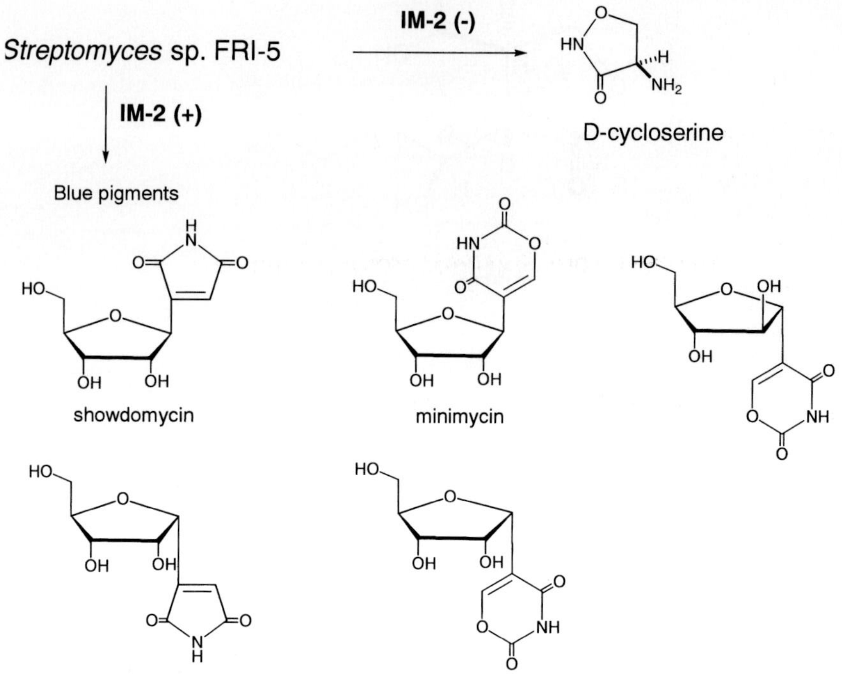

Fig. 4. Function of IM-2.

their structures as 2-(1'-hydroxyalkyl)-3-hydroxymethyl butyrolactones. They proposed a *trans*-configuration for Factor I and assigned a *cis*-configuration to other autoregulators. The configuration of hydroxyl groups on C6 was not determined. Later, Sakuda and Yamada revised their structures to those shown in Fig. 5 (Sakuda & Yamada, 1991). Factor I was confirmed to be *trans*- and the configuration of the hydroxyl group on C6 as β. Other of the Gräfe's factors' stereochemistry was revised from *cis* to *trans* and the orientation of hydroxyl group on C6 changed to α. Their absolute configurations have not been determined as yet. Gräfe's factors are actually exogenous ones and those from *S. vikiniensis* and *S. cyaneofuscatus* were obtained as a mixture. Their activities are about ten-fold less than those of endogenous factors, such as A-factor, VBs and IM-2. However, they are still effective enough to be regarded as microbial hormone-like substances.

BIOSYNTHESIS OF VIRGINIAE BUTANOLIDE A

At present, studies on the biosynthesis of butyrolactone autoregulators are being conducted only with the virginiae butanolide A (VB A). Since the quantity of VBs and other butyrolactone autoregulators produced in culture broth is very low, it is very difficult to carry out the incorporation

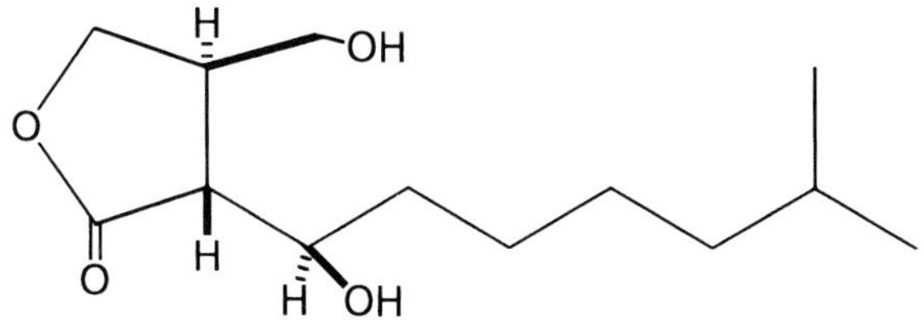

Factor I from *S. viridochromogenes*

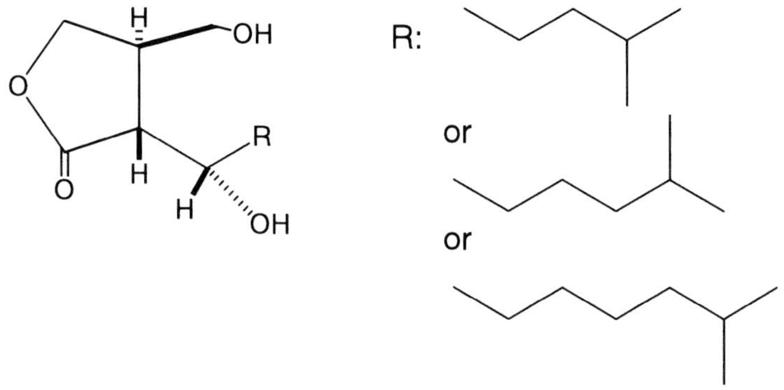

Factors from *S. bikiniensis* and *S. cyaneofuscatus*

Fig. 5. Structure of Gräfe's factors.

experiments using plausible isotope labelled precursors to obtain the labelling pattern of incorporated isotopes. Ohashi *et al.* (1989) obtained a very effective VB A producer: a spontaneous mutant of *Streptomyces antibioticus*, which was found during their studies on the distribution of VBs among *Streptomyces*. This strain produced a few mg/l of VB A. Therefore, it was used throughout these biosynthetic studies of VB A based on bioorganic chemical methodologies. The building blocks of the VB A molecule were identified by incorporation experiments of possible ^{13}C-labelled precursors. The constituents of the VB A molecule turned out to be one glycerol C_3 unit, two acetate C_2 units and one isovalerate C_5 unit (Sakuda *et al.*, 1990). The key intermediate: 3-oxo-7-methyloctanoyl CoA, was also confirmed by intact incorporation of the *N*-acetylcysteamine thioester of [2,3-^{13}C]-3-oxo-7-methyloctanoic acid (Sakuda *et al.*, 1992). The simplified biosynthetic pathway is shown in Scheme 1. The fate of glycerol-derived protons during butyrolactone ring formation was also chased by using deuterium-labelled glycerol for these incorporation studies. It was shown that the proton at the C2 position in the glycerol molecule is replaced and one proton on C1 is also

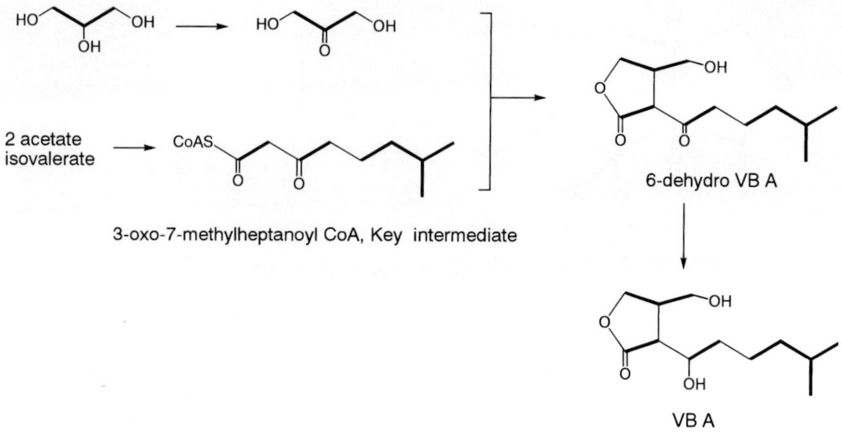

Scheme 1. Biosynthetic building blocks of virginiae butanolide A.

stereoselectively replaced during lactone formation. On the basis of these results, the two possible biosynthetic pathways A and B from dihydroxy-acetone and 3-oxo-7-methyloctanoyl CoA to VB A were proposed as shown in Scheme 2. Plausible intermediates 1, 2, 3, 4, 5 were chemically synthesized and were used as the substrates for the crude enzyme extracts from *S. antibioticus* cells. The simple dihydroxyacetone ester of 3-oxo-7-methyl-octanoic acid (4) was effectively converted to VB A whereas the phosphate derivatives, (1), (2), (3) proved to be poor substrates. Therefore, pathway B was deduced to be the main route to VB A. The intramolecular condensation reaction to form 6-dehydro VB A(5) from the ester (4) is NADH dependent and reduction from (5) to VB A is NADPH dependent (Sakuda *et al.*, 1993). In Scheme 3, the final three steps which involve an aldol-type condensation and reduction, leading to the lactone ring formation are shown in detail. Hydride delivery from NADH to the intermediate (6) is stereoselective. This biosynthetic pathway should be basically common to other autoregulators, VBs, A-factor and IM-2. Minor differences between them include: the key starter intermediates, 3-oxo-acyl CoA and the characteristics of the NADPH-dependent ketone reductase which catalyses the final steps. As branched-chain fatty acids are common in the cells of streptomycetes as primary metabolites, the key intermediates 3-oxo-acyl CoA are thought to come directly from this pool (Wallace *et al.*, 1995).

MODE OF ACTIONS OF BUTYROLACTONE AUTOREGULATORS

Virginiae butanolide receptor protein

Since VBs are secreted into the culture broth and are able to diffuse to other cells leading to a trigger in the expression of enzymes for virginiamycin

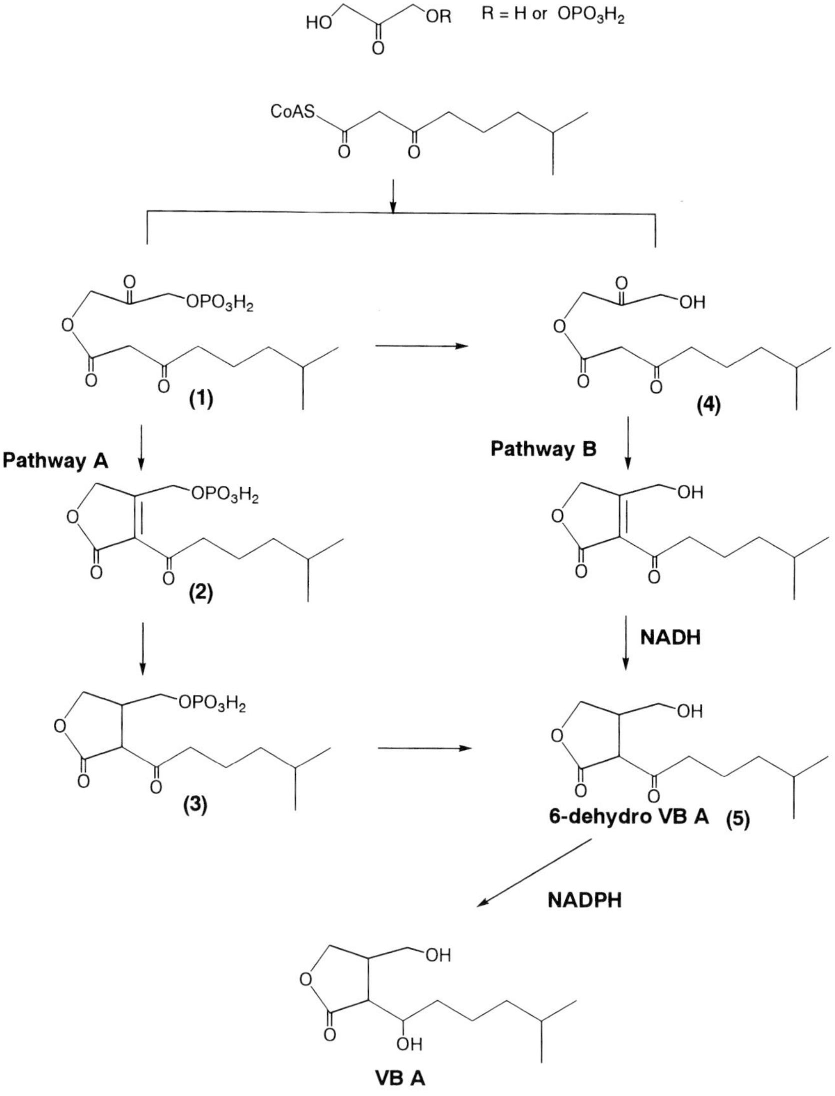

Scheme 2. Biosynthetic pathways of virginiae butanolide A.

biosynthesis, there should be some receptor protein which mediates the signal to facilitate the transcription of these enzyme-encoding genes. With this hypothesis in mind, the VB receptor protein in *S. virginiae* cells was searched for. VBs are compact molecules and every functional group in it is essential for their activity (Nihira *et al.*, 1988). Considering these conditions, a radioactive probe, $[^3H_2]$-VB C_7 was prepared and used as a ligand to identify

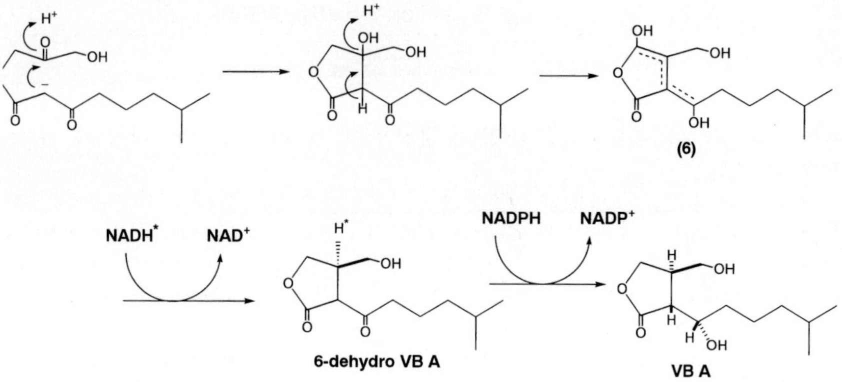

Scheme 3. Mechanism of butyrolactone forming reaction and its stereochemistry in final two steps of virginiae butanolide A biosynthesis.

and purify a VB receptor protein from a crude cell extract (Kim *et al.*, 1989) (see Fig. 6). The VB receptor was identified as a cytoplasmic protein and the numbers of this molecule present was estimated at about 30 to 40 per genome. The K_d value (dissociation constant) obtained in a crude state turned out to be 1.1 nM (Kim *et al.*, 1990). After laborious elimination of one protein named VbrA which always co-purified with the genuine VB receptor protein and was at first misidentified as the receptor (Okamoto *et al.*, 1992); the VB receptor protein was purified and named BarA (*b*utyrolactone *a*utoregulator *r*eceptor *A* (Okamoto *et al.*, 1995)). The molecular weight of BarA is 26 kDa and it forms a homodimer with a M.W. of 52 kDa. Based on the partial amino acid sequences obtained from the purified BarA protein, the *barA* gene was cloned and sequenced. *BarA* encodes 232 amino acid residues and at this time there was no homologous gene from other organisms reported. BarA is rich in Alanine residues (38 Ala/232 amino acids, 16.4%) and so quite hydrophobic in nature. The recombinant BarA

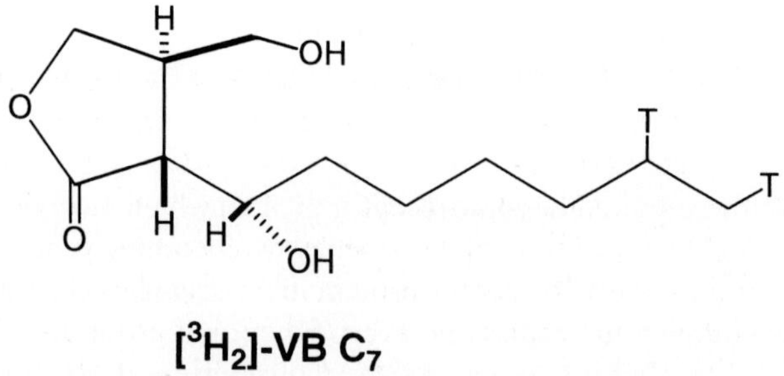

$[^3H_2]$-VB C_7

Fig. 6. Structure of tritium-labelled virginiae butanolide C_7.

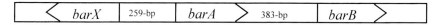

Fig. 7. Gene arrangement of *barA* flanking region.

which was expressed in *E. coli* was observed to be very unstable at room temperature.

Southern blot analysis of DNA from *Streptomyces griseus* IFO 13350 (A-factor producer), *Streptomyces* sp. FRI-5 (IM-2 producer), *Streptomyces coelicolor* and *Streptomyces lividans* with *barA* gene as the probe gave negative results. These results suggested that genes of autoregulator receptor proteins do not have high homology at the DNA level. A helix-turn-helix motif was found at the N-terminus of BarA and this fact strongly suggested that BarA is a regulatory protein binding to DNA. Located 383 bp downstream of the *barA* gene is *barB*, which is transcribed in the same direction as *barA*. The *barB* gene encodes 216 amino acids and also has a helix-turn-helix motif at the N-terminus which suggests that the product, BarB, is a DNA binding regulatory protein. Located 259 bp upstream of *barA*, the *barX* gene was found to be transcribed divergently to *barA* (Fig. 7) (Kinoshita *et al.*, 1997). BarX possesses a high homology (39.8% identity and 74.6% similarity) with AfsA of *S. griseus* which is thought to be involved in A-factor biosynthesis (Horinouchi *et al.*, 1989). Northern blot analysis of *barB* mRNA indicated that it appears at 12 h cultivation. When VB was added at 8 h cultivation, *barB* mRNA was detected at 10 h, two hours earlier. This result suggested that VB induces *barB* transcription. In order to study this mechanism *in vivo*, the promoter-probe plasmid pIJ4083 which has *xylE* as the reporter gene in *Streptomyces lividans* was utilized. *xylE* gene encodes cathecol 2,3-dioxygenase and its expression is detectable by colour formation. Promoter regions of *barA* and *barB* were inserted into the empty promoter site of the reporter gene, both in the presence of *barA* and absence of *barA*. The promoter regions of *barA* and *barB* were able to function and *xylE* was effectively expressed in *S. lividans*, but expression was not affected by VB addition. By contrast, when the *barA* gene, including its promoter region, was added to the upstream region of *barB* in promoter-probe plasmid pIJ4083, *xylE* was poorly expressed in *S. lividans*. However, following the addition of VB, *xylE* was much more effectively expressed (nine-fold). For the promoter region of the *barA* gene, a similar trait was seen although the effect observed was less (three-fold). These *in vivo* results showed that BarA regulates *barB* expression in conjunction with VB. These results were confirmed by *in vitro* experiments using a BIAcore system, which detects interaction between high molecular weight biomolecules such as proteins or protein and DNA by SPR (surface plasmon resonance). The fragments of the *barA* and *barB* promoter regions were immobilized on the sensor chips in flow cells of the BIAcore detector and recombinant BarA protein was applied to them.

A

B

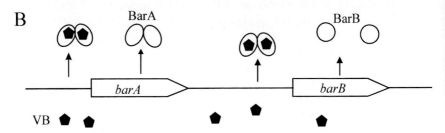

Scheme 4. Regulatory mechanisms of barA and barB transcription by BarA in combination with virginiae butanolide (VB).

Thus the binding activities of BarA to specific DNA fragments of the *barB* and *barA* promoter regions were confirmed by SPR. The results of gel shift assay also coincided with those of SPR. Those DNA fragments which showed positive SPR also showed the retardation of shifts in the presence of recombinant BarA and were restored by addition of VB. This effect is specific to VB and other butyrolactone autoregulators such as A-factor and IM-2 were unable to restore the retardation in the gel shift assay. The regulation of *barA* and *barB* transcripts by BarA in combination with VB is illustrated in Scheme 4. At present, the function of BarB is not clear. When VB is added at 0 h of cultivation, *S. virginiae* does not produce virginiamycin. In this case, BarB appeared at the early stages of exponential growth. Gene disruption of *barA* at the helix-turn-helix motif lead to a phenotype of early production of antibiotics (Nakano *et al.*, 1997). This result indicates a role for BarA in the pleiotropic down regulation of virginiamycin production by binding with plural target sites in the genome including the *barA* and *barB* promoter sites. Other hot spot target sites for BarA and BarB binding are observed in this study. In common with expression, the regulation of expression of secondary metabolite production genes in *Streptomyces* the VB control of virginiamycin is also controlled by many regulatory loops.

A-factor receptor protein

A-factor receptor protein was identified in the cytoplasm of *S. griseus* with a radioactive ligand as shown in Fig. 8 by the Tokyo University group (Miyake

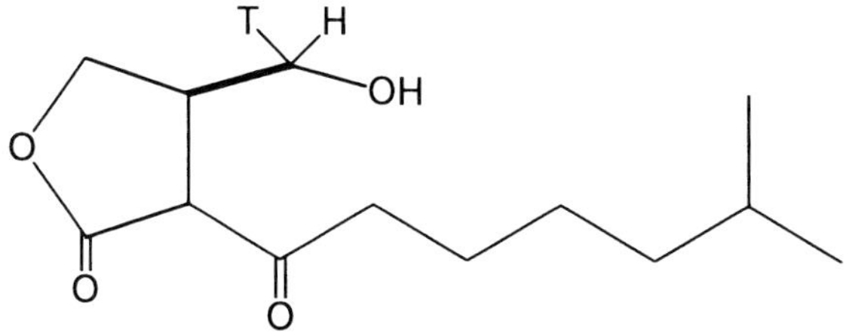

[5-³H]-A-factor

Fig. 8. Structure of tritium-labelled A-factor.

et al., 1989). The number of receptor molecules per genome is also very low at 30 to 40 and its Kd value was estimated at 0.7 nM. The A-factor independent streptomycin producer, *S. griseus* 2247, was found to be a receptor deficient strain (Miyake *et al.*, 1990). Therefore, the A-factor receptor was regarded as pleiotropic in nature, able to down-regulate the production of streptomycin and morphogenesis with repression restored when combined with A-factor. A-factor receptor protein was purified and its gene, *arpA*, cloned using partial amino acid sequences of the purified receptor protein (Onaka *et al.*, 1995). *arpA* encodes 276 amino acid residues and its product, ArpA, forms a homodimer. ArpA has a Gly rich C-terminus and shows 39.1% identity in amino acid sequence compared to BarA. At its N-terminus, there is also helix-turn-helix motif indicating that ArpA has DNA binding activity. The group of enzymes which are involved in streptomycin biosynthesis are well identified and clarified in terms of their genes and in some cases their functions (Piepersberg, 1997). Among them, a regulator gene, *strR*, has been characterized as an activator of the transcription of biosynthetic genes and a self-resistance gene, *aphD* which encodes streptomycin-6-phospho-transferase. *strR* and *aphD* are mainly cotranscribed via the *strR* promoter site. Vujaklija *et al.* verified the role of *strR* (Vujaklija *et al.*, 1991) and also identified a protein named A-factor dependent DNA binding protein, Adp, which was supposed to mediate between A-factor and ArpA, and expression of *strR* and *aphaD* as shown in Scheme 5 (Vujaklija-Horinouchi & Beppu, 1993).

AfsA, the gene product of *afsA*, is considered to be an enzyme involved in A-factor biosynthesis. The A-factor molecule generated (Horinouchi *et al.*, 1989) then combines with the receptor ArpA. This binding triggers the expression of the *adp* gene leading to downstream transcription of *strR*, *aphD* and activation of transcription of the streptomycin biosynthetic genes. Onaka and Horinouchi have determined the DNA sequence of the ArpA binding sites (Onaka & Horinouchi, 1997). Recently, Horinouchi and co-workers purified the key protein Adp and identified its gene *adp* and binding of ArpA at its promoter site (Horinouchi, personal communication).

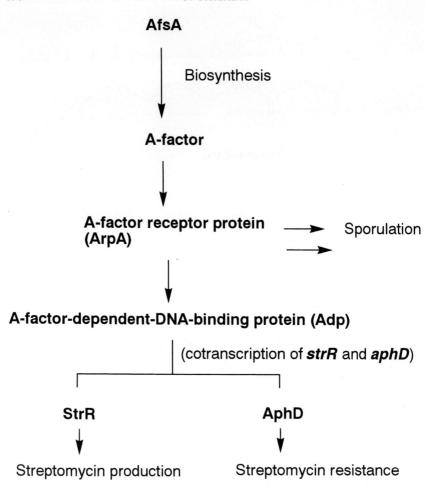

Scheme 5. Pleiotropic regulatory system of streptomycin production, self-resistance and sporulation by A-factor.

    Consequently the most up-to-date signal transduction mechanism can be described as follows. ArpA binds to the promoter site of *adp* and represses its transcription. ArpA releases from this promoter site by binding with A-factor and the expressed Adp activates cotranscription of *strR* and *aphD*. StrR further activates the gene cluster of streptomycin biosynthetic genes including *strB* and others, leading to the production of streptomycin. As for the triggering mechanism of morphological differentiation, there are no reports in the literature at present.

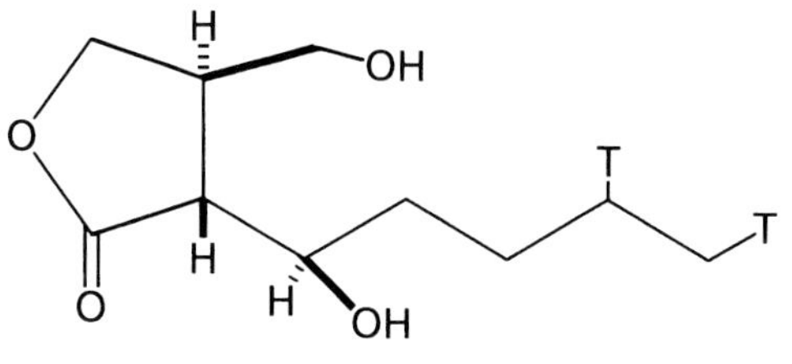

[³H₂]-IM-2 C₅

Fig. 9. Structure of tritium-labelled IM-2 C$_5$.

IM-2 receptor protein

As previously described the function of IM-2 is unique in comparison with VB or A-factor. IM-2 is produced when the carbon and nitrogen source is poor and it switches the pattern of secondary metabolite biosynthesis from D-cycloserine to nucleoside-type antibiotics and also induces blue pigment production in *Streptomyces* sp. FIR-5. The cytoplasmic IM-2 receptor protein has also been identified and purified using a radioactive ligand, [³H₂]-2-(1′-hydroxypentyl)-3-hydroxymethylbutyrolactone ([³H₂]-IM-2-C₅) (Fig. 9) (Ruengjitchatchawalya *et al.*, 1995). The molecular size of the IM-2 receptor was estimated at about 27 kDa by SDS–Page and 54 kDa in non-denatured form using a gel filtration column. The amino acid sequence of the purified receptor was determined and oligonucleotide probes were prepared on the basis of this sequence. However, Southern blot analysis of DNA fragments from *Streptomyces* sp. FIR-5 chromosome using these probes was unsuccessful. As the *barA* gene could not be used as the probe to clone the IM-2 receptor gene, *barX* gene was used instead, to probe and detect the corresponding gene in *Streptomyces* sp. FRI-5. Consequently the *farX* gene corresponding to *barX* was isolated, and at 184 bp downstream of it, the *farA* gene, which encodes a 221 amino acid protein, FarA, was found. This was proved to be the IM-2 receptor protein by heterologous expression in *S. lividans* and over expression in *E. coli* (Waki *et al.*, 1997). *farX* and *farA* are transcribed in the same direction as shown in Fig. 10. Recombinant FarA specifically binds with IM-2 and not with other butyrolactone autoregulators such as VB or A-factor types. FarX and BarX have 53.2% identity, and the homology between AfsA of *S. griseus* and FarX or BarX is 40.6 and 39.8%, respectively. Homology between FarA and BarA is 48.5% and that between FarA and ArpA (A-factor receptor) is 38.5%. FarA has a helix-turn-helix motif on its N-terminus and this feature strongly suggests that FarA is a DNA binding regulatory protein whose function is mediated in combination with IM-2. Northern blot analysis of the *farA* transcript showed that it

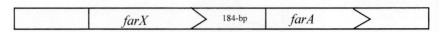

Fig. 10. Gene arrangement of *farX* and *farA*.

appeared 5 h into cultivation and addition of IM-2 at this time enhanced the amount of transcript at 8 h after start of cultivation. These data indicate that expression of FarA is also regulated by itself and IM-2 as in the case of BarA.

CONCLUSIONS

In Fig. 11, structures of the three characterized types of streptomycetes butyrolactone autoregulator are shown with their respective receptor proteins. They all have a 2-(1'-oxo or 1'-hydroxyalkyl)-3-hydroxymethylbutyrolactone skeleton in common. The configuration of the two substituents on the lactone ring is *trans* in all types of autoregulator and their absolute configuration is 2R, 3R. The different points are the length and branching states of 2-alkyl side chains, the redox states of C6 (oxo or hydroxy) and the orientation of hydroxyl groups on C6 (α for VB and β for IM-2). The distribution of these autoregulators among streptomycetes has been studied by several groups using each assay system and about 60% of *Streptomyces* were estimated to have autoregulators, although their roles in each *Streptomyces* were not determined (Ohashi *et al.*, 1989; Hashimoto *et al.*, 1992; Hara & Beppu, 1982*a*; Eritt *et al.*, 1984). Therefore it might be expected that there should be a number of different butyrolactone autoregulator congeners including enantiomers or stereoisomers. When we consider the role of these signal substances from the view point of microbial ecology, the survival strategy of microorganisms in nutrient rich environments may become apparent. After rapid propagation on a rich nutrient source, streptomycetes sense the deficiency of carbon or nitrogen sources and this leads to synchronized sporulation and/or production of antibiotic secondary metabolites by autoregulators, to both eliminate other microorganisms and to help their proliferated progeny to take advantage of the nutrients available. On the other hand, in poor nutrient environments, such as in soil or water, actinomycetes suppress secondary metabolite production to reduce energy consumption. And so no one has successfully isolated antibiotics from soils in nature.

As for practical applications of butyrolactone autoregulators, VB C has been used to enhance the production of virginiamycin. When added after 7.0–10.5 h of cultivation shortly after a decrease in CO_2 efflux, the yield of virginiamycin was much improved (Yang *et al.*, 1995, 1996). Since the structures of butyrolactone autoregulators are rather simple and their chemical syntheses are accessible, their application as microbial pesticides,

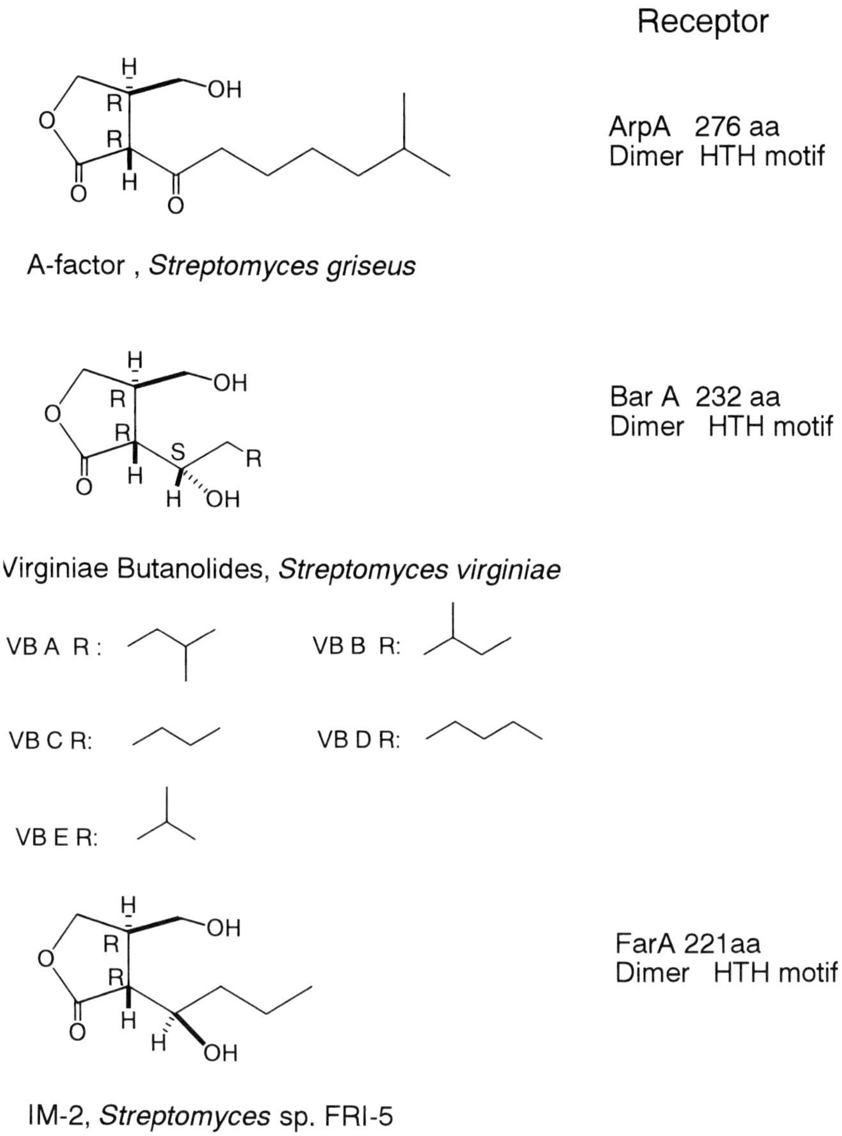

Receptor

A-factor , *Streptomyces griseus*

ArpA   276 aa
Dimer HTH motif

Virginiae Butanolides, *Streptomyces virginiae*

Bar A 232 aa
Dimer   HTH motif

VB A R:

VB B R:

VB C R:

VB D R:

VB E R:

IM-2, *Streptomyces* sp. FRI-5

FarA 221aa
Dimer   HTH motif

Fig. 11. Structures of three types of butyrolactone autoregulators and their receptor proteins.

as a sort of drug delivery system in combination with an antibiotic producer is also promising. The number of examples of autoregulators and respective receptor proteins will doubtless increase and studies in more detail of their regulatory mechanisms might open new avenues of research, including production of new secondary metabolites by awakening expression of dormant genes in streptomycetes.

REFERENCES

Eritt, I., Gräfe, U. & Fleck, W. F. (1984). Inducers of both cytodifferentiation and anthracycline biosynthesis of *Streptomyces griseus* and their occurrence in actinomycetes and other microorganisms. *Zeitschrift für Allgemeine Mikrobiologie*, **24**, 3–12.

Gräfe, U., Reinhardt, G., Schade, W., Eritt, I., Fleck, W. F. & Radics, L. (1983). Interspecific inducers of cytodifferentiation and anthracycline biosynthesis from *Streptomyces bikiniensis* and *S. cyaneofuscatus*. *Biotechnology Letters*, **5**, 591–6.

Gräfe, U., Schade, W., Eritt, I., Fleck, W. F. & Radics, L. (1982). A new inducer in anthracycline biosynthesis from *Streptomyces viridochromogenes*. *Journal of Antibiotics*, **35**, 1722–3.

Hara, H. & Beppu, T. (1982*a*). Mutants blocked in streptomycin production in *Streptomyces griseus* – the role of A-factor. *Journal of Antibiotics*, **35**, 349–58.

Hara, H. & Beppu, T. (1982*b*). Induction of streptomycin-inactivating enzyme by A-factor in *Streptomyces griseus*. *Journal of Antibiotics*, **35**, 1208–15.

Hashimoto, K., Nihira, T. & Yamada, Y. (1992*a*). Distribution of virginiae butanolides and IM-2 in the genus *Streptomyces*. *Journal of Fermentation and Bioengineering*, **73**, 61–5.

Hashimoto, K., Nihira, T., Sakuda, S. & Yamada, Y. (1992*b*). IM-2, a butyrolactone autoregulator, induces production of several nucleoside antibiotics in *Streptomyces* sp. FRI-5. *Journal of Fermentation and Bioengineering*, **73**, 449–55.

Horinouchi, S., Suzuki, H., Nishiyama, M. & Beppu, T. (1989). Nucleotide sequence and transcriptional analysis of the *Streptomyces griseus* gene (*afsA*) responsible for A-factor biosynthesis. *Journal of Bacteriology*, **171**, 1206–10.

Khokhlov, A. S. (1980). Problems of studies of specific cell autoregulators (on the example of substances produced by some actinomycetes). In *(IUPAC) Frontiers of Bioorganic Chemistry and Molecular Biology*, ed. S. N. Ananchenko, pp. 201–10. New York and Oxford: Pergamon Press.

Kim, H. S., Nihira, T., Tada, H., Yanagimoto, M. & Yamada, Y. (1989). Identification of binding protein of virginiae butanolide C, an autoregulator in virginiamycin production, from *Streptomyces virginiae*. *Journal of Antibiotics*, **42**, 769–78.

Kim, H. S., Tada, H., Nihira, T. & Yamada, Y. (1990). Purification of virginiae butanolide C-binding protein, a possible pleiotropic signal-transducer in *Streptomyces virginiae*. *Journal of Antibiotics*, **43**, 692–706.

Kinoshita, H., Ipposhi, H., Okamoto, S., Nakano, H., Nihira, T. & Yamada, Y. (1997). Butyrolactone autoregulator receptor protein (BarA) as a transcriptional regulator in *Streptomyces virginiae*. *Journal of Bacteriology*, **179**, 6986–93.

Kondo, K., Higuchi, Y., Sakuda, S., Nihira, T. & Yamada, Y. (1989). New virginiae butanolides from *Streptomyces virginiae*. *Journal of Antibiotics*, **42**, 1873–6.

Miyake, K., Horinouchi, S., Yoshida, M., Chiba, N., Mori, K., Nogawa, I., Morikawa, N. & Beppu, T. (1989). Detection and properties of A-factor-binding protein from *Streptomyces griseus*. *Journal of Bacteriology*, **171**, 4298–302.

Miyake, K., Kuzuyama, T., Horinouchi, S. & Beppu, T. (1990). The A-factor-binding protein of *Streptomyces griseus* negatively controls streptomycin production and sporulation. *Journal of Bacteriology*, **172**, 3003–8.

Mizuno, K., Sakuda, S., Nihira, T. & Yamada, Y. (1994). Enzymatic resolution of 2-acyl-3-hydroxymethyl-4-butanolide and preparation of optically active IM-2, the autoregulator from *Streptomyces* sp. FRI-5. *Tetrahedron*, **50**, 10 849–58.

Mori, K. (1983). Revision of the absolute configuration of A-factor. The inducer of streptomycin biosynthesis, basing on the reconfirmed (R)-configuration of (+)-paraconic acid. *Tetrahedron*, **39**, 3107–9.

Mori, K. & Chiba, N. (1990). Synthesis of optically active virginiae butanolide A, B, C and other autoregulators from streptomycetes. *Liebig's Annals Chemie*, 1990, 31–7.

Mori, K. & Yamane, K. (1982). Synthesis of optically active forms of A-factor, the inducer of streptomycin biosynthesis in inactive mutants of *Streptomyces griseus*. *Tetrahedron*, **38**, 2919–21.

Nakano, H., Takehara, E., Nihira, T. & Yamada, Y. (1998). Gene replacement analysis of the *Streptomyces virginiae barA* gene encoding the butyrolactone autoregulator receptor reveals that BarA acts as a repressor in virginamycin biosynthesis. *Journal of Bacteriology*, **180**, 3317–22.

Nihira, T., Shimizu, Y., Kim, H. S. & Yamada, Y. (1988). Structure–activity relationships of virginiae butanolide C, an inducer of virginiamycin production in *Streptomyces virginiae*. *Journal of Antibiotics*, **41**, 1828–37.

Ohashi, H., Zheng, Y. H., Nihira, T. & Yamada, Y. (1989). Distribution of virginiae butanolides in antibiotic-producing actinomycetes, and identification the inducing factor from *Streptomyces antibioticus* as virginae butanolide A. *Journal of Antibiotics*, **42**, 1191–5.

Okamoto, S., Nakamura, K., Nihira, T. & Yamada, Y. (1995). Virginiae butanolide binding protein from *Streptomyces virginiae*. Evidence that VbrA is not the virginiae butanolide binding protein and reidentification of the true binding protein. *Journal of Biological Chemistry*, **270**, 12 319–26.

Okamoto, S., Nihira, T., Kataoka, H., Suzuki, A. & Yamada, Y. (1992). Purification and molecular cloning of a butyrolactone autoregulator receptor from *Streptomyces virginiae*. *Journal of Biological Chemistry*, **267**, 1093–8.

Onaka, H. & Horinouchi, S. (1997). DNA-binding activity of the A-factor receptor protein and its recognition DNA sequence. *Molecular Microbiology*, **25**, 991–1000.

Onaka, H., Ando, N., Nihira, T., Yamada, Y., Beppu, T. & Horinouchi, S. (1995). Cloning and characterization of the A-factor receptor gene from *Streptomyces griseus*. *Journal of Bacteriology*, **177**, 6083–92.

Piepersberg, W. (1997). Molecular biology, biochemistry, and fermentation of aminoglycoside antibiotics. In *Biotechnology of Antibiotics*, 2nd edn., revised and expanded, ed. W. R. Strohl, pp. 81–163. New York: Marcel Dekker, Inc.

Ruengjitchatchawalya, M., Nihira, T. & Yamada, Y. (1995). Purification and characterization of the IM-2 binding protein from *Streptomyces* sp. strain FRI-5. *Journal of Bacteriology*, **177**, 551–7.

Sakuda, S. & Yamada, Y. (1991). Stereochemistry of butyrolactone autoregulators from *Streptomyces*. *Tetrahedron Letters*, **32**, 1817–20.

Sakuda, S., Higashi, A., Nihira, T. & Yamada, Y. (1990). Biosynthesis of virginiae butanolide A. *Journal of the American Chemical Society*, **112**, 898–9.

Sakuda, S., Higashi, A., Tanaka, S., Nihira, T. & Yamada, Y. (1992). Biosynthesis of virginiae butanolide A, a butyrolactone autoregulator from *Streptomyces*. *Journal of the American Chemical Society*, **114**, 663–8.

Sakuda, S., Tanaka, S., Mizuno, K., Sukcharoen, O., Nihira, T. & Yamada, Y. (1993). Biosynthetic studies on virginiae butanolide A, a butyrolactone autoregulator from *Streptomyces*. Part 2. Preparation of possible biosynthetic intermediates and conversion experiments in cell-free system. *Journal of the Chemical Society Perkin Transactions*, **1**, 2309–15.

Sato, K., Nihira, T., Sakuda, S., Yanagimoto, M. & Yamada, Y. (1989). Isolation and structure of a new butyrolactone autoregulator from *Streptomyces* sp. FRI-5. *Journal of Fermentation and Bioengineering*, **68**, 170–3.

Strohl, W. R. (1997). Industrial antibiotics: today and the future. In *Biotechnology of Antibiotics*, 2nd edn., revised and expanded, ed. W. R. Strohl, pp. 1–45. New York: Marcel Dekker, Inc.

Vujaklija, D., Horinouchi, S. & Beppu, T. (1993). Detection of an A-factor-responsive protein that binds to the upstream activation sequence of *strR*, a regulatory gene for streptomycin biosynthesis in *Streptomyces griseus*. *Journal of Bacteriology*, **175**, 2652–61.

Vujaklija, D., Ueda, K., Hong, S. K., Beppu, T. & Horinouchi, S. (1991). Identification of an A-factor-dependent promoter in the streptomycin biosynthesis gene cluster of *Streptomyces griseus*. *Molecular and General Genetics*, **229**, 119–28.

Waki, M., Nihira, T. & Yamada, Y. (1997). Cloning and characterization of the gene (*farA*) encoding the receptor for an extracellular regulatory factor (IM-2) from *Streptomyces* sp. strain FRI-5. *Journal of Bacteriology*, **179**, 5131–7.

Wallace, K. K., Zhao, B., McArthur, H. A. I. & Reynolds, K. A. (1995). *In vivo* analysis of straight-chain and branched-chain fatty acid biosynthesis in three actinomycetes. *FEMS Microbiology Letters*, **131**, 227–34.

Yamada, Y., Sugamura, K., Kondo, K., Yanagimoto, M. & Okada, H. (1987). The structure of inducing factors for virginiamycin production in *Streptomyces virginiae*. *Journal of Antibiotics*, **40**, 496–504.

Yanagimoto, M. & Terui, G. (1971). Physiological studies on staphylomycin production (II) Formation of a substance effective in inducing staphylomycin production. *Journal of Fermention Technology*, **49**, 611–18.

Yanagimoto, M. & Enatu, T. (1983). Regulation of a blue pigment production by γ-nonalactone in *Streptomyces* sp. *Journal of Fermention Technology*, **61**, 545–50.

Yanagimoto, M., Yamada, Y. & Terui, G. (1979). Extraction and purification of inducing material produced in staphylomycin fermentation. *Hakkokogaku*, **57**, 6–14.

Yang, K. Y., Morikawa, M., Shimizu, H., Shioya, S., Suga, K., Nihira, T. & Yamada, Y. (1996). Maximum virginiamycin production by optimization of cultivation conditions in batch culture with autoregulator addition. *Biotechnology and Bioengineering*, **49**, 437–44.

Yang, K. Y., Shimizu, H., Shioya, S., Suga, K., Nihira, T. & Yamada, Y. (1995). Optimum autoregulator addition strategy for maximum virginiamycin production in batch culture of *Streptomyces virginiae*. *Biotechnology and Bioengineering*, **46**, 437–42.

HOST/PATHOGEN INTERACTIONS DURING INFECTION BY ENTEROPATHOGENIC *ESCHERICHIA COLI*: A ROLE FOR SIGNALLING

GAD M. FRANKEL[1], BRENDAN WREN[2], GORDON DOUGAN[1], ROBIN M. DELAHAY[1], MIRANDA BATCHELOR[1], CHRISTINE HALE[1], ELIZABETH L. HARTLAND[1], MARK J. PALLEN[2], ILAN ROSENSHINE[3] AND STUART KNUTTON[4]

[1]*Department of Biochemistry, Imperial College of Science, Technology and Medicine, London SW7 2AZ, UK,* [2]*Microbial Pathogenesis Research Group, Department of Medical Microbiology, St Bartholomew's and the Royal London Schools of Medicine and Dentistry, London EC1A 7BE, UK,* [3]*Departments of Molecular Genetics and Biotechnology, and Clinical Microbiology, The Hebrew University, Faculty of Medicine, POB 12272, Jerusalem 9112, Israel and* [4]*Institute of Child Health, University of Birmingham, Birmingham B4 6NH, UK*

INTRODUCTION

Subversion of host cell functions is now recognized as a common theme in the pathogenesis of many bacterial infections. Such subversion can be displayed by bacterially induced cytoskeletal reorganization within host target cells, stimulated by bacterial activation of eukaryotic signal-transduction pathways. Stunning examples of this phenomenon are the interactions of enteropathogenic *Escherichia coli* (EPEC) and enterohaemorrhagic *E. coli* (EHEC) with mammalian intestinal enterocytes. EPEC, an established aetiological agent of human diarrhoea, remains an important cause of mortality amongst young infants in developing countries (Nataro & Kaper, 1998) and EHEC are an emerging cause of acute gastroenteritis and haemorrhagic colitis, often associated with severe/fatal renal and neurological complications (Nataro & Kaper, 1998). Subversion of intestinal epithelial cell function by EPEC and EHEC leads to the formation of distinctive 'attaching and effacing' (A/E) lesions on gut enterocytes which are characterized by localized destruction (effacement) of brush border microvilli, intimate attachment of the bacillus to the host cell membrane and the formation of an actin-rich underlying pedestal-like structure in the host cell (Fig. 1); similar A/E lesions are induced by EPEC and EHEC on a variety of tissue culture cell lines (Knutton *et al.*, 1989). *In vitro* studies employing cultured epithelial cells and defined EPEC mutants support a three-stage model of A/E lesion formation: (i) initial non-intimate attachment; (ii) signal transduction and cytoskeletal rearrangements in host cells; (iii) intimate bacterial adhesion, actin accumulation and pedestal formation (for review see Frankel *et al.*, 1998*a*). Although EPEC was the first *E. coli* to be associated with human disease in the 1940s and 1950s, it was not until the late 1980s and early 1990s that the pathogenic mechanisms and bacterial gene products employed to induce this complex brush border membrane lesion and diarrhoeal disease were identified. Recently, there has been a burst of new data that has started to unveil, at the molecular level, the mechanism underlying intercellular cross talk between bacteria and host cells, the nature of signalling in pathogenic bacteria and the targets within the mammalian cell of some of the bacterial effector proteins.

Rather than superficially reviewing the entire subject of signalling and pathogenic bacteria, this chapter will focus on EPEC and EHEC. Readers are referred to recent excellent reviews on other aspects of this subject (e.g. signal transduction, Cornelis, this volume; Cornelis & Wolf-Watz, 1997; Donnenberg *et al.*, 1997; Finlay & Cossart, 1997; Finlay & Falkow, 1997).

THE ACTIN PEDESTAL

One of the most striking early stage effects of EPEC on gut cells is the induced degradation of the highly organized actin cytoskeleton of brush border microvilli; actin filament breakdown results in effacement of brush border microvilli through vesiculation of the microvillous membrane. Efface-ment of brush border microvilli is followed by intimate bacterial attachment and the formation of so-called pedestals beneath bacteria composed of a complex including actin filaments and several actin-binding proteins (Finlay *et al.*, 1992; Knutton *et al.*, 1989). The typical morphology of the actin-rich structures evolves during the infection from flat, cup-like structures into more elongated pedestals on which the bacterium rests; highly extended pedestals are frequently seen on infected cultured epithelial cells (Fig. 1). The base of the pedestals is rich in myosin and tropomyosin, while actin filaments, villin, and α-actinin are distributed uniformly along it (Finlay *et al.*, 1992). High concentrations of ezrin, plastin, and talin may also be present within the pedestals (Finlay *et al.*, 1992), but the distribution of these proteins along the stalk is currently undefined.

THE GENETIC BASIS OF A/E LESION AND PEDESTAL FORMATION

The first gene to be associated with A/E lesion and pedestal formation was the *eae* gene, encoding intimin (Jerse *et al.*, 1990), an outer membrane protein required for intimate attachment to EPEC to host cells and for reorganiza-tion of polymerized actin within the pedestal-like structures (Donnenberg *et*

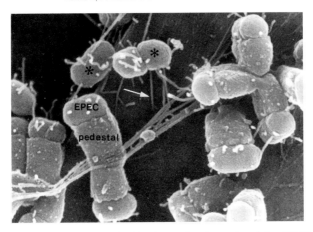

Fig. 1. Scanning electron micrograph showing pedestals produced by EPEC on cultured epithelial cells. EspA filaments (arrow) are seen to connect these bacteria and the cell surface × 60 000. (Reproduced by courtesy of Oxford University Press.)

al., 1997). The *eae* gene is encoded on a large (~ 35 kb) pathogenicity island, the LEE (McDaniel *et al.*, 1995) which is necessary and sufficient for the A/E phenotype (McDaniel & Kaper, 1997). The LEE region encodes 41 open reading frames (Elliott *et al.*, 1998; Perna *et al.*, 1998) and has inserted into at least two different loci on the *E. coli* chromosome: at minute 82 adjacent to the *selC* locus and at minute 94 at the *pheU* (gene loci correspond to the *E. coli* K-12 chromosomal genetic map which does not encode LEE). There is at least one more unmapped insertion site for the LEE and more may exist in wild type *E. coli* strains (for review see Kaper *et al.*, 1998). These observations indicate that the LEE has inserted into the *E. coli* genome at multiple times during the evolution of the EPEC/EHEC family. These pathogenic *E. coli* have acquired other virulence-associated factors in addition to LEE. Some factors, such as the Shiga toxins, are encoded by bacteriophages in EHEC, whereas EPEC have acquired a large *c.* 90 kb plasmids encoding bundle forming pili and PerABC, a transcriptional activator. EHEC have acquired a different plasmid encoding haemolysin production and EspP/ PssA, the primary sequence of which is highly related to an auto-transporter protein family found in different Gram-negative pathogens (Brunder *et al.*, 1997; Djafari *et al.*, 1997).

BACTERIAL PROTEIN SECRETION SYSTEMS

Most bacterial virulence-associated determinants are either surface located or are secreted outside the bacterium. There are only a limited number of ways by which Gram-negative bacteria can transport proteins across their unique double membrane. In Gram-negative bacteria four pathways of protein secretion have been described thus far (Finlay & Falkow, 1997). Proteins destined for secretion by type II and type IV pathways, belonging to the sec-dependent pathway, are synthesized with an amino-terminal signal peptide composed predominantly of hydrophobic amino acids. These proteins bind an export specific chaperone, SecB. Transport across the cytoplasmic membrane occurs in association with a multi-subunit organelle, the translocase, comprising a number of inner membrane proteins including a cytoplasmic membrane associated ATPase (SecA) and periplasmic signal peptidase. Type II secretion from the periplasm and across the outer membrane requires an additional set of proteins some of which (PulS and PulD) are localized at the outer membrane. Proteins secreted by the type IV system, also known as auto-transporter proteins, are found in different Gram-negative bacterial pathogens. In this system proteins are secreted across the inner membrane by the sec-dependent pathway and pass the outer membrane following an auto-proteolytic cleavage and release of the amino terminal to the extracellular space. In contrast, Type I and Type III secretion pathways are sec-independent and proteins are transported across the two membrane layers simultaneously, α-Haemolysin is the prototype polypeptide secreted by the Type I pathway, while the Type III secretion system is found in many Gram-negative pathogens and is responsible for delivery of virulence-associated factors involved in subversion of host cell signal transduction pathways required for bacterial adhesion, invasion and disease.

Type III secretion systems

Many pathogens of animals (for example, *Yersinia, Salmonella, Shigella* and *Bordetella*) and plants (for example, *Pseudomonas syringae, Ralstonia solanaceum, Xanthomonas campestris*) utilize a complex specialized secretion system termed the Type III secretion system (Hueck, 1998; Lee, 1997). Type III secretion systems deliver effector proteins across the bacterial cell envelope and eukaryotic cell membrane into the host cell cytosol (Hueck, 1998; Lee, 1997). Some, but not all the proteins secreted via the Type III secretion system require specific secretion chaperones (Cornelis & Wolf-Watz, 1997). The components responsible for secretion of proteins across the bacterial cell envelope are broadly conserved in all Type III secretion systems, so that one Type III secretion system can often export proteins usually secreted by another Type III system (Rosqvist *et al.*, 1995). Many components are also shared with the flagellar export apparatus (Hueck, 1998; Lee, 1997). Electron microscopical studies of the *Salmonella* pathogenicity island 1 (SPI1) type III secretion system indicates the existence of a macromolecular complex which spans both bacterial membranes and consists of a basal-body-like structure with two outer membrane rings, two inner membrane rings and a connecting cylindrical structure (Kubori *et al.*, 1998).

The proteins secreted across the bacterial cell wall vary from system to system: in *Yersinia* they are termed Yops (Cornelis & Wolf-Watz, 1997) in *Shigella flexneri* Ipas (Menard *et al.*, 1996a) and in *Salmonella* (*Salmonella* has at least two Type III secretion systems, SPI1 and SPI2; Hueck, 1998) Sips and Sops (Collazo & Galan, 1997; Wood *et al.*, 1996). Some proteins secreted by Type III systems show no obvious similarity to other secreted proteins; in other cases, there are clear sequence similarities between secreted proteins from one system or another (e.g. between the Sips and the Ipas) (Collazo & Galan, 1997). However, even when secreted proteins from one Type III secretion system are clearly homologous to those from another, they may have different functions (e.g. IpaB of *S. flexneri* facilitates bacterial escape from the phagolysosome, whereas *Salmonella*, although possessing an IpaB homologue, SipB, remains predominantly within the phagolysosome) (Strauss & Falkow, 1997).

The secreted proteins can be divided into translocases responsible for the translocation of effector proteins from the bacterial cell into the eukaryotic cell and effector proteins that subvert host cell functions. While in some cases a subversive functions of effector proteins has been proposed, for example, YopH is a protein–tyrosine phosphatase (Persson *et al.*, 1997), YpkA (YopO) a protein serine/threonine kinase (Galyov *et al.*, 1994; Hakansson *et al.*, 1996a), and SopE an activator of Rho GTPases (Hardt *et al.*, 1998), the functions of many proteins secreted by Type III systems remain obscure. It is often unclear whether a secreted protein is an effector protein, part of the translocation apparatus or both (e.g. SipB, Collazo & Galan, 1997).

Type III secretion systems differ not just in the associated secreted proteins but also in the manner in which these proteins are exported. When co-cultivated with eukaryotic cells under certain conditions, *Salmonella* liberates the effector proteins secreted by the SPI1 type III systems into the culture medium, whereas in *Yersinia* the secretion of Yops is polarized, with secreted proteins apparently transferred directly from bacterial to host cells, without being released into the external milieu (Persson *et al.*, 1995). Translocation of Yops into the host cell cytosol also appears to require contact between the bacterial and eukaryotic cells, whereas protein translocation in other systems can, at least under some growth conditions, occur without cell-to-cell contact, for example, IpaB and IpaC are able to enter eukaryotic cells even when adsorbed to inert micro particles (Menard *et al.*, 1996b).

The structural basis for cell-to-cell contact and protein translocation has yet to be fully elucidated for any Type III secretion system. Interactions between various secreted proteins (e.g. YopB/YopD, YopN/TyeA, IpaB/IpaC) have been described (Hakansson *et al.*, 1993; Iriarte *et al.*, 1998; Menard *et al.*, 1996a), but nothing is known about the mechanisms of protein–protein binding. Ginocchio *et al.* (1994) have shown that contract with epithelial cells induces the formation of a transient surface appendage (the invasome) on *S. typhimurium* and, although the origin and biological function of these extracellular organelles remains controversial, evidence showing that the formation of this structure requires the SPI1 Type III secretion system has been described. However, a recent study has questioned whether these *Salmonella* structures are Type III dependent and involved in protein translocation (Reed *et al.*, 1998). Roine *et al.* (1997) have recently described a pilus-like surface structure in the plant pathogen *Pseudomonas syringae* composed of a protein secreted by a Type III system, but they have yet to show a role for this appendage in cell-to-cell contact or protein translocation.

The EPEC Type III secretion system

Based on experimental data and homology with other Type III secretion systems, 12 LEE genes, including *esc* R, S, T, U, C, J, V, N, D, F, and *sep* Q and Z, are proposed to encode components of a Type III secretion system apparatus (Elliott *et al.*, 1998; Jarvis *et al.*, 1995). In addition to secretion, the *esc* and *sep* genes appear to be required for the delivery (translocation) of proteins by EPEC into the host cytosol and cell membrane. Two proteins have thus far been shown to be translocated during EPEC infection into epithelial cells: EspB and the translocated intimin receptor, Tir also known as EspE (see below). The *escV, N* and *U* genes were confirmed experimentally to be required for translocation of Tir and EspB.

EscC is a major component of the EPEC Type III secretion system exposed at the bacterial surface and may form one of the outer membrane rings in EPEC's Type III secretion system. EscC is homologous to YscC and PulD, that form ring-shaped oligomeric complexes in the outer bacterial membrane with an ~20 nm diameter central pore (Koster *et al.*, 1997). Other members of the PulD-like proteins are involved in morphogenesis of filamentous bacteriophages, biogenesis of Type IV pili, and Type II and Type III protein secretion. Thus, EscC may also be a channel-forming protein in the EPEC outer membrane. EscN appears to be an ATPase that energizes the process of protein secretion/translocation, a requirement for many of the secretion pathways. Since EspB is translocated to the host cell membrane and has partial structural similarity to YopB of *Yersinia*, a protein involved in forming the translocation-channel in the host cell membrane (Hakansson *et al.*, 1996*b*), it is possible that EspB serves a similar function in EPEC.

Although the structural basis for protein translocation is not fully elucidated for any Type III secretion system, integrating data from studies of different Type III secretion systems suggests the presence of channel-forming proteins of bacterial origin in both the bacterial outer membrane and in the plasma membrane of the infected host cells, and a filamentous structure connecting bacterial and host cell membranes. A model for protein translocation was suggested to comprise at least three components: (i) an energized basal-body-like structure (ii) a channel in the host cell membrane and (iii) a hollow filamentous structure connecting the bacterial basal-body-like structure and the host cell membrane.

THE EPEC SECRETED PROTEINS AND THE TRANSLOCON

Three major EPEC-secreted proteins (or Esps) are known to be exported by the LEE-encoded Type III secretion system (Elliott *et al.*, 1998; Jarvis *et al.*, 1995) EspA, EspB and EspD (for review see Frankel *et al.*, 1998). All three Esps are required for signal transduction and for the formation of the A/E lesion although very little was known about their biological function. By employing EspA antiserum to visualize the EspA protein on EPEC during growth and A/E lesion formation, EspA was shown to be a structural protein and a major component of a large (∼ 50 nm diameter) transiently expressed filamentous surface organelle which, prior to intimin-mediated intimate attachment, forms a direct link between the bacterium and the host cell (Knutton *et al.*, 1998). Immunogold-labelling showed that EspA-filaments are constructed from a small number of smaller fibrils which resemble pili, and, although the number and arrangement of such pili in the intact organelle have yet to be defined, preliminary data are consistent with a hollow cylindrical structure. EspD may be a component of the EspA filaments because an *espD* mutant was found to secrete only low levels of EspA and produced barely detectable filaments. EspB, on the other hand, is unlikely to be a component of the structure, because antiserum to EspB did not stain EspA-filaments and intact EspA filaments were observed on the surface of an *espB* mutant EPEC.

The role of the EspA filaments in bringing bacterial and host cells into close association fulfils an essential first step in the molecular cross-talk between bacterium and host cell. This first step might, by analogy with the role of pili in conjugation, be followed by the creation of another structure that mediates macromolecular transfer between cells.

An alternative to the above model is that the EspA filament may play a direct role in the translocation process. It is tempting to speculate that it could act as a 'molecular go-between', carrying signals and/or proteins from the bacterium to the host cell and vice versa. Two lines of evidence support a direct role for the EspA filament in protein translocation.

(i) Esp secretion by EPEC in the absence of epithelial cells was shown independently by several groups. However, this probably represents basal levels of secretion since, upon contact with host cells, there is an immediate burst of EspB secretion and this secretion burst is strongly enhanced by intimate cell binding (Wolff *et al.*, 1998). Following bacterial attachment, EspB is translocated into the host cell (Wolff *et al.*, 1998). In contrast, an *espA* mutant was able to secrete EspB, and an EspB-adenylate cyclase (CyaA) fusion protein but was unable to trans-

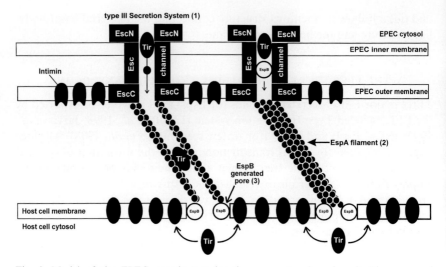

Fig. 2. Model of the EPEC protein translocation apparatus (translocon). The proposed translocon consists of three components: a type III secretion system pore in the bacterial envelope (1), a pore in the host cell membrane generated by translocated EspB (3) and a hollow EspA filament (2) connecting the two pores to provide a continuous channel from the bacterial to the host cell cytosol. The translocon is used to translocate Tir into the host cell, where it becomes inserted into the host cell membrane to act as a receptor for surface expressed intimin.

locate EspB or EspB-CyaA suggesting that a functional EspA is required for translocation of EspB.

(ii) One of the more significant recent findings is the realization that a host cell intimin receptor is, in fact, a LEE-encoded bacterial protein first described by Kenny *et al.* (1997) in EPEC as Tir and shortly after by Deibel *et al.* (1998) in an O26:NM EHEC strain as EspE. The 78–80 kDa Tir/EspE proteins were shown to be secreted by the Type III secretion system and translocated into the host cell, where they are localized to the plasma membrane. Again, translocation of Tir is dependent on expression of functional EspA filaments.

Integration of these, and other data, were used as a basis for a proposed model of protein translocation in EPEC (Fig. 2) in which effector proteins are carried from pathogenic bacteria to the host cell through a pore in hollow EspA filaments bridging between bacterial and host cell membrane channels.

EspA shows sequence similarity to other filament-forming proteins

The structural similarities between the EspA-containing filaments and flagellar filaments (which rely on a similar export system to EspA) prompted a search for sequence similarities between flagellin and EspA. A search using

```
                *         *              *         *         *
EspA        AAISAKANNLTTTVNNSQLEIQQMSNTLNLLTSARSDMQSLQYRT 184

496298      KAVDTQRSVLGASQNRFESTITNLNNTVNNLTSARSRIQDADYST 330

FLIC_SALCH  AQVDALRSDLGAVQNRFNSAITNLGNTVNNLSSARSRIEDSDYAT 455

                aa    a     L      N        I      NT N LtSARS   q     Y T
```

Fig. 3. Alignment of similar regions of EspA, a flagellin from *Yersinia enterocolitica* (Entrez UID 496298) and FliC from *Salmonella choleraesuis*. Conserved heptad repeats of hydrophobic residues are indicated by asterixes. The predicted coiled coil region identified by COILS is underlined. Residues present in all three sequences are shown beneath the alignment in upper case; residues shared between EspA and one other sequence are shown in lower case.

the program BlastP 1.0 reported 62 sequences giving high-scoring segment pairs with EspA. Eighteen of these were flagellins from Gram-negative bacteria (*Yersinia, Salmonella, Campylobacter, Pseudomonas putida*), although all had high P values (0.15–0.99992). Similar results were obtained with a BlastP 2.0 search, where an alignment between a flagellin from *Y. enterocolitica* and *EspA* gave a P value of 0.016.

The region of EspA that showed most similarity to flagellins spanned 45 amino acids at the C terminus of EspA (residues 139–184), and this aligned with a stretch of sequence that ended around 30–35 residues from the C terminus of the flagellins. This region of EspA showed 33% identity to a flagellin from *Y. enterocolitica* (Entrez UID 496298), and contained a stretch of six contiguous residues (LTSARS) that are shared with some other flagellins (Fig. 3).

PropSearch is a program, which aims to find weak similarities based on protein sequence properties independent of sequence alignments (Hobohm & Sander, 1995). When PropSearch was used to search for distant homologues of EspA, six of the ten highest scoring proteins were flagellins or flagellar-hook proteins. The highest scoring protein was a flagellin from Listeria monocytogenes (Swissprot entry flaa_lismo). This sequence scored 9.42, which equates, according to the program's authors, with a probability of 87% that EspA and this flagellin adopt the same fold. The second highest score (10.85 or 80% probability of adopting the same fold) was for the protein VirB5 from *Agrobacterium tumefaciens*, which is almost certainly a structural component of the pilus involved in T-DNA transfer (Fullner *et al.*, 1996).

EspA contains a coiled coil domain

The structural similarities between the EspA-containing filaments and fimbrial and flagellar filaments suggested that the assembly of the EspA-containing filaments might involve a mechanism similar to that employed by these other filaments, namely an interaction between coiled-coil domains of protein monomers. Analysing the sequence of EspA with the COILS program (Lupas *et al.*, 1991) strongly predicted (>80% probability) a coiled coil domain spanning residues 140–172, within the region showing similarity to flagellins. This region still scored highly when, as recommended in the COILS documentation, an alternative weighting option and scoring matrix were used. Scrutiny of the putative coiled-coil region of EspA sequence revealed a run of heptad repeats of hydrophobic residues, characteristic of coiled coils (Fig. 3). However, although widespread in Type III secretion systems in general, the importance of a coiled-coil domain in protein–protein interaction was demonstrated experimentally only for YopN (Iriarte *et al.*, 1998).

EspA-associated filaments mediate initial EPEC attachment to target cells

A characteristic feature of EPEC interaction with epithelial cells is the formation of microcolonies on the host cell surface and A/E lesions beneath the attached bacteria. Immunofluorescence staining of monolayers during the late stages of EPEC infection revealed only scanty staining of bacteria with the EspA antiserum, although stained EspA filaments could sometimes be seen on the surface of bacteria that had formed A/E lesions (Fig. 1). However, these structures are only seen on the side of the bacterium that is not in contact with the cell and they appear to be excluded from the site of intimate bacterial association. Accordingly, formation of EspA filaments might precede intimate attachment and A/E lesion formation and thus occur at a time when the bacteria adhere only weakly to host cells. Gentle washing revealed two populations of cell-associated bacteria. One population had induced A/E lesion formation and expressed few EspA filaments; the second and larger population consisted of bacteria that had not induced A/E lesions. Bacteria in this second group were covered with filaments, staining with anti-EspA, that appeared to have mediated attachment of bacteria to the eukaryotic cell surface. EspA filament mediating EPEC attachment to target cells was also visualized by scanning electron microscopy. Moreover, the anti-EspA antiserum reduced the level of adhesion to cells by 100-fold. These observations suggest that the EspA filaments might have a dual function: they may function as adhesins during the first stage of the infection and as part of the Type III translocon, be involved in activation of host cell signal transduction pathways and A/E lesion formation during the second stage of the infection.

HOST-CELL SIGNALLING

Subversion of host cell signal transduction pathways by EPEC leads to localized actin polymerization, bundling of the newly polymerized actin filaments, and organization of the polymerized bundles beneath attached bacteria. Since mutants in *eae* or *tir* can induce unorganized actin polymerisation but not bundling and pedestals, it seems that these events are dependent upon translocation of effector EPEC proteins. It is possible, although not yet proven, that the cytoplasmic localization of a subpopulation of EspB may be involved in this process.

Several signal transduction pathways in the host cell have been shown to be activated following EPEC infection, including phosphorylation and dephosphorylation of several host cell proteins, release of second messengers, and activation of host cell transcriptional factors. Although the signal(s) and the mechanisms responsible for EPEC-induced cytoskeletal reorganization and pedestal formation are currently unknown, Rac, Rho and Cdc42-dependent pathways do not appear to be involved (Ben-Ami *et al.*, 1998). However, several other signal transduction pathways appear to be stimulated in epithelial cells following infection with EPEC, including tyrosine phosphorylation of several host cell proteins, serine/threonine phosphorylation of myosin light chain, and dephosphorylation of a 240 kDa host cell protein and activation of the transcriptional factor NFkB (associated with increased IL-8 production and transmigration of polymorphonuclear cells) (for review see Frankel *et al.*, 1998*a,b*). The significance of these events to A/E lesion formation remains controversial.

Tyrosine phosphorylation of PLC-γ1 was detected in cells infected with wild-type EPEC following A/E lesion formation (Kenny & Finlay, 1997). This event was reported to follow intimin–Tir interaction. Tyrosine phosphorylation of PLC-γ1 leads to generation of two second messengers: phosphatidylinositol (IP) metabolites, and diacylglycerol (DAG). However, as recent results have shown no increase in intracellular Ca^{2+} levels at early stages of infection and during A/E lesion formation (Bain *et al.*, 1998), stimulation of the classical phospholipase C pathway seems to represent a late event, perhaps related to induction of apoptosis in infected cells.

It was shown that Tir becomes tyrosine phosphorylated immediately upon translocation, causing the translocated polypeptide to appear as a 90 kDa protein on SDS–PAGE (hence Tir was initially referred to as Hp90, and thought to be a host cell-derived intimin receptor) (Rosenshine *et al.*, 1992). It is believed that Tir consists of at least three functional regions, an extracellular domain that interacts with intimin, trans membrane domain/s and cytoplasmic domain/s that can interact, directly or indirectly, with the host cell cytoskeleton. However, the physiological significance of Tir phosphorylation is at present not clear for several reasons. Intimin can bind the unphosphorylated form of Tir, and Tir from O157:H7 EHEC does not appear to become tyrosine phosphorylated in the host cell. Furthermore, experiments with tyrosine protein kinase inhibitors did not inhibit A/E lesion formation. Several reports have shown that purified intimin can bind mammalian cells in the absence of Tir, which suggests there may be an additional intimin receptor of cellular origin; *in vitro* studies have shown that intimin, like *Yersinia* invasin, has the ability to bind $\beta 1$ integrins (Frankel *et al.*, 1996). However, whether intimin–integrin interactions play any role in colonization and A/E lesion formation has yet to be determined.

Intimin, is homologous to the invasins, proteins which promote eukaryotic cell invasion by *Yersinia*. Studying the intimin family of proteins showed that, like invasin, their cell binding activity is localized to the C-terminal 280 amino acids (Int280) (Frankel *et al.*, 1994), and that within this domain lies a 76-amino acid loop formed by a disulphide bridge between two cysteines at positions 862 and 937 (Kelly *et al.*, 1998). This loop is required for intimin-mediated intimate attachment and invasion into cultured mammalian cells. Recent studies have shown that introducing small in-frame mutations at the C terminus of EPEC intimin could dramatically reduce intimin-mediated cell invasion without affecting A/E lesion formation. In particular, deletion of the last amino acid (Lys 939) from the intimin C terminus segregated intimin-mediated A/E lesion formation from intimin-mediated HEp-2 cell invasion (Frankel *et al.*, 1998*b*). These results show that intimin can modulate signal transduction pathways leading to cell invasion.

A recent study has shown that EPEC induces transient proliferation and elongation of microvillous-like processes (MLP) on the eukaryotic cell surface. HEp-2 cells responded to infection with UMD864 (intimin positive, EspB negative) by increasing the MLP network during early stages which, in later stages, developed to 'cage-like' structures surrounding the bacteria. Similar remodelling of the eukaryotic cell surface was observed following inoculation with Int280-coated covaspheres. These results suggest that (i) MLP production is an intimin-dependent event; (ii) that intimin can induce cytoskeletal reorganisation, but not A/E lesion formation, even in the absence of Esp-mediated cell signalling and translocation of Tir; and (iii) that two sequential signal transduction pathways are activated during EPEC infection; intimin-mediated production of MLP and MLP retraction and A/E lesion and pedestal formation (A. D. Phillips, G. Dougan & G. Frankel, unpublished results).

BACTERIAL CELL SIGNALLING

Previous reports using polyclonal intimin and EspA antisera showed, following incubation of HEp-2 cells with EPEC strain E2348/69, strong intimin and EspA expression by all the attached bacteria. Intimin was uniformly distributed over the bacterial surface, apart from at the site of intimate bacterial attachment to the HEp-2 cells and EspA was seen as part of a filamentous structure. However, intimin and EspA expression were greatly reduced or not detected after A/E lesion were formed (Knutton *et al.*, 1997, 1998). The fact that the EspA filaments and intimin are eliminated after

A/E lesion formation implies that the bacterial cells may have responded to some changes in the host cell. Contact-dependent control of bacterial gene expression has recently been described in two settings: synthesis of Yops after Yersinia-host-cell contact (Pettersson *et al.*, 1996) and up-regulation of gene expression after binding of P-pili of uropathogenic *E. coli* to erythrocytes (Zhang & Normark, 1996). Based on these precedents, it was speculated that the EspA filaments might also, on contacting host cells, transmit a signal from the host cell to the bacterium, altering bacterial gene expression. Intimin and EspA have been shown to be highly immunogenic and patients with EPEC infection develop immune response to these proteins. Accordingly, if the observed down-regulation of intimin expression following A/E lesion formation occurs *in vivo*, it could be an important EPEC regulatory mechanism for overcoming host immune responses.

CONCLUSIONS

An emerging theme in the pathogenesis of bacterial infections is subversion by bacterial pathogens of host cell functions including signal-transduction pathways and cytoskeletal organization. A striking example of this is shown by EPEC and EHEC. These pathogens colonize the intestinal mucosa and, by subverting intestinal epithelial cell function, produce the characteristic A/E histopathological lesion. Recent studies on EPEC, EHEC and other bacterial pathogens have started to unveil the molecular mechanisms underlying intercellular cross-talk between bacteria and host cells, the nature of signalling in pathogenic bacteria and the mammalian cell targets of some of the bacterial effector proteins. Furthermore, detailed analysis of the host proteins targeted by effector proteins will enable us to learn more about the cellular processes that bacterial pathogens choose to sabotage. Extensive investigation of this fascinating topic in many laboratories world-wide is guaranteed to produce new and exciting insights into the subversive nature of this evolving population of microbial pathogens. From a practical standpoint, the ability to deliver specific proteins into eukaryotic cells could open new areas for research in medicine and industry.

REFERENCES

Bain, C., Keller, R., Collington, G. K., Trabulsi, L. R. & Knutton, S. (1998). Increased levels of intracellular calcium are not required for the formation of attaching and effacing lesions by enteropathogenic and enterohemorrhagic *Escherichia coli*. *Infection and Immunity*, **66**, 3900–8.

Ben-Ami, G., Ozeri, V., Hanski, E., Hofmann, F., Aktories, K., Hahn, K. M., Bokoch, G. M. & Rosenshine, I. (1998). Agents that inhibit Rho, Rac, and Cdc42 do not block formation of actin pedestals in HeLa cells infected with enteropathogenic *Escherichia coli*. *Infection and Immunity*, **66**, 1755–8.

Brunder, W., Schmidt, H. & Karch, H. (1997). EspP, a novel extracellular serine protease of enterohaemorrhagic *Escherichia coli* O157:H7 cleaves human coagulation factor V. *Molecular Microbiology*, **24**, 767–78.

Collazo, C. M. & Galan, J. E. (1997). The invasion-associated type III system of *Salmonella typhimurium* directs the translocation of Sip proteins into the host cell. *Molecular Microbiology*, **24**, 747–56.

Cornelis, G. R. & Wolf-Watz, H. (1997). The *Yersinia* Yop virulon: a bacterial system for subverting eukaryotic cells. *Molecular Microbiology*, **23**, 861–7.

Deibel, C., Kramer, S., Chakraborty, T. & Ebel, F. (1998). EspE, a novel secreted protein of attaching and effacing bacteria, is directly translocated into infected host cells, where it appears as a tyrosine-phosphorylated 90 kDA protein. *Molecular Microbiology*, **28**, 463–74.

Djafari, S., Ebel, F., Deibel, C., Kramer, S., Hudel, M. & Chakraborty, T. (1997). Characterization of an exported protease from Shiga toxin-producing *Escherichia coli*. *Molecular Microbiology*, **25**, 771–84.

Donnenberg, M. S., Kaper, J. B. & Finlay, B. B. (1997). Interactions between enteropathogenic *Escherichia coli* and host epithelial cells. *Trends in Microbiology*, **5**, 109–14.

Elliott, S. J., Wainwright, L. A., McDaniel, T. K., Jarvis, K. G., Deng, Y. K., Lai, L. C., McNamara, B. P., Donnenberg, M. S. & Kaper, J. B. (1998). The complete sequence of the locus of enterocyte effacement (LEE) from enteropathogenic *Escherichia coli* E2348/69. *Molecular Microbiology*, **28**, 1–4.

Finlay, B. B., Rosenshine, I., Donnenberg, M. S. & Kaper, J. B. (1992). Cytoskeletal composition of attaching and effacing lesions associated with enteropathogenic *Escherichia coli* adherence to HeLa cells. *Infection and Immunity*, **60**, 2541–3.

Finlay, B. B. & Cossart, P. (1997). Exploitation of mammalian host cell functions by bacterial pathogens. *Science*, **276**, 718–25.

Finlay, B. B. & Falkow, S. (1997). Common themes in microbial pathogenicity revisited. *Microbiology and Molecular Biology Reviews*, **61**, 136–69.

Frankel, G., Candy, D. C. A., Everest, P. & Dougan, G. (1994). Characterization of the C-terminal domains of intimin-like proteins of enteropathogenic and enterohemorrhagic *Escherichia coli, Citrobacter freundii*, and *Hafnia alvei*. *Infection and Immunity*, **62**, 1835–42.

Frankel, G., Lider, O., Hershkoviz, R., Mould, A. P., Kachalsky, S. G., Candy, D. C. A., Cahalon, L., Humphries, M. J. & Dougan, G. (1996). The cell-binding domain of intimin from enteropathogenic *Escherichia coli* binds to β1 integrins. *Journal of Biological Chemistry*, **271**, 20359–64.

Frankel, G., Phillips, A. D., Rosenshine, I., Dougan, G., Kaper, J. B. & Knutton, S. (1998a). Enteropathogenic and enterohaemorrhagic *Escherichia coli*: more subversive elements. *Molecular Microbiology*, in press.

Frankel, G., Phillips, A. D., Novakova, M., Batchelor, M., Hicks, S. & Dougan, G. (1998b). Generation of *Escherichia coli* intimin-derivatives with differing biological activities using site-directed mutagenesis of the intimin C-terminus domain. *Molecular Microbiology*, **29**, 559–70.

Fullner, K. J., Lara, J. C. & Nester, E. W. (1996). Pilus assembly by Agrobacterium T-DNA transfer genes. *Science*, **273**, 1107–9.

Galyov, E. E., Hakansson, S. & Wolf-Watz, H. (1994). Characterization of the operon encoding the YpkA Ser/Thr protein kinase and the YopJ protein of *Yersinia pseudotuberculosis*. *Journal of Bacteriology*, **176**, 4543–8.

Ginocchio, C. C., Olmsted, S. B., Wells, C. L. & Galan, J. E. (1994). Contact with epithelial cells induces the formation of surface appendages on *Salmonella typhimurium*. *Cell*, **76**, 717–24.

Hakansson, S., Bergman, T., Vanooteghem, J. C., Cornelis, G. & Wolf-Watz, H. (1993). YopB and YopD constitute a novel class of *Yersinia* Yop proteins. *Infection and Immunity*, **61**, 71–80.

Hakansson, S., Galyov, E. E., Rosqvist, R. & Wolf-Watz, H. (1996a). The *Yersinia* YpkA Ser/Thr kinase is translocated and subsequently targeted to the inner surface of the HeLa cell plasma membrane. *Molecular Microbiology*, **20**, 593–603.

Hakansson, S., Schesser, K., Persson, C., Galyov, E. E., Rosqvist, R., Homble, F. & Wolf-Watz, H. (1996b). The YopB protein of *Yersinia pseudotuberculosis* is essential for the translocation of Yop effector proteins across the target cell plasma membrane and displays a contact-dependent membrane disrupting activity. *EMBO Journal*, **15**, 5812–23.

Hardt, W. D., Chen, L. M., Schuebel, K. E., Bustelo, X. R. & Galan, J. E. (1998). *S. typhimurium* encodes an activator of Rho GTPases that induces membrane ruffling and nuclear responses in host cells. *Cell*, **93**, 815–26.

Hobohm, U. & Sander, C. (1995). A sequence property approach to searching protein databases. *Journal of Molecular Biology*, **251**, 390–9.

Hueck, C. J. (1998). Type III protein secretion systems in bacterial pathogens of animals and plants. *Microbiology and Molecular Biology Reviews*, **62**, 379–433.

Iriarte, M., Sory, M. P., Boland, A., Boyd, A. P., Mills, S. D., Lambermont, I. & Cornelis, G. R. (1998). TyeA, a protein involved in control of Yop release and in translocation of *Yersinia* Yop effectors. *EMBO Journal*, **17**, 1907–18.

Jarvis, K. G., Giron, J. A., Jerse, A. E., McDaniel, T. K., Donnenberg, M. S. & Kaper, J. B. (1995). Enteropathogenic *Escherichia coli* contains a putative type III secretion system necessary for the export of proteins involved in attaching and effacing lesion formation. *Proceedings of the National Academy of Sciences, USA*, **92**, 7996–8000.

Jerse, A. E., Yu, J., Tall, B. D. & Kaper, J. B. (1990). A genetic locus of enteropathogenic *Escherichia coli* necessary for the production of attaching and effacing lesions on tissue cultures cells. *Proceedings of the National Academy of Sciences, USA*, **87**, 7839–43.

Kaper, J. B., Elliott, S., V., S., Perna, T. P., F., M. G. & Blatner, F. R. (1998). Attaching-and-effacing intestinal histopathology and locus of enterocyte effacement. In *Escherichia coli and other Shiga Toxin-producing E. coli Strains*, ed. J. B. Kaper & A. D. O'Brien, pp. 163–182. Washington, DC; AMS Press.

Kelly, G., Prasannan, S., Daniel, S., Frankel, G., Dougan, G., Connerton, I. & Mathews, S. (1998). Sequential assignment of the triple labelled 30.1 kDa cell adhesion domain of intimin from enteropathogenic *E. coli*. *Journal of Biomolecular NMR*, **12**, 189–91.

Kenny, B., DeVinney, R., Stein, M., Reinscheid, D. J., Frey, E. A. & Finlay, B. B. (1997). Enteropathogenic *E. coli* (EPEC) transfers its receptor for intimate adherence into mammalian cells. *Cell*, **91**, 511–20

Kenny, B. & Finlay, B. B. (1997). Intimin-dependent binding of enteropathogenic *Escherichia coli* to host cells triggers novel signaling events, including tyrosine phosphorylation of phospholipase C-γ1. *Infection and Immunity*, **65**, 2528–36.

Knutton, S., Adu-Bobie, J., Bain, C., Phillips, A. D., Dougan, G. & Frankel, G. (1997). Down regulation of intimin expression during attaching and effacing enteropathogenic *Escherichia coli* adhesion. *Infection and Immunity*, **65**, 1644–62.

Knutton, S., Baldwin, T., Williams, P. H. & McNeish, A. S. (1989). Actin accumulation at sites of bacterial adhesion to tissue culture cells: basis of a new diagnostic test for enteropathogenic and enterohaemorrhagic *Escherichia coli*. *Infection and Immunity*, **57**, 1290–8.

Knutton, S., Rosenshine, I., Pallen, M. J., Nisan, I., Neves, B. C., Bain, C., Wolff, C., Dougan, G. & Frankel, G. A. (1998). Novel EspA-associated surface organelle of enteropathogenic *Escherichia coli* involved in protein translocation into epithelial cells. *EMBO Journal*, **17**, 2166–76.

Koster, M., Bitter, W., de Cock, H., Allaoui, A., Cornelis, G. R. & Tommassen, J. (1997). The outer membrane component, YscC, of the Yop secretion machinery of *Yersinia enterocolitica* forms a ring-shaped multimeric complex. *Molecular Microbiology*, **26**, 789–97.

Kubori, T., Matsushima, Y., Nakamura, D., Uralil, J., Lara-Tejero, M., Sukhan, A., Galan, J. E. & Aizawa, S. I. (1998). Supramolecular structure of the *Salmonella typhimurium* type III protein secretion system. *Science*, **280**, 602–5.

Lee, C. A. (1997). Type III secretion systems: machines to deliver bacterial proteins into eukaryotic cells? *Trends in Microbiology*, **5**, 148–56.

Lupas, A., Van Dyke, M. & Stock, J. (1991). Predicting coiled coils from protein sequences. *Science*, **252**, 1162–4.

McDaniel, T. K., Jarvis, K. G., Donnenberg, M. S. & Kaper, J. B. (1995). A genetic locus of enterocyte effacement conserved among diverse enterobacterial pathogens. *Proceedings of the National Academy of Sciences, USA*, **92**, 1664–8.

McDaniel, T. K. & Kaper, J. B. (1997). A cloned pathogenicity island from enteropathogenic *Escherichia coli* confers the attaching and effacing phenotype on *E. coli* K-12. *Molecular Microbiology*, **23**, 399–407.

Menard, R., Dehio, C. & Sansonetti, P. J.(1996*a*). Bacterial entry into epithelial cells: the paradigm of *Shigella*. *Trends in Microbiology*, **4**, 220–6.

Menard, R., Prevost, M. C., Gounon, P., Sansonetti, P. & Dehio, C. (1996*b*). The secreted Ipa complex of *Shigella flexneri* promotes entry into mammalian cells. Proceedings of the National Academy of Sciences, USA, **93**, 1254–8.

Nataro, J. & Kaper, J. B. (1998). Diarrheagenic *Escherichia coli*. *Clinical Microbiology Review*, **11**, 143–210.

Perna, N. T., Mayhew, G. F., Posfai, G., Elliott, S., Donnenberg, M. S., Kaper, J. B. & Blattner, F. R. (1998). Molecular evolution of a pathogenicity island from enterohemorrhagic *Escherichia coli* O157:H7. *Infection and Immunity*, **66**, 3810–17.

Persson, C., Carballeira, N., Wolf-Watz, H. & Fallman, M. (1997). The PTPase YopH inhibits uptake of *Yersinia*, tyrosine phosphorylation of p130Cas and FAK& the associated accumulation of these proteins in peripheral focal adhesions. *EMBO Journal*, **16**, 2307–18.

Persson, C., Nordfelth, R., Holmstrom, A., Hakansson, S., Rosqvist, R. & Wolf-Watz, H. (1995). Cell-surface-bound *Yersinia* translocate the protein tyrosine phosphatase YopH by a polarized mechanism into the target cell. *Molecular Microbiology*, **18**, 135–50.

Pettersson, J., Nordfelth, R., Dubinina, E., Bergman, T., Gustafsson, M., Magnusson, K. E. & Wolf-Watz, H. (1996). Modulation of virulence factor expression by pathogen target cell contact. *Science*, **273**, 1231–3.

Reed, K. A., Clark, M. A., Booth, T. A., Hueck, C. J., Miller, S. I., Hirst, B. H. & Jepson, M. A. (1998). Cell-contact-stimulated formation of filamentous appendages by *Salmonella typhimurium* does not depend on the type III secretion system encoded by *Salmonella* pathogenicity island 1. *Infection and Immunity*, **66**, 2007–17.

Roine, E., Wei, W., Yuan, J., Nurmiaho-Lassila, E. L., Kalkkinen, N., Romantschuk, M. & He, S. Y. (1997). Hrp pilus: an hrp-dependent bacterial surface appendage produced by *Pseudomonas syringae* pv. tomato DC3000. *Proceedings of the National Academy of Sciences, USA*, **94**, 3459–64.

Rosenshine, I., Donnenberg, M. S., Kaper, J. B. & Finlay, B. B. (1992). Signal transduction between enteropathogenic *Escherichia coli* (EPEC) and epithelial cells: EPEC induces tyrosine phosphorylation of host cell proteins to initiate cytoskeletal rearrangements and bacterial uptake. *EMBO Journal*, **11**, 3551–60.

Rosqvist, R., Hakansson, S., Forsberg, A. & Wolf-Watz, H. (1995). Functional conservation of the secretion and translocation machinery for virulence proteins of yersiniae, salmonellae and shigellae. *EMBO Journal*, **14**, 4187–95.

Strauss, E. J. & Falkow, S. (1997). Microbial pathogenesis: genomics and beyond. *Science*, **276**, 707–12.

Wolff, C., Nisan, I., Hanski, E., Frankel, G. & Rosenshine, I. (1998). Protein translocation into HeLa cells by infecting enteropathogenic *Escherichia coli*. *Molecular Microbiology*, **28**, 143–55.

Wood, M. W., Rosqvist, R., Mullan, P. B., Edwards, M. H. & Galyov, E. E. (1996). SopE, a secreted protein of *Salmonella dublin*, is translocated into the target eukaryotic cell via a sip-dependent mechanism and promotes bacterial entry. *Molecular Microbiology*, **22**, 327–38.

Zhang, J. P. & Normark, S. (1996). Induction of gene expression in *Escherichia coli* after pilus-mediated adherence. *Science*, **273**, 1234–6.

SURVIVAL STRATEGY OF *YERSINIA* IN ITS HOST

CÉCILE NEYT AND GUY R. CORNELIS

Microbial Pathogenesis Unit, Christian de Duve Institute of Cellular Pathology and Faculté de Médecine, Université Catholique de Louvain, B-1200 Brussels, Belgium

INTRODUCTION

Introduction: the Yersinia *lifestyle*

Invasive pathogenic bacteria have in common the capacity to overcome the defence mechanisms of their animal host and to proliferate in its tissues. Recent data reveal the existence of major virulence mechanisms in various pathogenic bacteria. One of these systems involves the delivery of bacterial proteins inside eukaryotic cells by extracellularly located bacteria that are in close contact with the target cell surface. The Yop system of *Yersinia* spp., which will be described here, represents an archetype for this new mechanism. The other animal pathogens sharing related systems are *Salmonella* spp., *Shigella* spp., enteropathogenic *E. coli* (EPEC) and *Pseudomonas aeruginosa*. Related systems are also found in the plant pathogens such as *Erwinia amylovora*, *P. syringae*, *Xanthomonas campestris* and *Ralstonia solanacearum* (for review, see Alfano & Collmer, 1997; Van den Ackerveken & Bonas, 1997).

The genus *Yersinia* includes three species that are pathogenic for rodents and humans; *Yersinia pestis* is the agent of black death, *Yersinia pseudotuberculosis* is an agent of mesenteric adenitis and septicaemia and *Yersinia enterocolitica*, the most prevalent in humans, causes gastrointestinal syndromes, ranging from an acute enteritis to mesenteric lymphadenitis (Cover & Aber, 1989). *Y. pestis* is generally inoculated by a flea bite, while the two others are food-borne pathogens. In spite of these differences in the infection routes, all three share a common tropism for lymphoid tissues and a common capacity to resist the non-specific immune response, in particular phagocytosis and killing by macrophages and polymorphonuclear leukocytes (PMNs). They are extracellular pathogens, and their survival strategy is based on this ability to escape the non-specific immune response.

Virulence mechanism of Yersinia *spp.*

The *Yersinia* virulence mechanism consists of the injection of bacterial proteins into the cytosol of eukaryotic cells, from bacteria adhering to the

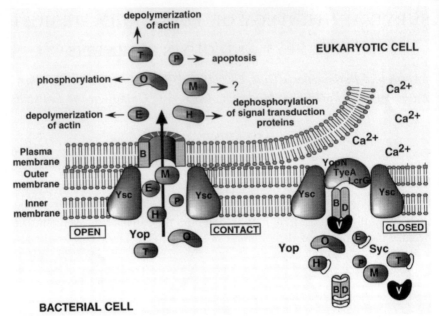

Fig. 1. Tentative model of the interaction between *Yersinia* and a eukaryotic cell. When *Yersinia* is placed at 37 °C in a rich environment, the Ysc secretion apparatus is installed and a stock of Yop proteins is synthesized. Some of these proteins are capped with their specific Syc chaperones, which presumably prevent premature associations. As long as there is no contact with a eukaryotic cell, the YopN-TyeA plug blocks the Ysc secretion channel. Upon Ca^{2+}-depletion or contact with the eukaryotic target cell, the secretion channel opens up and the YopB translocator inserts in the eukaryotic cell with the help of YopD and LcrV. The Yop effectors (YopE, YopH, YopM, YopO/YpkA, YopP/YopJ, YopT) are then transported through the secretion channel and translocated across the plasma membrane, guided by the translocators. YopE and YopT act on the cytoskeleton, while YopP induces apoptosis.

cell surface. The bacteria will first secrete a set of 12 proteins called Yops across its two membranes. Some of these Yops will then form a translocation apparatus, allowing the others, called the effectors, to cross the cell plasma membrane. The injection through the bacterial and cell membranes is thought to occur in one step. The secretion across the bacterial membrane occurs via a type III secretion pathway and requires a specific apparatus (called Ysc for Yop secretion). The crossing of the cell plasma membrane requires some Yops called translocators, amongst which are YopB and YopD. The Yops thus form two distinct groups: some Yops, such as YopE, YopH, ... are intracellular effectors delivered inside eukaryotic cells, while other Yops, the translocators, form a delivery apparatus. A model of the system is presented in Fig. 1.

This complete anti-host system comprising the effectors, the delivery apparatus and the specialized secretion system, as well as its regulation, is encoded by a virulence plasmid of 70 kb called pYV (Laroche *et al.*, 1984). This virulence plasmid is well conserved among the three species.

The fate of the Yops will first be reviewed, from secretion to delivery and action in eukaryotic cells. Then the effects of this virulence apparatus on eukaryotic cells will be analysed. Finally, the regulation of gene expression will be discussed briefly.

YOP SECRETION

The Ysc secretion apparatus

Effector and translocator Yops are transported across the two bacterial membranes and the bacterial cell wall by a specialized secretion system (Ysc) that represents the first recognized member of the type III family of secretion systems. The Ysc apparatus comprises about 20 proteins. Several proteins share sequence homology with components of other type III secretion systems of plant and animal pathogens and also with proteins involved in the assembly of flagella (He, 1997). Recently, the secretion apparatus from *Salmonella* has been purified and visualized by electron microscopy and it appears to be almost identical to the basis of the flagellum assembly secretion machinery (Kubori *et al.*, 1998). Some of the proteins involved in the secretion apparatus have been characterized. One of them (YscN) has essential ATP-binding sites typical of ATPases and some are inner membrane proteins (YscR, YscU, LcrD). YscC is an outer membrane protein that appears as an insoluble complex (Plano & Straley, 1995). It belongs to the 'secretins' family as protein PIV involved in the extrusion of the filamentous phages (Russel, 1994). YscC assembles in very stable multimers, forming a ring-shaped structure with an external diameter of about 200 Å and an apparent central pore of about 50 Å (Koster *et al.*, 1997). It requires the lipoprotein VirG for its stability and insertion in the membrane (Allaoui *et al.*, 1995*a*; Koster *et al.*, 1997).

Secretion signal of the Yops

The signal required to secrete a Yop protein is located in the N-terminal region of the protein, but it does not have the features of a classical signal peptide and it is not cleaved off during secretion. This signal is contained within the first 15 and 17 residues of YopE and YopH, respectively (Sory *et al.*, 1995; Schesser *et al.*, 1996). The YopN secretion signal is localized in the 15 first codons of the gene (Anderson & Schneewind, 1997). The minimal domain of YopM sufficient for secretion of YopM-Cya was found to be shorter than 40 residues (Boland *et al.*, 1996). For YopO/YpkA and YopP/YopJ, it is shorter than 77 and 43 residues, respectively (Sory *et al.*, unpublished observation).

There is no similarity between the secretion domains of the Yops, with respect to amino acid sequence, hydrophobicity profile, distribution of charged residues or prediction of secondary structure, which suggested recognition of a conformational motif of the nascent protein (Michiels & Cornelis, 1991). A systematic mutagenesis of the secretion signal by Anderson and Schneewind (1997) led to doubts about the proteic nature of this signal. No point mutation could be identified that specifically abolished secretion of YopE or YopN. Moreover, frameshift mutations that completely altered the peptide sequences of the signals also failed to prevent secretion. Anderson and Schneewind (1997) concluded that the signal that leads to the secretion of Yops could be in their messenger RNA rather than in their peptide sequence. Recently, Cheng *et al.* (1997) showed that there is a second secretion signal in YopE and they showed that this second, and weaker, secretion signal corresponds to the SycE-binding site (see further). Not surprisingly, it is only functional in the presence of the SycE chaperone (Cheng *et al.*, 1997), in agreement with the pilot hypothesis of Wattiau and Cornelis (1993) for SycE. There are thus two different signals driving the export of YopE by the type III secretion apparatus. The first one would be the structure of the 5′ mRNA and the second one, built into the protein would use the chaperone as a pilot. The same could apply to effectors YopH and YopT. Some other effector Yops do not seem to have a chaperone, in which case, they would only be recognized by their N- or 5′-terminal signal.

Finally, less is known about secretion of the translocators. No signal sequence is removed from YopB, YopD and LcrV, but their secretion signal has not yet been identified. Some observations tend to suggest that secretion of YopB and YopD could proceed by a mechanism slightly different from that used by the effector Yops. First, Sarker *et al.* (1998*a*) have shown that LcrV is required for the secretion of both YopB and YopD and that LcrV has the capacity to bind to both of them. Secondly, mutations in some genes such as *virG* (Allaoui *et al.*, 1995*a*), *yscF* (Allaoui *et al.*, 1995*b*) or *yscM*/*lcrQ* (Rimpiläinen *et al.*, 1992; Stainier *et al.*, 1997) lead to phenotypes in which YopB, YopD and LcrV are secreted differently from the other Yops. Recent work suggests that YopB and YopD are secreted as an heterodimer (Neyt & Cornelis, 1998). A well-formed secretion apparatus is thus needed for the secretion of this bulky complex, which explains the observations concerning the different secretion phenotype of YopB and YopD in the *virG*, *yscF* and *yscM* mutants.

The Syc cytosolic chaperones SycE, SycH, SycT and SycD

The correct functioning of the system requires the presence of small cytosolic proteins, called Syc for 'specific Yop chaperone' (for a review, see Wattiau *et al.*, 1996). Each chaperone is specific to one or two Yops. Five Syc proteins have been described so far: SycE for YopE (Wattiau & Cornelis, 1993), SycH for YopH (Wattiau *et al.*, 1994), SycT for YopT (Iriarte & Cornelis, 1998), SycN for YopN (Iriarte & Cornelis, 1999) and SycD for YopB and YopD (Wattiau *et al.*, 1994; Neyt & Cornelis, 1998). Each chaperone is necessary, either directly or indirectly for the secretion of its cognate Yop(s), and it is associated with it in the bacterial cytoplasm. The gene encoding the chaperone is located next to the gene encoding the corresponding Yop. The Syc chaperones are weakly, or even not, related in terms of amino acid sequence. However, they share some common features: an acidic pI, a size in the range of 15–19 kDa and a C-terminal amphiphilic α-helix.

Concerning the roles and the mode of action of the Syc chaperones, it appears that SycD is quite different from the others. It will thus be discussed separately. The role of SycT and SycN will not be described since they have not yet been completely analysed.

Role of SycE and SycH

Wattiau *et al.* (1994) suggested that the Syc chaperones could act as a form of secretion pilots to drive nascent Yops to the secretion machinery. This hypothesis is actually well accepted since YopE deleted from its N-terminal domain can only be secreted if SycE is present. But, this is presumably not the only role for the Syc chaperones.

In the next section it will be seen that YopH and YopE have a discrete domain (residues 15–50 for YopE and residues 20–70 for YopH) that is specifically required for their translocation into eukaryotic cells (Sory *et al.*, 1995). Woestyn *et al.* (1996) showed that it is the very same region of YopE and YopH that binds the cognate chaperone. In addition, they showed that, in a *sycH* mutant, YopH secretion is more efficient in the absence of YopB and YopD than in their presence (Woestyn *et al.*, 1996). This result suggests that SycH could prevent the association of YopH with YopB and/or YopD, but this hypothesis still awaits an experimental confirmation.

SycE also has an anti-degradation role. The half-life of YopE is longer in wild-type bacteria than in *sycE* mutant bacteria (Frithz-Lindsten *et al.*, 1995; Cheng *et al.*, 1997). SycH does not have such a clear anti-degradation role because YopH can be detected in the cytosol of *sycH* mutant bacteria (Wattiau *et al.*, 1994).

Finally, SycE and SycH could also have an anti-folding role, in order to maintain the Yop proteins in a conformation which is adequate for secretion. However, the fact that these chaperones only bind to a small defined domain argues against that hypothesis.

In summary, it appears that the SycE and SycH chaperones fulfil several roles concerning YopE and YopH stability, secretion and possibly conformation and interaction between translocators and effectors. One could imagine that the chaperone first acts as a secretion pilot leading the Yop protein to the secretion locus and simultaneously preventing premature association with the translocators. By binding the Yop, the Syc chaperone could ensure stability and proper conformation of the protein. At the secretion stage, the chaperone is released from the partner Yop, and the translocation domain would be free to interact with the translocation machinery, which then leads the Yop to the host cell cytoplasm.

Role of SycD

SycD is a chaperone for the YopD and YopB proteins. It is somewhat different from SycE and SycH, according to its role and the way it acts. First, it serves two Yops rather than one; secondly, it serves translocators, while SycE and SycH serve effector Yops. Thirdly, SycD binds to several domains on YopB (Neyt & Cornelis, 1998), while SycE and SycH bind their cognate Yop at a unique site. The last property evokes SecB, a molecular chaperone in *E. coli*, which is dedicated to the export of newly synthesized proteins (Kumamoto & Beckwith, 1985), and also has multiple binding sites on the maltose binding protein (Khisty *et al.*, 1995).

The SycD chaperone appears to be associated, in the bacterial cytoplasm, to a heteromer consisting of YopB and YopD rather than to each protein separately (Neyt & Cornelis, 1998). There appear to be two roles of SycD in this complex. First, it seems to be a protective element. C. Neyt and G. R. Cornelis (unpublished data) showed that overexpression of YopB and YopD in *E. coli* leads to cell lysis and death, while simultaneous overexpression of SycD prevents this toxicity. This suggests that SycD protects the bacteria against the toxicity of these proteins. Secondly SycD would also be an anti-association factor, preventing a premature association of the YopB-YopD complex with LcrV (see below). In the absence of SycD, YopB and YopD are degraded in the bacterial cytoplasm, probably by a bacterial 'house-keeping' protease.

Conclusion

In conclusion, the Syc chaperones probably constitute two different families. SycE, SycH and presumably SycT appear to be both secretion/ translocation pilots and anti-association factors, while SycD would rather have a protective role covering the YopB–YopD cytoplasmic association until secretion and could prevent premature associations between the translocators. Concerning the anti-degradation role of the Syc chaperones, it is unclear whether the Yops are degraded in the absence of their chaperone because they are not secreted or because the chaperone has a direct antidegradation role.

DELIVERY OF EFFECTOR YOPS INTO EUKARYOTIC CELLS

Translocation across the eukaryotic cell plasma membrane

Characterization of the translocation phenomenon

The evidence for YopD-mediated translocation of the YopE protein was essentially genetical. In 1994, this elegant hypothesis was confirmed by two different approaches. The first one was based on immunofluorescence and confocal laser scanning microscopy examinations. Rosqvist *et al.* (1994) showed that the YopE protein appeared in the cytosol of HeLa cells infected with wild-type *Y. pseudotuberculosis*. In contrast, when cells were infected with a mutant strain of *Y. pseudotuberculosis* unable to produce YopD, YopE was no longer internalized, showing that the YopD protein was essential for the translocation of YopE across the target cell membrane (Rosqvist *et al.*, 1994). The second approach was based on a reporter enzyme strategy introduced by Sory and Cornelis (1994). The reporter system consisted of the calmodulin-activated adenylate cyclase domain (called Cya) of the *Bordetella pertussis* cyclolysin (Glaser *et al.*, 1988). Because the catalytic domain of cyclolysin is unable to enter eukaryotic cells by itself, accumulation of cyclic AMP (cAMP) would essentially reflect Yop internalization. Infection of HeLa cells with recombinant *Y. enterocolitica* producing a hybrid YopE-Cya protein resulted in a marked increase in cAMP even when internalization of the bacteria themselves was prevented by cytochalasin D. Infection with a *Y. enterocolitica* mutant unable to produce both the YopD and YopB proteins did not lead to cAMP accumulation, confirming the involvement of YopD and/or YopB in translocation of the YopE protein across eukaryotic membranes (Sory & Cornelis, 1994). The same methods were applied to demonstrate translocation of YopH and YopM across the plasma membrane of epithelial cells and macrophages (Persson *et al.*, 1995; Sory *et al.*, 1995; Boland *et al.*, 1996). Delivery into eukaryotic cells of less abundant Yop proteins such as YopO/YpkA, YopP/YopJ and YopT turned out to be more difficult to monitor. Håkansson *et al.* (1996b) constructed a mutant of *Y. pseudotuberculosis* unable to produce the more abundant Yop effectors (YopE, YopH, YopM) as well as YopK (see below), and this strain allowed the visualization of translocation of YpkA/YopO into HeLa cells, by confocal microscopy (Håkansson *et al.*, 1996b; Holmström *et al.*, 1997; Fällman *et al.*, 1997). A similar multiple-*yop*-mutant was constructed in *Y. enterocolitica* (Boland & Cornelis, 1998), and this mutant allowed the demonstration of the delivery of YopO-Cya, YopP-Cya and YopT-Cya into macrophages (Iriarte *et al.*, 1998; M. P. Sory *et al.*, unpublished observations; Iriarte & Cornelis, 1998), by a system dependent on both YopD and YopB.

The translocation requires living bacteria adhering to their target and is the fate of extracellular bacteria. Selective killing of extracellular bacteria by gentamicin inhibits cytotoxicity (Rosqvist *et al.*, 1990). In agreement with that, Sory and Cornelis (1994) showed that the use of cytochalasin-D, which inhibits the entry of bacteria into cells, does not greatly affect the amount of YopE that is internalized.

This translocation phenomenon is 'polarized' in the sense that the majority of the Yop effector molecules produced are directed into the cytosol of the eukaryotic cell and not to the outside environment (Rosqvist *et al.*, 1994). However, there is some discrepancy about the degree of 'directionality'. As will be seen later, proteins YopN, TyeA and LcrG are probably involved in this 'contact-oriented' phenomenon (Forsberg *et al.*, 1994; Rosqvist *et al.*, 1994; Persson *et al.*, 1995; Boland *et al.*, 1996; Sarker *et al.*, 1998b; Iriarte *et al.*, 1998).

A translocation signal on Yop effectors?

Taking advantage of the Yop-Cya strategy, Sory *et al.* (1995), identified a domain required for the internalization of YopE and YopH into murine PU5–1.8 macrophages. Starting from hybrids that were readily translocated, they engineered gradual deletions into the *yop* gene, starting from a restriction site at the hinge between *yopE* or *yopH* and *cyaA*. Internalization into macrophages, revealed by cAMP production, required the 50 N-terminal amino acids of YopE, and the 71 N-terminal amino acids of YopH, while secretion required only 15 and 17 residues, respectively. Sory *et al.* (1995) concluded that YopE and YopH are modular proteins composed of a secretion domain, a translocation domain and an effector domain. The same experiments, carried out on YopE from *Y. pseudotuberculosis*, confirmed these observations (Schesser *et al.*, 1996). The domain required for translocation of the other effectors has also been shown to reside in the N-terminal domain. For YopM, it is localized within the first 100 residues and it extends further than the 41 first residues (Boland *et al.*, 1996). For YopO/YpkA and YopP/YopJ, the signal is localized within the 77 N-terminal residues and the 99 N-terminal residues, respectively (Iriarte *et al.*, 1998; M. P. Sory *et al.*, unpublished data). For YopT, it is within the 124 N-terminal residues (Iriarte & Cornelis, 1998).

The fact that efficient translocation requires a domain longer than that required for secretion does not necessarily mean that secretion and translocation occurs in two distinct steps. Since the 'translocation' domain appears to be the second secretion signal, it could be rather seen as an affinity domain that will determine the order of translocation of the Yops rather than a real translocation domain necessary for the injection. It appears that the crossing of the bacterial and cell membranes *in vivo*, would rather occur in one step (Lee *et al.*, 1998).

The delivery apparatus

Role of YopB and YopD

YopD is the first element that was shown to be required for the translocation of the effector Yops (Rosqvist *et al.*, 1994; Hartland *et al.*, 1994; Sory & Cornelis, 1994). It is encoded, together with two other Yops, i.e. LcrV and YopB and the chaperone SycD, by the large *lcrGVHyopBD* operon. Analysis of non-polar *yopB* and *yopD* mutants showed that YopB is also individually required for translocation of the effectors across the eukaryotic cell plasma membrane (Håkansson *et al.*, 1996*a,b*; Boland *et al.*, 1996).

YopD consists of 306 amino acids and has a molecular mass of 33.3 kDa. Its calculated pI is 7.0 in *Y. pseudotuberculosis* and 6.6 in *Y. enterocolitica* (Håkansson *et al.*, 1993). Analysis of YopD with the Lupas algorithm (Lupas *et al.*, 1991) suggests the presence of a domain that could form coiled coils, which are common structures involved in protein-protein interactions. This putative coiled coil is C-terminal in YopD, spanning residues 249–292. The same domain could also form an amphipathic helix. The hydropathy analysis identifies a 31-amino acid length hydrophobic region in the middle of YopD (Håkansson *et al.*, 1993). The Eisenberg plot analysis (Eisenberg, 1984) suggests that YopD is a transmembrane protein (Håkansson *et al.*, 1993). However, its function is unknown.

YopB is a 401-residue protein, having a molecular mass of 41.8 kDa (Håkansson *et al.*, 1993). The calculated isoelectric point is 7.3 in *Y. pseudotuberculosis* and 6.7 in *Y. enterocolitica*. Analysis of YopB with the Lupas algorithm (Lupas *et al.*, 1991) predicts the presence of two putative coiled coils, spanning residues 103–165 and residues 330–385. The central part of YopB contains two hydrophobic regions, separated by only 15 amino acids and as for YopD, the Eisenberg plot of YopB suggests that it is a transmembrane protein (Eisenberg, 1984; Håkansson *et al.*, 1993). YopB has a moderate level of similarity with proteins of the RTX family of α-haemolysins and leukotoxins such as LktA of *Pasteurella haemolytica* (Strathdee & Lo, 1989) and HlyA of *E. coli* (Felmlee & Welch, 1988; Bhakdi *et al.*, 1986). The homology between YopB and the RTX proteins is limited to the hydrophobic regions. Since, in the RTX proteins, these hydrophobic regions are believed to be involved in disrupting the target cell membrane (Welch, 1991), it seems logical to assume that they play the same role in YopB suggesting that the translocation apparatus could be some kind of a pore, where YopB would be the main element. The observation of Håkansson *et al.* (1996*b*) that *Y. pseudotuberculosis* has a YopB- and contact-dependent lytic activity on sheep erythrocytes supports this hypothesis. Moreover, purified YopB has the ability to disrupt lipid bilayers (Håkansson *et al.*, 1996*b*). This YopB-dependent lytic activity is higher when the effector *yop* genes are deleted, suggesting that the pore is normally filled with effectors during contact (Håkansson *et al.*, 1996*b*). The presence of sugar molecules of a given size in the medium can inhibit YopB-mediated sheep erythrocyte lysis, which allowed an approximate determination of the size of the putative pore: since dextran 4 has an inhibitory effect, while raffinose has no significant effect, the inner diameter of the pore would be between 12 Å and 35 Å.

The fact that YopB and YopD are both hydrophobic proteins needed for translocation, suggests that they could associate at some stage to fulfill their function. This idea is reinforced by the presence of hypothetical coiled coils in both proteins. In good agreement with this hypothesis, YopB and YopD appear to be associated in the bacterium prior to their secretion (Neyt & Cornelis, 1998). In an attempt to localize the domain of YopB that is involved in this interaction with YopD, C. Neyt and G. N. Cornelis (unpublished data) analysed the capacity of a set of truncated YopB proteins to bind to YopD. The outcome of this analysis is that the binding does not occur at one precise site on YopB but rather at different sites, along the protein. These observations suggest that YopB and YopD could insert together in the eukaryotic membrane, and that the putative pore described above could consist of YopB and YopD, but this has not been shown yet. Until now, the pore has been neither purified nor observed on eukaryotic target cells by electron microscopy.

Role of YopQ/YopK

Yop translocation through the putative pore seems to be controlled by the 21-kDa YopK/YopQ (Holmström *et al.*, 1995). A *yopK* mutant of *Y. pseudotuberculosis* delivers more YopE and YopH into HeLa cells than the wild-type strain, whereas a strain overexpressing YopK is impaired in translocation. Overproduction of YopK also leads to a reduction of the YopB-dependent lytic effect on infected HeLa cells and sheep erythrocytes, probably by influencing the size of the pore as shown by the protective effect of differentially sized sugar moieties (Holmström *et al.*, 1997).

Role of LcrV

The *lcrGVsycDyopBD* operon also encodes the LcrV protein, known since the mid-1950s as a protective antigen of plague (Burrows & Bacon, 1956). LcrV has been described as a regulatory protein involved in the calcium-response since an in-frame deletion mutant in *lcrV* was found to be Ca^{2+}-independent and downregulated in transcription of *yop* genes (Perry *et al.*, 1986; Bergman *et al.*, 1991; Price *et al.*, 1991; Straley *et al.*, 1993; Skrzypek & Straley, 1995). However, recent data from Sarker *et al.* (1998*a*) indicates that LcrV could be a functional element of the translocation apparatus since it is necessary for the secretion of YopB and YopD. Moreover, LcrV interacts with both YopB and YopD proteins (Sarker *et al.*, 1998*a*) as well as with LcrG (Nilles *et al.*, 1997; Sarker *et al.*, 1998*a*). From these results, it was suggested that LcrV constitutes a third component of an organized delivery apparatus. It could form some kind of a short pilus underneath YopB and YopD, but this is still pure speculation.

Role of LcrG

LcrG is a 96-amino acid protein (11.0 kDa) (Skrzypek & Straley, 1993) that is required for efficient translocation of the Yop effectors (Sarker *et al.*, 1998*b*). Translocation of YopE-, YopH-, YopO-, YopM- and YopP-Cya hybrids is strongly decreased in an *lcrG* non polar mutant. LcrG has been shown to bind to LcrV (Sarker *et al.*, 1998*b*; Nilles *et al.*, 1997), suggesting that they could act together at some stage. LcrG has also been shown to bind heparan sulphate proteoglycans on the surface of HeLa cells (see below) but the role of this interaction in translocation still remains to be investigated (Boyd *et al.*, 1998).

Model of assembly of the translocation machinery

Inside the bacterium, YopB and YopD are associated, and this complex is capped with the chaperone SycD (see above). At the moment of secretion, the chaperone would be released in the bacterial cytoplasm and LcrV would bind to the YopB-YopD heteromer, promoting the secretion of the complex and possibly its polymerization. The putative pore in the eukaryotic cell plasma membrane would thus comprise both YopB and YopD, but this is pure speculation. This pore could resemble the one formed by perforin from the cytotoxic T lymphocytes.

Control of Yop release

The proteins involved in control of Yop release: YopN, LcrG, TyeA

In vitro, *Yersinia* only secrete Yops in the absence of Ca^{2+} but they deliver into eukaryotic cells grown in the presence of Ca^{2+}. Thus, the real signal triggering Yop secretion must be the contact between the bacteria and the eukaryotic cell. It is likely that the absence of Ca^{2+} mimics the contact but this could not be shown yet. The interpretation of the phenotype of mutants affected in the Ca^{2+} response has thus to be cautious.

The isolation of Ca^{2+}-blind mutants (Yother & Goguen, 1985) allowed the identification of three genes involved in the control of Yop release by Ca^{2+}-chelation: *yopN* (Forsberg *et al.*, 1991), *lcrG* (Skrzypek & Straley, 1993; Sarker *et al.*, 1998*b*) and *tyeA* (Iriarte *et al.*, 1998). These mutants are deregulated for Yop secretion in the sense that they secrete Yops even in the presence of Ca^{2+}. YopN is a 32.6-kDa protein encoded by the first gene of a locus also containing *tyeA*. In low Ca^{2+} conditions, most of the YopN produced is released in the culture supernatant, while in the presence of Ca^{2+}, the protein is not released but it is exposed at the bacterial surface.

TyeA is a 92-amino acid protein (10.8 kDa) encoded immediately down-stream of *yopN* (Viitanen *et al.*, 1990; Forsberg *et al.*, 1991). It is not secreted, but it is loosely associated with the membrane and it binds to the second coiled-coil of YopN (Iriarte *et al.*, 1998). Surprisingly, this protein plays a role in translocation of some Yop effectors (Iriarte *et al.*, 1998) (see below): a *tyeA* mutant is impaired in translocation of YopE and YopH, but not of YopM, YopO, YopP and YopT (Iriarte *et al.*, 1998; Iriarte & Cornelis, 1998).

LcrG is also involved in the control of Yop release, since the *lcrG* mutants are Ca^{2+} blind (Skrzypek & Straley, 1993; Sarker *et al.*, 1998*b*) but it turned out to be required for efficient translocation of all the known Yop effectors into macrophages (Sarker *et al.*, 1998*b*).

Contact control

It has been suggested that YopN could function as a sensor and a stop valve controlling Yop secretion. After contact with the eukaryotic cell, the YopN sensor would interact with a ligand on the target cell surface, be removed and allow Yop secretion and delivery to the target cell (Rosqvist *et al.*, 1994). However, YopN has never been shown to interact either with Ca^{2+} or with a cell receptor. The fact that *lcrG* and *tyeA* mutants are also deregulated for Yop secretion in the presence of Ca^{2+} or depolarized in the presence of eukaryotic cells (Skrzypek & Straley, 1993; Sarker *et al.*, 1998*b*; Iriarte *et al.*, 1998) suggests that the control of delivery of the effectors requires more than YopN, but rather a complex system comprising at least these three proteins. One can speculate that YopN-LcrG-TyeA form a recognition complex at the bacterial surface that interacts with a receptor on the surface of eukaryotic cells. Information concerning this hypothetical receptor is still scarce but a first element appeared recently: LcrG is able to bind directly to heparin-agarose beads and, in agreement with this, it presents heparin-binding motifs. Nevertheless, it is still difficult to establish a model that integrates the dual function of YopN, TyeA and LcrG.

The Yop effectors and their targets

The outcome of the injection of the Yop effectors into the eukaryotic cell is the ability of *Yersinia* to obstruct a cellular immune response. Six effectors have been identified: YopE, YopH, YopO, YopM, YopP and YopT. The effectors and their targets will be reviewed here.

YopE

YopE is a 23-kDa protein. As mentioned previously, YopE contributes to the ability of *Yersinia* to resist phagocytosis (Rosqvist *et al.*, 1990). YopE induces the disruption of the microfilament structure of the host cells

(Rosqvist *et al.*, 1991). However, YopE does not act directly on actin (Rosqvist *et al.*, 1991). The actual enzyme activity and the target of YopE remain thus to be identified.

YopH

YopH, originally described as Yop51 (Michiels & Cornelis, 1988) and Yop2b (Bölin & Wolf-Watz, 1988), is probably the best characterized Yop. It is a protein tyrosine phosphatase of 51 kDa. The protein tyrosine phosphorylation process forms part of signal transduction pathways that control many cellular functions, including fundamental processes such as phagocytosis, mitogenesis and cell division (Lodish *et al.*, 1995). After infection of HeLa cells with *Yersinia*, YopH leads to inhibition of bacterial uptake, dephosphorylation of p130^{Cas} and FAK and disruption of peripheral focal complexes (Rosqvist *et al.*, 1988; Persson *et al.*, 1997; Black & Bliska, 1997). Focal adhesions are sites where integrin receptors serve as a transmembrane bridge between extracellular matrix proteins and intracellular signalling proteins. FAK is involved in the early steps of the integrin-mediated signalling cascade and is therefore believed to function as a transmitter/ amplifier (Lodish *et al.*, 1995). This effect of YopH leads to the inhibition of phagocytosis by PMNs and macrophages.

YopM

YopM is an acidic 41 kDa protein that contains a succession of 12 repeated structures (Leung & Straley, 1989) related to the very common leucine-rich repeat (LRR) motifs (Kobe & Deisenhofer, 1994). Because of these LRR motifs, YopM exhibits a weak similarity with a large number of proteins, including the α chain of the platelet membrane glycoprotein Ib. This observation suggested to Leung and Straley (1989) that YopM could bind thrombin and interfere with platelet-mediated events of the inflammatory response. In agreement with this hypothesis, *in vitro* studies showed that YopM-containing culture supernatants of *Y. pestis* inhibit platelet aggregation, whereas culture supernatants of a *yopM* mutant do not (Leung *et al.*, 1990). However, this hypothesis is questioned by two recent observations. First, the domains of the GPI-bα, that are involved in the interaction with thrombin lie outside the region sharing homology with YopM (De Marco *et al.*, 1994). Secondly, Boland *et al.* (1996) showed that YopM is delivered inside eukaryotic cells in a similar manner to YopE, YopH and YpkA. YopM thus belongs to the group of intracellular effectors, but its action remains unknown.

YopO/YpkA

YpkA is a 81 kDa serine/threonine kinase that shows sequence similarity to eukaryotic counterparts (Galyov *et al.*, 1993). It is targeted to the inner surface of the plasma membrane of the eukaryotic cell (Håkansson *et al.*,

1996*a*). Given the kinase activity of YpkA and its localization, it is reasonable to suggest that YpkA also interferes with some signal-transduction pathway of the eukaryotic cell.

YopP/YopJ

This Yop was first described in *Y. enterocolitica* as Yop30 (Cornelis *et al.*, 1987), later named YopP in *Y. enterocolitica* and YopJ in *Y. pseudotuberculosis* (Galyov *et al.*, 1994). YopP, encoded by the same operon as YopO (Cornelis *et al.*, 1987), is a 32.5-kDa protein. Both YopP and YopJ have been shown to induce apoptosis in murine macrophages (Mills *et al.*, 1997; Monack *et al.*, 1997). YopP also inhibits TNFα release. Thereby, *Yersinia* might eliminate macrophages without inducing an inflammatory response, and favor extracellular proliferation in lymphoid tissues (see above).

YopP and YopJ share a high level of similarity with AvrRxv from *Xanthomonas campestris* (Whalen *et al.*, 1993), AvrA from *Salmonella* (Hardt & Galan, 1997) and y410 of *Rhizobium* (Freiberg *et al.*, 1997). Until now, no function is known for AvrA and y410. However, AvrRxv is one of many avirulence proteins identified in plant pathogens that mediate the hypersensitive response, a process that is likely to result from the activation of a programmed cell death pathway (Whalen *et al.*, 1993; Mittler & Lam, 1996). However, no cytotoxic effect has been described for AvrA so far. It is striking to observe that animal and plant pathogens share a type III secretion-dependent effector to elicit programmed cell death in their respective hosts.

YopT

YopT is a 35.5-kDa Yop protein that induces a cytotoxic effect in HeLa cells and macrophages (Iriarte & Cornelis, 1998). The effect on HeLa cells consists of the disruption of the actin filaments and the alteration of the cell cytoskeleton.

EFFECTS ON THE HOST CELLS

The cell types that are actual targets of the Yop effector proteins *in vivo* are not known for the moment. As was said above, *Yersinia* spp. have the capacity to resist phagocytosis and killing and to inhibit the inflammatory response, suggesting that macrophages and PMNs are *in vivo* targets. Endothelial cells and epithelial cells of the gastrointestinal tract may also be relevant targets for the Yop virulon. Endothelial cells have an important role in the development of the immune and inflammatory responses, by recruiting PMNs through expression of adhesion molecules and epithelial cells also synthesize and secrete a number of cytokines. In this section, what is known about the effects of *Yersinia* infection on the host cells will be reviewed.

Macrophages

The contact between *Yersinia* and a macrophage leads to different effects: the bacteria has the capacity to trigger apoptosis, to suppress the normal release of TNFα, to impair phagocytosis and to inhibit the respiratory burst. Each of these four aspects will be treated individually below.

Induction of apoptosis

Three groups, two working with *Y. enterocolitica* (Mills *et al.*, 1997; Ruckdeschel *et al.*, 1997*b*) and one working with *Y. pseudotuberculosis* (Monack *et al.*, 1997) showed independently that *Yersinia* triggers apoptosis of cultured macrophages. Infected macrophages displayed general features of apoptosis, such as membrane blebbing (apoptotic body formation), cellular shrinkage (Mills *et al.*, 1997; Ruckdeschel *et al.*, 1997*b*) and DNA fragmentation. Infection of macrophages with secretion and translocation mutants of *Y. enterocolitica* did not lead to apoptosis, showing that a translocated Yop effector is involved. Screening of a library of *yop* mutants showed that the YopE cytotoxin is not involved, and identified YopP as the effector responsible for apoptosis (Mills *et al.*, 1997). In an independent study, Monack *et al.* came to the conclusion that YopJ, the *Y. pseudotuberculosis* homolog of YopP, is required for the induction of the cell death process (Monack *et al.*, 1997). The mechanism by which *Yersinia* induces macrophage apoptosis remains to be elucidated. Recently, Ruckdeschel *et al.* (1998) showed that *Y. enterocolitica* inhibits NF-κB activation in murine J774A.1 and peritoneal macrophages; analysis of different *Y. enterocolitica* mutants revealed a striking correlation between the abilities to inhibit NF-κB activation and to trigger apoptosis. Several reports showed that apoptosis can be prevented by the expression of NF-κB, suggesting that the induction of NF-κB may be part of a survival mechanism (Liu *et al.*, 1996; Beg & Baltimore, 1996; Wang *et al.*, 1996; Barinaga, 1996; Taglialatela *et al.*, 1997). These results suggest that *Yersinia* could trigger apoptosis by suppressing the cellular activation of NF-κB (Ruckdeschel *et al.*, 1998).

Inhibition of TNFα release

TNFα is a pro-inflammatory cytokine playing a central role in the development of the immune and inflammatory responses to infection. Secreted mainly by macrophages, TNFα acts on various cell types involved in the host's defence mechanisms. Ruckdeschel *et al.* (1997*a*) working with the mouse monocyte-macrophage cell line J774A.1 and *Y. enterocolitica* showed that a functional type III secretion machinery is required for the phenomenon to occur, and they suggested a correlation between this inhibition of TNFα release and the inhibition of the ERK1/2, p38 and JNK mitogen-activated protein kinases (MAPKs) activities. It has been shown recently both in *Y. enterocolitica* (Boland & Cornelis, 1998) and in *Y. pseudotubercu-*

losis (Palmer *et al.*, 1998) that the *Yersinia*-induced inhibition of TNFα release requires not only the type III secretion apparatus but also a functional Yop translocation apparatus and the effector YopP (*Y. enterocolitica*)/YopJ (*Y. pseudotuberculosis*). Since YopP is also involved in the triggering of apoptosis (see above), it may well be that both phenomena are linked. One can thus speculate that YopP could act upstream or at the junction of cascades leading to apoptosis on one hand and to the inhibition of TNFα on the other hand; alternatively, the initial role of YopP could be to induce the death of the macrophage by triggering apoptosis, thereby impairing the synthesis and the release of TNFα.

Inhibition of phagocytosis

Working *in vitro* with *Y. pseudotuberculosis* and resident mouse peritoneal macrophages, Rosqvist *et al.* (1988, 1990) showed by a double immunofluorescence technique (Heeseman & Laufs, 1985) that YopE and YopH could act in concert to enable *Yersinia* to inhibit their own uptake by macrophages. Hence they are able to proliferate in the Peyer's patches as extracellular microcolonies (Hanski *et al.*, 1989).

Inhibition of the respiratory burst

Yersinia spp. are able to impair the oxidative burst of the macrophages and, so far, the only Yop effector protein that has been shown to be possibly involved in this phenomenon is YopH. However, the role of YopH in the inhibition of the respiratory burst remains a matter of debate (Hartland *et al.*, 1994; Green *et al.*, 1995).

Polymorphonuclear leukocytes

Polymorphonuclear leukocytes (PMNs) constitute the second group of professional phagocytes that are encountered by *Yersinia* invading the lymphoid tissues of their host. pYV$^+$ *Y. enterocolitica* impede to some extent their phagocytosis by PMNs. However, when ingested, the pYV$^+$ bacteria are hardly killed, while pYV$^-$ bacteria are almost instantly killed (Visser *et al.*, 1995; Ewald *et al.*, 1994), thus implying that plasmid encoded factors can interfere with killing mechanisms. These involve oxygen-dependent mechanisms (oxidative burst) and oxygen-independent mechanisms which include acidification of the phagosome and attack by antimicrobial polypeptides. The inhibition of the oxidative burst will first be described and then the resistance to antimicrobial peptides.

Inhibition of the PMN oxidative burst

The interaction between *Yersinia enterocolitica* and PMNs leads to the inhibition of the PMN oxidative burst and this phenomenon is dependent on pYV-encoded proteins (Lian & Pai, 1985). It seems that the adhesin YadA acts in concert with YopE and YopH to resist antibacterial activities of PMNs under opsonizing conditions (Ruckdeschel *et al.*, 1996; China *et al.*, 1994). The mechanism by which YadA enables *Y. enterocolitica* to resist phagocytosis could involve a reduction of complement-mediated opsonization (China *et al.*, 1994). YadA binds the complement factor H (China *et al.*, 1993) and thus reduces the opsonization by C3b molecules (China *et al.*, 1993). There is a correlation between the lack of oxidative burst and the reduction of opsonization by C3b molecules (Tertti *et al.*, 1987). YopE and YopH, however, would rather inhibit the bactericidal functions of the cells.

Resistance to antimicrobial peptides

Antimicrobial polypeptides present in azurophilic granules of human granulocytes include bactericidal/permeability-increasing protein, cathepsin G, elastase, proteinase 3, azurocidin, lysozyme and defensins. These antimicrobial polypeptides are released into the phagolysosome through fusion of cytoplasmic granules with the phagosomes. YadA seems to be involved in the resistance of *Y. enterocolitica* to the antimicrobial activity of polypeptides from human granulocytes, although the involvement of other plasmid-encoded factors could not be completely ruled out (Visser *et al.*, 1996).

Epithelial cells

HeLa cells have been of great importance in the discovery of injection of Yop effectors into eukaryotic cells by extracellular adhering bacteria, since they are very sensitive to the cytotoxic effect of YopE (Rosqvist *et al.*, 1994; Sory & Cornelis, 1994; Persson *et al.*, 1995; Håkansson *et al.*, 1996*b*). This cytotoxic effect consists of rounding up of the cells and detachment from the extracellular matrix (Goguen *et al.*, 1986; Rosqvist *et al.*, 1990). Different effector Yops are involved in this cytotoxicity, namely YopE, YopH, YopO and YopT (see section describing the effectors for details).

Colon epithelial cells also appear to be programmed to provide a set of chemotactic and activating signals to adjacent and underlying immune and inflammatory cells in the earliest phases after microbial infection (Jung *et al.*, 1995). Virulent *Y. enterocolitica* induce a significantly lower level of IL-8 secretion by T84 cells than non-virulent *Y. enterocolitica*, and the YopB and YopD proteins are required for this suppressive effect (Schulte *et al.*, 1996). It is easily conceivable that this effect favours *Yersinia*, especially during the early phase of infection, by avoiding massive attraction of PMNs to the site of infection.

REGULATION OF THE SYSTEM

Most of the genes involved in Yop synthesis and delivery are organized as a single regulon under a dual transcriptional control. The first level of regulation is temperature-dependent. Activation of the transcription of the genes results from the temperature-influenced interplay between a transcriptional activator, VirF and chromatin structure (Cornelis *et al.*, 1989, 1991; Lambert de Rouvroit *et al.*, 1992). The second regulation prevents full expression of *yop* genes as long as the secretion apparatus is closed. By analogy with the secreted anti-σ factor involved in regulation of flagellum synthesis (Hughes *et al.*, 1993), the most likely hypothesis is that this feedback inhibition is mediated by an inhibitor that is normally expelled via the Yop secretion apparatus. Genetic evidence suggested that the secreted LcrQ protein of *Y. pseudotuberculosis* could be this hypothetical regulator (Rimpiläinen *et al.*, 1992). In *Y. enterocolitica*, the pYV plasmid carries two copies of the *yscM* gene (the counterpart of lcrQ), called *yscM1* and *yscM2*. These two different YscM proteins behave like LcrQ in *Y. pseudotuberculosis*. LcrQ and YscM are secreted and their overexpression leads to a complete inhibition of Yop synthesis but how they exert their inhibitory activity is still unknown but it appears that they rather act indirectly (Pettersson *et al.*, 1996; Stainier *et al.*, 1997). Hence, up-regulation of *yop* expression and polarized translocation of effector Yops are triggered by the opening of the secretion apparatus in response to the signal generated upon interaction of the pathogen with its target cell.

CONCLUSIONS AND FUTURE PERSPECTIVES

The Yop virulon constitutes a new and sophisticated type of bacterial weapon. The general mechanism is now well understood, but many interesting questions remain. In particular, the exact structure of the delivery apparatus is still unknown and the mechanism of the control of Yop release is still more intriguing. Why do some Yop need TyeA to be translocated? And why do some Yop have a chaperone and other not? What is the role of LcrV in the delivery apparatus? The mode of secretion of the Yop effectors appears to be also a matter of debate: are they secreted post- or co-translationnaly? The fact that two parallel secretion pathways could exist would explain some results but is nevertheless difficult to include in the present model.

Actually, it appears also that secretion and translocation would be coupled; the injection of the Yops from the bacterial cytoplasm to the eukaryotic cytosol would thus occur in one step, without intermediate between bacteria and cell. The translocation signal, which also corresponds to the binding domain of the Syc chaperones, would thus rather be an affinity domain that will determine the order of translocation of the Yops.

The study of the targets of the intracellular Yops is also very appealing, since it would lead to a better understanding of these cellular processes and would also help to clarify why *Yersinia* uses so many Yops.

Besides answering all these fascinating basic questions, one could also envision developing possible medical applications of the new concepts that came into sight. One can envision that such a sophisticated virulence apparatus could be an appropriate target for 'antipathogenicity drugs'. One may also consider engineered *Yersinia* as vectors to deliver antigens when a CTL response is desirable. This second application is probably close to realization, at least in the laboratory.

ACKNOWLEDGEMENTS

C.N. is a research assistant funded by the Belgian 'Fonds National de la Recherche Scientifique'. The *Yersinia* project is supported by the Belgian 'Fonds National de la Recherche Scientifique Médicale' (Convention 3.4595.97), the 'Direction Générale de la Recherche Scientifique-Communauté Française de Belgique' (Action de Recherche Concertée 94/99–172) and by the 'Interuniversity Poles of Attraction Program – Belgian State, Prime Minister's Office, Federal Office for Scientific, Technical and Cultural affairs' (PAI 4/03).

REFERENCES

Alfano, J. R. & Collmer, A. (1997). The type III (Hrp) secretion pathway of plant pathogenic bacteria: trafficking harpins, Avr proteins, and death. *Journal of Bacteriology*, **179**, 5655–62.

Allaoui, A., Scheen, R., Lambert de Rouvroit, C. L. & Cornelis, G. R. (1995*a*). VirG, a *Yersinia enterocolitica* lipoprotein involved in Ca^{2+} dependency, is related to ExsB of *Pseudomonas aeruginosa*. *Journal of Bacteriology*, **177**, 4230–7.

Allaoui, A., Schulte, R. & Cornelis, G. R. (1995*b*). Mutational analysis of the *Yersinia enterocoliticia virC* operon: characterization of *yscE, F, G, I, J, K* required for Yop secretion and *yscH* encoding YopR. *Molecular Microbiology*, **18**, 343–55.

Anderson, D. M. & Schneewind, O. (1997). A mRNA signal for the type III secretion of Yop proteins by *Yersinia enterocolitica*. *Science*, **278**, 1140–3.

Barinaga, M. (1996). Life-death balance within the cell. *Science*, **274**, 724.

Beg, A. A. & Baltimore, D. (1996). An essential role for NF-κB in preventing TNF-α-induced cell death. *Science*, **274**, 782–4.

Bergman, T., Håkansson, S., Forsberg, A., Norlander, L., Macellaro, A., Backman, A., Bölin, I. & Wolf-Watz, H. (1991). Analysis of the V antigen *lcrGVH-yopBD* operon of *Yersinia pseudotuberculosis*: evidence for a regulatory role of LcrH and LcrV. *Journal of Bacteriology*, **173**, 1607–16.

Bhakdi, S., Mackman, N., Nicaud, J. M. & Holland, I. B. (1986). *Escherichia coli* hemolysin may damage target cell membranes by generating transmembrane pores. *Infection and Immunity*, **52**, 63–9.

Black, D. S. & Bliska, J. B. (1997). Identification of p130^{Cas} as a substrate of *Yersinia* YopH (Yop51), a bacterial protein tyrosine phosphatase that translocates into mammalian cells and targets focal adhesions. *EMBO Journal*, **16**, 2730–44.

Boland, A. & Cornelis, G. R. (1998). Suppression of macrophage TNFα release during *Yersinia* infection: role of YopP. *Infection and Immunity*, **66**, 1878–84.

Boland, A., Sory, M-P., Iriarte, M., Kerbourch, C., Wattiau, P. & Cornelis, G. R. (1996). Status of YopM and YopN in the *Yersinia* Yop virulon: YopM of *Y. enterocolitica* is internalized inside the cytosol of PU5–1.8 macrophages by the YopB, D, N delivery apparatus. *EMBO Journal*, **15**, 5191–201.

Bölin, I. & Wolf-Watz, H. (1988). The plasmid encoded Yop2b protein of *Yersinia pseudotuberculosis* is a virulence determinant regulated by calcium and temperature at the level of transcription. *Molecular Microbiology*, **2**, 237–45.

Boyd, A. P., Sory, M-P., Iriarte, M. & Cornelis, G. R. (1998). Heparin interferes with translocation of Yop proteins into HeLa cells and binds to LcrG, a regulatory component of the *Yersinia* Yop apparatus. *Molecular Microbiology*, **27**, 425–36.

Burrows, T. W. & Bacon, G. A. (1956). The basis of virulence in *Pasteurella pestis*: an antigen determining virulence. *British Journal of Experimental Pathology*, **37**, 481–93.

Cheng, L. W., Anderson, D. M. & Schneewind, O. (1997). Two independent type III secretion mechanisms for YopE in *Yersinia enterocolitica*. *Molecular Microbiology*, **24**, 757–65.

China, B., N'Guyen, B. T., de Bruyere, M. & Cornelis, G. R. (1994). Role of YadA in resistance of *Yersinia enterocolitica* to phagocytosis by human polymorphonuclear leukocytes. *Infection and Immunity*, **62**, 1275–81.

China, B., Sory, M-P., N'Guyen, B. T., de Bruyere, M. & Cornelis, G. R. (1993). Role of the YadA protein in prevention of opsonization of *Yersinia enterocolitica* by C3b molecules. *Infection and Immunity*, **61**, 3129–36.

Cornelis, G. R., Sluiters, C., Delor, I., Geib, D., Kaniga, K., Lambert de Rouvroit, C. L., Sory, M-P., Vanooteghem, J-C. & Michiels, T. (1991). *ymoA*, a *Yersinia enterocolitica* chromosomal gene modulating the expression of virulence functions. *Molecular Microbiology*, **5**, 1023–34.

Cornelis, G. R., Sluiters, C., Lambert de Rouvroit, C. L. & Michiels, T. (1989). Homology between VirF, the transcriptional activator of the *Yersinia* virulence regulon, and AraC, the *Escherichia coli* arabinose operon regulator. *Journal of Bacteriology*, **171**, 254–62.

Cornelis, G. R., Vanooteghem, J-C. & Sluiters, C. (1987). Transcription of the *yop* regulon from *Y. enterocolitica* requires trans acting pYV and chromosomal genes. *Microbial Pathogenesis*, **2**, 367–79.

Cover, T. L. & Aber, R. C. (1989). *Yersinia enterocolitica*. *New England Journal of Medicine*, **321**, 16–24.

De Marco, L., Mazzucato, M., Masotti, A. & Ruggeri, Z. M. (1994). Localization and characterization of an α-thrombin-binding site on platelet glycoprotein Ibα. *Journal of Biological Chemistry*, **269**, 6478–84.

Eisenberg, D. (1984). Three-dimensional structure of membrane and surface proteins. *Annual Review of Biochemistry*, **53**, 595–623.

Ewald, J. H., Heesemann, J., Rudiger, H. & Autenrieth, I. B. (1994). Interaction of polymorphonuclear leukocytes with *Yersinia enterocolitica*: role of the *Yersinia* virulence plasmid and modulation by the iron-chelator desferrioxamine B. *Journal of Infectious Diseases*, **170**, 140–50.

Fällman, M., Persson, C. & Wolf-Watz, H. (1997). *Yersinia* proteins that target host cell signalling pathways. *Journal of Clinical Investigations*, **99**, 1153–7.

Felmlee, T. & Welch, R. A. (1988). Alterations of amino acid repeats in the *Escherichia coli* hemolysin affect cytolytic activity and secretion. *Proceedings of the National Academy of Sciences, USA*, **85**, 5269–73.

Forsberg, A., Rosqvist, R. & Wolf-Watz, H. (1994). Regulation and polarized transfer of the *Yersinia* outer proteins (Yops) involved in antiphagocytosis. *Trends in Microbiology*, **2**, 14–19.

Forsberg, A., Viitanen, A. M., Skurnik, M. & Wolf-Watz, H. (1991). The surface-located YopN protein is involved in calcium signal transduction in *Yersinia pseudotuberculosis*. *Molecular Microbiology*, **5**, 977–86.

Freiberg, C., Fellay, R., Bairoch, A., Broughton, W. J., Rosenthal, A. & Perret, X. (1997). Molecular basis of symbiosis between *Rhizobium* and legumes. *Nature*, **387**, 394–401.

Frithz-Lindsten, E., Rosqvist, R., Johansson, L. & Forsberg, A. (1995). The chaperone-like protein YerA of *Yersinia pseudotuberculosis* stabilizes YopE in the cytoplasm but is dispensible for targeting to the secretion loci. *Molecular Microbiology*, **16**, 635–47.

Galyov, E. E., Håkansson, S., Forsberg, A. & Wolf-Watz, H. (1993). A secreted protein kinase of *Yersinia pseudotuberculosis* is an indispensable virulence determinant. *Nature*, **361**, 730–2.

Galyov, E. E., Håkansson, S. & Wolf-Watz, H. (1994). Characterization of the operon encoding the YpkA Ser/Thr protein kinase and the YopJ protein of *Yersinia pseudotuberculosis*. *Journal of Bacteriology*, **176**, 4543–8.

Glaser, P., Sakamoto, H., Bellalou, J., Ullmann, A. & Danchin, A. (1988). Secretion of cyclolysin, the calmodulin-sensitive adenylate cyclase-haemolysin bifunctional protein of *Bordetella pertussis*. *EMBO Journal*, **7**, 3997–4004.

Goguen, J. D., Walker, W. S., Hatch, T. P. & Yother, J. (1986). Plasmid-determined cytotoxicity in *Yersinia pestis* and *Yersinia pseudotuberculosis*. *Infection and Immunity*, **51**, 788–94.

Green, S. P., Hartland, E. L., Robins Browne, R. M. & Phillips, W. A. (1995). Role of YopH in the suppression of tyrosine phosphorylation and respiratory burst activity in murine macrophages infected with *Yersinia enterocolitica*. *Journal of Leukocyte Biology*, **57**, 972–7.

Håkansson, S., Bergman, T., Vanooteghem, J. C., Cornelis, G. & Wolf-Watz, H. (1993). YopB and YopD constitute a novel class of *Yersinia* Yop proteins. *Infection and Immunity*, **61**, 71–80.

Håkansson, S., Galyov, E. E., Rosqvist, R. & Wolf-Watz, H. (1996*a*). The *Yersinia* YpkA Ser/Thr kinase is translocated and subsequently targeted to the inner surface of the HeLa cell plasma membrane. *Molecular Microbiology*, **20**, 593–603.

Håkansson, S., Schesser, K., Persson, C., Galyov, E. E., Rosqvist, R., Homblé, F. & Wolf-Watz, H. (1996*b*). The YopB protein of *Yersinia pseudotuberculosis* is essential for the translocation of Yop effector proteins across the target cell plasma membrane and displays a contact dependent membrane disrupting activity. *EMBO Journal*, **15**, 5812–23.

Hanski, C., Kutschka, U., Schmoranzer, H. P., Naumann, M., Stallmach, A., Hahn, H., Menge, H. & Riecken, E. O. (1989). Immunohistochemical and electron microscopic study of interaction of *Yersinia enterocolitica* serotype 0:8 with intestinal mucosa during experimental enteritis. *Infection and Immunity*, **57**, 673–8.

Hardt, W-D. & Galan, J. E. (1997). A secreted *Salmonella* protein with homology to an avirulence determinant of plant pathogenic bacteria. *Proceedings of the National Academy of Sciences, USA*, **94**, 9887–92.

Hartland, E. L., Green, S. P., Phillips, W. A. & Robins Browne, R. M. (1994). Essential role of YopD in inhibition of the respiratory burst of macrophages by *Yersinia enterocolitica*. *Infection and Immunity*, **62**, 4445–53.

He, S. Y. (1997) Hrp-controled interkingdom protein transport: learning from flagellar assembly? *Trends in Microbiology*, **5**, 489–95.

Heesemann, J. & Laufs, R. (1985). Double immunofluorescence microscopic technique for accurate differentiation of extracellularly and intracellularly located bacteria in cell culture. *Journal of Clinical Microbiology*, **22**, 168–75.

Holmström, A., Pettersson, J., Rosqvist, R., Håkansson, S., Tafazoli, F., Fällman, M., Magnusson, K. E., Wolf-Watz, H. & Forsberg, A. (1997). YopK of *Yersinia pseudotuberculosis* controls translocation of Yop effectors across the eukaryotic cell membrane. *Molecular Microbiology*, **24**, 73–91.

Holmström, A., Rosqvist, R., Wolf-Watz, H. & Forsberg, A. (1995). Virulence plasmid-encoded YopK is essential for *Yersinia pseudotuberculosis* to cause systemic infection in mice. *Infection and Immunity*, **63**, 2269–76.

Hughes, K. T., Gillen, K. L., Semon, M. J. & Karlinsey, J. E. (1993). Sensing structural intermediates in bacterial flagellar assembly by export of a negative regulator [see comments]. *Science*, **262**, 1277–80.

Iriarte, M. & Cornelis, G. R. (1998). YopT, a new *Yersinia* Yop effector protein, affects the cytoskeleton of host cells. *Molecular Microbiology*, **29**, 915–29.

Iriarte, M., Sory, M-P., Boland, A., Boyd, A. P., Mills, S. D., Lambermont, I. & Cornelis, G. R. (1998). TyeA, a protein involved in control of Yop release and in translocation of *Yersinia* Yop effectors. *EMBO Journal*, **17**, 1907–18.

Iriarte, M. & Cornelis, G. R. (1999). Assignment of SycN, YscX and YscY, three new elements of the *Yersinia* Yop virulon. *Journal of Bacteriology*, in press.

Jung, H. C., Eckmann, L., Yang, S. K., Panja, A., Fierer, J., Morzycka Wroblewska, E. & Kagnoff, M. F. (1995). A distinct array of proinflammatory cytokines is expressed in human colon epithelial cells in response to bacterial invasion. *Journal of Clinical Investigation*, **95**, 55–65.

Khisty, V. J., Munske, G. R. & Randall, L. L. (1995). Mapping of the binding frame for the chaperone SecB within a natural ligand, galactose-binding protein. *Journal of Biological Chemistry*, **270**, 25920–7.

Kobe, B. & Deisenhofer, J. (1994). The leucine-rich repeat: a versatile binding motif. *Trends in Biochemical Science*, **19**, 415–20.

Koster, M., Bitter, W., de Cock, H., Allaoui, A., Cornelis, G. R. & Tommassen, J. (1997). The outer membrane component, YscC, of the Yop secretion machinery of *Yersinia enterocolitica* forms a ring-shaped multimeric complex. *Molecular Microbiology*, **26**, 789–98.

Kubori, T., Matsushima, Y., Nakamura, D., Uralil, J., Lara-Tejero, M., Sukhan, A., Galan, J. E. and Aizawa, S-I. (1998). Supramolecular structure of the *Salmonella typhimurium* Type III protein secretion system. *Science*, **280**, 602–5.

Kumamoto, C. A. & Beckwith, J. (1985). Evidence for specificity at an early step in protein export in *Escherichia coli*. *Journal of Bacteriology*, **163**, 267–74.

Lambert de Rouvroit, C. L., Sluiters, C. & Cornelis, G. R. (1992). Role of the transcriptional activator, VirF, and temperature in the expression of the pYV plasmid genes of *Yersinia enterocolitica*. *Molecular Microbiology*, **6**, 395–409.

Laroche, Y., Van Bouchaute, M. & Cornelis, G. (1984). A restriction map of virulence plasmid pVYE439–80 from a serogroup 9 *Yersinia enterocolitica* strain. *Plasmid*, **12**, 67–70.

Lee, V. T., Anderson, D. M. & Schneewind, O. (1998). Targeting of *Yersinia* Yop proteins into the cytosol of HeLa cells: one-step translocation of YopE across acterial and eukaryotic membranes is dependent on SycE chaperone. *Molecular Microbiology*, **28**, 593–601.

Leung, K. Y. & Straley, S. C. (1989). The *yopM* gene of *Yersinia pestis* encodes a released protein having homology with the human platelet surface protein GPIb α. *Journal of Bacteriology*, **171**, 4623–32.

Leung, K. Y., Reisner, B. S. & Straley, S. C. (1990). YopM inhibits platelet aggregation and is necessary for virulence of *Yersinia pestis* in mice. *Infection and Immunity*, **58**, 3262–71.

Lian, C. J. & Pai, C. H. (1985). Inhibition of human neutrophil chemiluminescence by plasmid-mediated outer membrane proteins of *Yersinia enterocolitica*. *Infection and Immunity*, **49**, 145–51.

Liu, Z. G., Hsu, H., Goeddel, D. V. & Karin, M. (1996). Dissection of TNF receptor 1 effector functions: JNK activation is not linked to apoptosis while NF-κB activation prevents cell death. *Cell*, **87**, 565–76.

Lodish, H., Baltimore, D., Berk, A., Zipursky, S. L., Matsudaira, P. & Darnel, J. E. (1995). In *Integrative and Specialized Cellular Activities*, ed. J. E. Darnel, pp. 850–1342. New York: Scientific American Books Inc.

Lupas, A., van Dyke, M. & Stock, J. (1991). Predicting coiled coils from protein sequences. *Science*, **252**, 1162–4.

Michiels, T. & Cornelis, G. (1988). Nucleotide sequence and transcription analysis of *yop51* from *Yersinia enterocolitica* W22703. *Microbial Pathogenesis*, **5**, 449–59.

Michiels, T. & Cornelis, G. R. (1991). Secretion of hybrid proteins by the *Yersinia* Yop export system. *Journal of Bacteriology*, **173**, 1677–85.

Mills, S. D., Boland, A., Sory, M-P., Van der Smissen, P., Kerbourch, C., Finlay, B. B. & Cornelis, G. R. (1997). *Yersinia enterocolitica* induces apoptosis in macrophages by a process requiring functional type III secretion and translocation mechanisms and involving YopP, presumably acting as an effector protein. *Proceedings of the National Academy of Sciences, USA*, **94**, 12 638–43.

Mittler, R. & Lam, E. (1996). Sacrifice in the face of foes: pathogen-induced programmed cell death in plants. *Trends in Microbiology*, **4**, 10–15.

Monack, D. M., J. Mecsas, J., Ghori, N. & Falkow, S. (1997). *Yersinia* signals macrophages to undergo apoptosis and YopJ is necessary for this cell death. *Proceedings of the National Academy of Sciences, USA*, **94**, 10 385–90.

Neyt, C. & Cornelis, G. R. (1998). Role of SycD, the chaperone of the *Yersinia* Yop translocators YopB and YopD. *Molecular Microbiology*, **31**, in press.

Nilles, M. L., Williams, A. W., Skrzypek, E. & Straley, S. C. (1997). *Yersinia pestis* LcrV forms a stable complex with LcrG and may have a secretion-related regulatory role in the Low-Ca^{2+} response. *Journal of Bacteriology*, **179**, 1307–16.

Palmer, L. E., Hobbie, S., Galan, J. E. & Bliska, J. B. (1998). YopJ of *Yersinia pseudotuberculosis* is required for the inhibition of macrophage TNFα production and downregulation of the MAP kinases p38 and JNK. *Molecular Microbiology*, **27**, 953–65.

Perry, R. D., Harmon, P. A., Bowmer, W. S. & Straley, S. C. (1986). A low-Ca^{2+} response operon encodes the V antigen of *Yersinia pestis*. *Infection and Immunity*, **54**, 428–34.

Persson, C., Carballeira, N., Wolf-Watz, H. & Fällman, M. (1997). The PTPase YopH inhibits uptake of *Yersinia*, tyrosine phosphorylation of p130^{Cas} and FAK, and the associated accumulation of these proteins in peripheral focal adhesions. *EMBO Journal*, **16**, 2307–18.

Persson, C., Nordfelth, R., Holmström, A., Håkansson, S., Rosqvist, R. & Wolf-Watz, H. (1995). Cell-surface-bound *Yersinia* translocate the protein tyrosine phosphatase YopH by a polarized mechanism into the target cell. *Molecular Microbiology*, **18**, 135–50.

Pettersson, J., Nordfelth, R., Dubinina, E., Bergman, T., Gustafsson, M., Magnusson, K. E. & Wolf-Watz, H. (1996). Modulation of virulence factor expression by pathogen target cell contact. *Science*, **273**, 1231–3.

Plano, G. V. & Straley, S. C. (1995). Mutations in *yscC*, *yscD*, and *yscG* prevent high-level expression and secretion of V antigen and Yops in *Yersinia pestis*. *Journal of Bacteriology*, **177**, 3843–54.

Price, S. B., Cowan, C., Perry, R. D. & Straley, S. C. (1991). The *Yersinia pestis* V antigen is a regulatory protein necessary for Ca^{2+}-dependent growth and maximal expression of low-Ca^{2+} response virulence genes. *Journal of Bacteriology*, **173**, 2649–57.

Rimpiläinen, M., Forsberg, A. & Wolf-Watz, H. (1992). A novel protein, LcrQ, involved in the low-calcium response of *Yersinia pseudotuberculosis* shows extensive homology to YopH. *Journal of Bacteriology*, **174**, 3355–63.

Rosqvist, R., Bölin, I. & Wolf-Watz, H. (1988). Inhibition of phagocytosis in *Yersinia pseudotuberculosis*: a virulence plasmid-encoded ability involving the Yop2b protein. *Infection and Immunity*, **56**, 2139–43.

Rosqvist, R., Forsberg, A., Rimpiläinen, M., Bergman, T. & Wolf-Watz, H. (1990). The cytotoxic protein YopE of *Yersinia* obstructs the primary host defence. *Molecular Microbiology*, **4**, 657–67.

Rosqvist, R., Forsberg, A. & Wolf-Watz, H. (1991). Intracellular targeting of the *Yersinia* YopE cytotoxin in mammalian cells induces actin microfilament disruption. *Infection and Immunity*, **59**, 4562–9.

Rosqvist, R., Magnusson, K-E. & Wolf-Watz, H. (1994). Target cell contact triggers expression and polarized transfer of *Yersinia* YopE cytotoxin into mammalian cells. *EMBO Journal*, **13**, 964–72.

Ruckdeschel, K., Harb, S., Roggenkamp, A., Hornef, M., Zumbihl, R., Kohler, S., Heesemann, J. & Rouot, B. (1998). *Yersinia enterocolitica* impairs activation of transcription factor NF-κB: involvement in the induction of programmed cell death and in the suppression of the macrophage TNF-α production. *Journal of Experimental Medicine*, **187**, 1069–79.

Ruckdeschel, K., Machold, J., Roggenkamp, A., Schubert, S., Pierre, J., Zumbihl, R., Liautard, J. P., Heesemann, J. & Rouot, B. (1997a). *Yersinia enterocolitica* promotes deactivation of macrophage mitogen-activated protein kinases extracellular signal-regulated kinase-1/2, p38, and c-Jun NH_2-terminal kinase. *Journal of Biological Chemistry*, **272**, 15920–7.

Ruckdeschel, K., Roggenkamp, A., Lafont, V., Mangeat, P., Heesemann, J. & Rouot, B. (1997b). Interaction of *Yersinia enterocolitica* with macrophages leads to macrophage cell death through apoptosis. *Infection and Immunity*, **65**, 4813–21.

Ruckdeschel, K., Roggenkamp, A., Schubert, S. & Heesemann, J. (1996). Differential contribution of *Yersinia enterocolitica* virulence factors to evasion of microbicidal action of neutrophils. *Infection and Immunity*, **64**, 724–33.

Russel, M. (1994). Phage assembly: a paradigm for bacterial virulence factor export? *Science*, **265**, 612–14.

Sarker, M. R., Neyt, C., Stainier, I. & Cornelis, G. R. (1998a). The *Yersinia* Yop virulon: LcrV is required for extrusion of the translocators YopB and YopD. *Journal of Bacteriology*, **180**, 1207–14.

Sarker, M. R., Sory, M-P., Boyd, A. P., Iriarte, M. & Cornelis, G. R. (1998b). LcrG controls internalization of *Yersinia* Yop effector proteins into eukaryotic cells. *Infection and Immunity*, **66**, 2976–9.

Schesser, K., Frithz-Lindsten, E. & Wolf-Watz, H. (1996). Delineation and mutational analysis of the *Yersinia pseudotuberculosis* YopE domains which mediate translocation across bacterial and eukaryotic cellular membranes. *Journal of Bacteriology*, **178**, 7227–33.

Schulte, R., Wattiau, P., Hartland, E. L., Robins Browne, R. M. & Cornelis, G. R. (1996). Differential secretion of interleukin-8 by human epithelial cell lines upon entry of virulent or nonvirulent *Yersinia enterocolitica*. *Infection and Immunity*, **64**, 2106–13.

Skrzypek, E. & Straley, S. C. (1993). LcrG, a secreted protein involved in negative regulation of the low-calcium response in *Yersinia pestis*. *Journal of Bacteriology*, **175**, 3520–8.

Skrzypek, E. & Straley, S. C. (1995). Differential effects of deletions in lcrV on secretion of V antigen, regulation of the low-Ca^{2+} response, and virulence of *Yersinia pestis*. *Journal of Bacteriology*, **177**, 2530–42.

Sory, M. P. & Cornelis, G. R. (1994). Translocation of a hybrid YopE-adenylate cyclase from *Yersinia enterocolitica* into HeLa cells. *Molecular Microbiology*, **14**, 583–94.

Sory, M. P., Boland, A., Lambermont, I. & Cornelis, G. R. (1995). Identification of the YopE and YopH domains required for secretion and internalization into the cytosol of macrophages, using the *cyaA* gene fusion approach. *Proceedings of the National Academy of Sciences, USA*, **92**, 11 998–12 002.

Stainier, I., Iriarte, M. & Cornelis, G. R. (1997). YscM1 and YscM2, two *Yersinia enterocolitica* proteins causing down regulation of *yop* transcription. *Molecular Microbiology*, **26**, 833–43.

Straley, S. C., Plano, G. V., Skrzypek, E., Haddix, P. L. & Fields, K. A. (1993). Regulation by Ca^{2+} in the *Yersinia* low-Ca^{2+} response. *Molecular Microbiology*, **8**, 1005–10.

Strathdee, C. A. & Lo, R. Y. (1989). Cloning, nucleotide sequence, and characterization of genes encoding the secretion fuction of the *Pasteurella haemolytica* leukotoxin determinant. *Journal of Bacteriology*, **171**, 916–28.

Taglialatela, G., Robinson, R. & Perez Polo, J. R. (1997). Inhibition of nuclear factor kappa B (NFkappaB) activity induces nerve growth factor-resistant apoptosis in PC12 cells. *Journal of Neuroscience Research*, **47**, 155–62.

Tertti, R., Eerola, E., Lehtonen, O. P., Stahlberg, T. H., Viander, M. & Toivanen, A. (1987). Virulence-plasmid is associated with the inhibition of opsonization in *Yersinia enterocolitica* and *Yersinia pseudotuberculosis*. *Clinical and Experimental Immunology*, **68**, 266–74.

Van den Ackerveken, G. & Bonas, U. (1997). Bacterial avirulence proteins as triggers of plant disease restistance. *Trends in Microbiology*, **5**, 394–8.

Viitanen, A. M., Toivanen, P. & Skurnik, M. (1990). The *lcrE* gene is part of an operon in the *lcr* region of *Yersinia enterocolitica* O:3. *Journal of Bacteriology*, **172**, 3152–62.

Visser, L. G., Annema, A. & van Furth, R. (1995). Role of Yops in inhibition of phagocytosis and killing of opsonized *Yersinia enterocolitica* by human granulocytes. *Infection and Immunity*, **63**, 2570–5.

Wang, C. Y., Mayo, M. W. & Baldwin, A. S. J. (1996). TNF- and cancer therapy-induced apoptosis: potentiation by inhibition of NF-κB. *Science*, **274**, 784–7.

Wattiau, P. & Cornelis, G. R. (1993). SycE, a chaperone-like protein of *Yersinia enterocolitica* involved in the secretion of YopE. *Molecular Microbiology*, **8**, 123–31.

Wattiau, P., Bernier, B., Deslee, P., Michiels, T. & Cornelis, G. R. (1994). Individual chaperones required for Yop secretion by *Yersinia*. *Proceedings of the National Academy of Sciences, USA*, **91**, 10 493–7.

Wattiau, P., Woestyn, S. & Cornelis, G. R. (1996). Customized secretion chaperones in pathogenic bacteria. *Molecular Microbiology*, **20**, 255–62.

Welch, R. A. (1991). Pore-forming cytolysins of gram-negative bacteria. *Molecular Microbiology*, **5**, 521–8.

Whalen, M. C., Wang, J. F., Carland, F. M., Heiskell, M. E., Dahlbeck, D., Minsavage, G. V., Jones, J. B., Scott, J. W., Stall, R. E. & Staskawicz, B. J. (1993). Avirulence gene avrRxv from *Xanthomonas campestris* pv. vesicatoria specifies resistance on tomato line Hawaii 7998. *Molecular Plant and Microbe Interactions*, **6**, 616–27.

Woestyn, S., Sory, M-P., Boland, A., Lequenne, O. & Cornelis, G. R. (1996). The cytosolic SycE and SycH chaperones of *Yersinia* protect the region of YopE and YopH involved in translocation across eukaryotic cell membranes. *Molecular Microbiology*, **20**, 1261–71.

Yother, J. & Goguen, J. D. (1985). Isolation and characterization of Ca^{2+}-blind mutants of *Yersinia pestis*. *Journal of Bacteriology*, **164**, 704–11.

MATHEMATICAL MODELLING OF SIGNALLING IN *DICTYOSTELIUM DISCOIDEUM*

JONATHAN A. SHERRATT[1], JOHN C. DALLON[1], THOMAS HÖFER[2] AND PHILIP K. MAINI[3]

[1]*Department of Mathematics, Heriot-Watt University, Edinburgh EH14 4AS, UK*
[2]*Institute of Biophysics, Humboldt University of Berlin, Invalidenstrasse 42, D-10115 Berlin, Germany*
[3]*Centre for Mathematical Biology, Mathematical Institute, 24–29 St Giles', Oxford OX1 3LB, UK*

INTRODUCTION

The cellular slime mould *Dictyostelium discoideum* (*D. discoideum*) has a remarkable life cycle, incorporating many key features of morphogenesis in higher organisms, including chemotaxis, cell differentiation and multicellular organization. In starvation conditions, the unicellular amoebae aggregate into a multicellular slug (containing about 10^5 cells), which is capable of coordinated movement towards chemical and light sources. Within the slug, cells begin to differentiate and sort into 'prestalk' (about 20%) and 'prespore' (about 80%); at some point the slug becomes stationary, and these cells form the stalk and spores of a 'fruiting body', from which individual spore cells are dispersed when conditions become more favourable. For detailed reviews of this life cycle, see Bonner (1982) and Devreotes (1982). The experimental accessibility of *D. discoideum*, along with the key features mentioned above, has led to its widespread adoption as a prototype morphogenetic system. What was once thought of as a simple system has become a rich source of information for the processes of gene regulation during development, signal transduction pathways, amoeboidal movement and chemotaxis. The aggregation stage has been particularly well studied. The key to this process is intercellular signalling by cyclic adenosine $3',5'$-monophosphate (cAMP). Starvation causes a small number of cells, distributed throughout a population, to act as pacemakers, emitting cAMP periodically (Raman *et al.*, 1976). Surrounding cells move towards these pacemakers because of a chemotactic response to cAMP; they also secrete cAMP themselves in an autocatalytic manner, propagating the signal across the spatial domain. The waves of cAMP take the form either of target patterns, concentric circles, or spirals.

Initial mathematical models for *D. discoideum* morphogenesis focused on the kinetics of cAMP, both intra- and extracellular, and the cAMP-receptor on the cell surface (Goldbeter & Segel, 1977). Crucially, investigation of the spatiotemporal dynamics of this reaction predicted spiral wave formation (Hagan & Cohen, 1981; Tyson *et al.*, 1989), in agreement with the spiral patterns seen in a field of aggregating *D. discoideum* amoebae. A number of other aspects of *D. discoideum* biology have also been extensively modelled mathematically, including slug migration (Odell & Bonner, 1986), cell sorting within the slug (Meinhardt, 1983), mound formation at the end of aggregation (Vasiev *et al.*, 1989; Levine *et al.*, 1997), and the formation of cellular streams during aggregation. This chapter will concentrate on this last issue, and review some of the mathematical models proposed during the past few years. The reasons for this focus are as follows. First, it is an excellent example of the ability of biochemical signalling in amoebae to generate coordinated behaviour, and secondly, it is one of a very few areas of cell biology to which a wide variety of different mathematical modelling approaches have been applied, and comparison of the results of these approaches provides valuable insight into the biological process.

The term 'streaming' refers to the fact that, as *D. discoideum* cells move into the aggregation centre, they do so not as a uniform field, but rather in discrete streams, separated by about 50 μm (Fig. 1). The wide variety of mathematical models for this phenomenon are all based on the inclusion of cell movement into existing models for the spatiotemporal dynamics of cAMP. This section briefly reviews these dynamics and their mathematical modelling.

The basic reason for cAMP wave propagation during *D. discoideum* aggregation is that cAMP dynamics fall into a category known as an 'excitable system'. This refers to the fact that behaviour occurs on two different time scales; the processes involved in the production and release of cAMP occurring more rapidly than the competing process which turns off cAMP production. The competing process eventually wins and 'resets' the system to the original state of very low cAMP production. However, the difference in time scales generates 'cycles' of cAMP concentration. To explain this, it is assumed that an initial moderate level of cAMP is added uniformly to a field of *D. discoideum* amoebae. This stimulates the amoebae to produce and release cAMP, causing cAMP levels to increase even further, until the much slower inhibition of the production 'catches up'. This returns the amoebae to a low cAMP producing state, so that the cAMP concentration returns to its original level. In reality, however, cAMP is added not uniformly, but at one point in space, by a pacemaker cell. The above cAMP-cycle then occurs in the immediate vicinity of the pacemaker. However, during the phase of this cycle at which cAMP levels are high, sufficient cAMP diffuses into the surrounding regions to stimulate other amoebae, initiating a cAMP cycle there, and this process is repeated throughout the field of cells. It is this series of phase-lagged cAMP cycles that constitutes the observed periodic and spiral waves of cAMP.

A simple mathematical representation of cAMP dynamics, which highlights its excitability, was developed by Martiel and Goldbeter (1984). The

Fig. 1. Aggregation of *Dictyostelium* (str NP377) on an agar plate, showing the formation of spiral cAMP waves which induce (*a*) cell movement, (*b*) the onset of cell streaming, and (*c*) the developed cell stream morphology in the whole aggregation territory. Pictures are taken *ca.* 15 min apart. The position of the cAMP waves in (*a*), (*b*) can be inferred from the different light scattering responses of elongated (moving) and rounded (stationary) cells; amoebae elongate under the influence of the cAMP waves and form the bright bands in the photograph. (Courtesy of P.C. Newell.)

model consists of differential equations for three variables, extracellular cAMP concentration (u), active cAMP-receptor concentration (v), and intracellular cAMP concentration (w):

$$\partial u/\partial t = k_1 w - k_2 u + D\nabla^2 u \tag{1a}$$

$$\partial v/\partial t = -f_1(u) + f_2(u) \cdot (1 - v) \tag{1b}$$

$$\partial w/\partial t = k_3 f_3(u, v) - k_4 w \tag{1c}$$

where

$$f_1(u) = \frac{k_5 + k_6 u}{1 + u} \quad f_2(u) = \frac{k_7 + k_8 u}{1 + k_9 u} \quad f_3(u, v) = \frac{k_{10}(1 + u)^2 + u^2 v^2}{k_{11}(1 + u)^2 + u^2 v^2}$$

Here f_1 and f_2 are kinetic rate functions for receptor desensitization and resensitization respectively, and f_3 describes the activation of adenylate cyclase by bound and active receptors; the ks are positive constants. Numerical simulations of these equations, illustrating spiral waves, were presented by Tyson *et al.* (1989). A more mathematical account of spiral wave formation in an excitable system such as this is given in the book of Grindrod (1991).

The representation of signal transduction in the Martiel–Goldbeter model is extremely simplistic. A great deal is known about the details of cAMP dynamics. Briefly, the binding of cAMP to cell surface receptors induces excitation and adaptation of guanylyl cyclase and adenylyl cyclase, on time scales of seconds and minutes respectively. The first of these controls the chemotactic response of the amoeba (see below), while the second causes synthesis and secretion of cAMP. These processes are regulated by the G proteins in a complicated manner which is not yet fully understood. A mathematical model reflecting this level of detail was proposed by Tang and Othmer (1994, 1995) for the pathway involving adenylyl cyclase; their later paper simplifies the system to five differential equations, of the form

$$\frac{dw_1}{d\tau} = \alpha_4 u_2 - w_1 - \alpha_4 u_2 w_1 \tag{2a}$$

$$\frac{dw_2}{d\tau} = \beta_2 \beta_3 c_2 u_4 - \beta_5 w_2 + \beta_6 c_3 w_3 - c_3 \beta_4 u_1 w_2 - \beta_2 \beta_3 c_2 u_4 (w_2 + c_3 w_3) \tag{2b}$$

$$\frac{dw_3}{d\tau} = -(\beta_5 + \beta_6) w_3 + \beta_4 u_1 w_2 \tag{2c}$$

$$\frac{dC_i}{d\tau} = \gamma_1 \gamma_2 w_1 + \gamma_5 (1 - w_1) - \gamma_4 \frac{C_i}{C_i + \gamma_3} - sr(C_i) \tag{2d}$$

$$\frac{\partial C_o}{\partial \tau} = \Delta_1 \nabla^2 C_o - \hat{\gamma}_9 \frac{C_o}{C_o + \gamma_8} + \frac{\rho}{1 - \rho} \left(sr(C_i) - \gamma_7 \frac{C_o}{C_o + \gamma_6} \right) \tag{2e}$$

where

$$u_1 = \frac{\alpha_0 C_o + (\beta_5 - \alpha_0 C_o) w_3}{\alpha_1 + \alpha_0 C_o + \beta_4 w_2} \qquad u_2 = \frac{\alpha_2 \alpha_3 c_1 u_1 (1 - w_1)}{1 + \alpha_4 + \alpha_2 \alpha_3 c_1 u_1 - \alpha_4 w_1} \qquad u_4 = \frac{\beta_0 C_o}{\beta_1 + \beta_0 C_o}$$

In these equations C_i represents internal cAMP, C_o external cAMP, u_1 (u_4) the fraction of stimulatory (inhibitory) receptors bound with cAMP, u_2 the fraction of the activated subunit of the stimulatory G protein, w_1 the activated adenylyl cyclase complex, w_2 a subunit of the hypothesized inhibitory G protein and w_3 represents a complex of w_2 and u_1. Further details of the signal transduction mechanism are reviewed by H. G. Othmer and P. Schaap (unpublished data).

A DISCRETE MODEL FOR *D. DISCOIDEUM* AGGREGATION

The phenomenon of cell streaming results from the interaction of the spatiotemporal dynamics of cAMP with the movement of *D. discoideum* cells up gradients of cAMP. Mathematical models for streaming differ in two basic ways. First, they use representations of cAMP kinetics with various levels of detail, ranging from caricatures to fairly accurate accounts of the current level of knowledge. Secondly, models differ in the way in which the cell populations are represented, varying between continuum models which average over the cells, and discrete models in which individual cells are represented as discrete objects. Mathematically this latter difference is the most fundamental, giving completely different types of equation system. Two detailed models at opposite ends of both of these spectra will be discussed. In this section the model of Dallon and Othmer (1997) is described, in which a realistic representation of cAMP kinetics is used (taken from Tang & Othmer, 1994), and in which the *D. discoideum* amoebae are represented as discrete objects. In the following section the model of Höfer *et al.* (1995*a*,*b*) is described, which uses a caricature of cAMP dynamics and represents the amoebae as a continuous population. Various other models are mentioned in the Discussion.

When representing each amoeba individually, the key mathematical assumptions are those made on cell movement. In the work of Dallon and Othmer (1997), the following movement rules were used:

(i) The cell moves if the time derivative of the extracellular cAMP concentration is greater than $0.02\,\mu M\,min^{-1}$. This ensures that a triangular wave of cAMP of duration 200 s above baseline and amplitude $0.1\,\mu M$ initiates movement.

(ii) All cells move for a fixed duration (100 s for wild-type cells) in the direction of the cAMP gradient at the cell when the motion started. This level of persistence is based on experimental observations of the time for which cells move during aggregation (Alcantara & Monk, 1974; Tomchick & Devreotes, 1981). However, it is somewhat arbitrary, and is an important parameter to vary in model simulations.

(iii) The cells move at a fixed speed of $30\,\mu m\,min^{-1}$, which is the maximum cell speed measured in experiments of Alcantara and Monk (1974). In reality, cells will move more slowly initially, speeding up as a result of successive stimulation by cAMP or when they form streams (Varnum *et al.*, 1985). However, such variations in speed would represent a significant increase in mathematical complexity.

Using these rules, Dallon and Othmer (1997) followed the movement of a fixed number of cells, varying between 120 000 and 160 000, in a 1 cm² area. In parallel with this cell tracking, the Tang–Othmer equations (2) were used to determine the concentration of cAMP and related variables. This combined system was solved numerically using techniques similar to the particle-in-cell method developed for combustion problems (details in Dallon & Othmer, 1997). Figure 2(*a*) shows a model simulation for parameter values corresponding to wild-type *D. discoideum*. The model

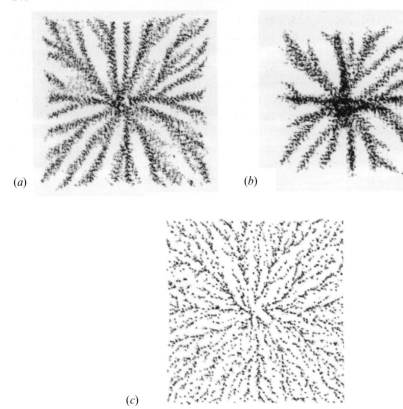

Fig. 2. Aggregation patterns for simulated wild-type (*a*) and mutant (*b*), (*c*) cells. (*a*) The predicted pattern for wild-type cells, with a movement duration of 100 s. (*b*) The pattern for *streamer F* mutants, with a movement duration of 500 s. (*c*) The pattern for *jittery* mutants, with movement duration 20 s. The solutions are shown after a model run of 95 min duration; cells in the centre are oscillatory, with parameter γ_2 ranging between 0.4 and 0.17.

predicts spiral waves of cAMP (not shown), with cell aggregation induced by the cAMP wave, and the formation of cell streams. Both this overall behaviour and the predicted space and time scales agree very closely with experimental observations.

The discrete nature of the model enables it to be used to predict the implications of mutations. Figure 2(*b*) shows a simulation corresponding to the *streamer F* mutant studied by Ross and Newell (1981); here, the time over which cells move between direction changes has been increased from 100 s (wild-type) to 500 s (*streamer F*). This change causes the cell streams to be fewer in number, but larger and more compact, again in close agreement with experimental observations. Figure 2(*c*) shows the results of reducing the duration of movement to 20 s; this is an artificial mutation, which Dallon and Othmer named *jittery*. In this case, thin and highly fragmented streams form so that aggregation to the pacemaker does not occurs. These differences can be explained by reference to the time scale of a cAMP wave pulse, which takes about two minutes for both the wave front and wave back to move past a fixed point in space. Therefore, *jittery* cells reorient several times during the passage of a single cAMP pulse. Since these global waves of cAMP are the sum of small cAMP bursts from each cell, as described above, they can have very rough profiles, with several local maxima and minima. Reorienting several times during a wave front will cause the cell to become caught at a local maximum. In contrast, in the wild type cells, and to an even greater extent for the *streamer F* mutant, cell direction is primarily set in the wave-front and not reset until the cAMP pulse has passed.

A CONTINUOUS MATHEMATICAL MODEL

The representation of the *D. discoideum* amoebae as discrete objects in the model of Dallon and Othmer (1997) discussed above is relatively unusual within mathematical models for cell biology. Other examples are few: for example, work of Weliky *et al.* (1991) on *Xenopus* gastrulation, and the study of juxtacrine signalling by Collier *et al.* (1996); there are also some other examples of application to *D. discoideum* signalling that will be discussed later. These various discrete models have all been proposed by individual investigators, on a somewhat *ad hoc* basis. In contrast, an established body of theory exists for models of a 'continuum' type. In these, cells are not represented as discrete objects, but rather via a 'cell density', which denotes the number of cells per unit area at a point in the domain. Models of this type have a long history of application to developmental biology and medicine; see Murray (1989) for review. Höfer *et al.* (1995*a*) studied streaming using such a continuum representation of *D. discoideum* amoebae, coupled to the simple Martiel–Goldbeter representation of cAMP kinetics. In this section the model of Höfer *et al.* (1995*a*) is described and the results compared with those of Dallon and Othmer (1997).

Within a continuum model, cellular dynamics are represented by a 'conservation equation', in which the various contributions to overall cell movement are represented by separate terms. In the case of *D. discoideum* amoebae, there is a background level of random migration, in addition to directed movement up gradients of cAMP. Mathematically, this gives the equation:

$$\partial n/\partial t = \underbrace{\nabla \cdot [\mu(n)\nabla n]}_{\text{random migration}} - \underbrace{\nabla \cdot [\chi(v)n\nabla u]}_{\text{chemotaxis}} \qquad (3)$$

where $n(\underline{x}, t)$ denotes the cell density; recall that $u(\underline{x}, t)$ and $v(\underline{x}, t)$ represent the concentrations of extracellular cAMP and active cAMP- receptor, respectively. It would be straightforward to include terms representing cell division and death in this equation, but these are omitted because the extent of these processes is essentially negligible during aggregation. Höfer *et al.* (1994) studied equation 3 coupled to the Martiel–Goldbeter equations 1 for cAMP kinetics. The most immediate outcome of these simulations was an explanation for the so-called 'chemotactic wave paradox' (Soll *et al.*, 1993). As a pulse of cAMP passes an amoeba, it is observed to move in the wave front, but not in the wave back, giving significant cell movement in the opposite direction to that of the pulse. This is intuitively suprising, since there are equal (but opposite) cAMP gradients in both the wavefront and waveback; because the gradient in the waveback would promote movement with the cAMP pulse, the cell should spend longer in the waveback than in the wavefront, causing a small net movement in the same direction as the pulse. This contradiction was resolved by Höfer *et al.* (1994), using simulations of a slightly simplified version of equations (1, 3). Briefly, because the time scales of cell movement and desensitization to cAMP are similar for *D. discoideum*, the cell becomes sensitized while they are moving in the wavefront of the cAMP pulse, and by the time the waveback reaches the cell, it is no longer sensitive to cAMP gradients, so that little further movement occurs (illustrated in Fig. 3).

Having confirmed that the model predicts the correct timecourse of cell movement in cAMP gradients, Höfer *et al.* (1995*a*) used the model (1,3) to simulate aggregation on an agar plate; a typical example is illustrated in Fig. 4. Initially, a rotating spiral wave pattern of cAMP develops from a disrupted wave front, inducing cell movement towards the wave core. The interaction of the cAMP waves and cell chemotaxis then causes initial inhomogeneities in cell density to grow, leading to the formation of cell streams. This in turn disrupts the cAMP wave, which reinforces the streaming pattern. Thus this model predicts that the observed streaming pattern is the result of an instability along the length of an advancing front of cells, with cells gradually sorting into clumps, via movement up small cAMP gradients, as successive arms of the cAMP spiral wave move past them.

There are two very notable points of difference between the Höfer *et al.* (1995*a*) model, and that of Dallon & Othmer presented in the previous section. The first is that Höfer *et al.* (1995*a*) are unable to predict initiation of the spiral wave of cAMP. Rather, the spiral is induced rather artificially in simulations such as that illustrated in Fig. 4. This is common practice: studies such as those of Tyson *et al.* (1989) showed that cAMP dynamics are able to support spiral waves, but do not explain their initiation. In contrast, Dallon and Othmer's (1997) model does predict spiral wave generation, caused by small asymmetries in cell locations around pacemaker cells (see above). Subsequently, spiral wave generation has been demonstrated in continuum models, based on desynchronization of cells on the developmental path (Lauzeral *et al.*, 1997).

The second key difference between the models concerns mathematical tractability. The differential equations in terms of which the model of Höfer

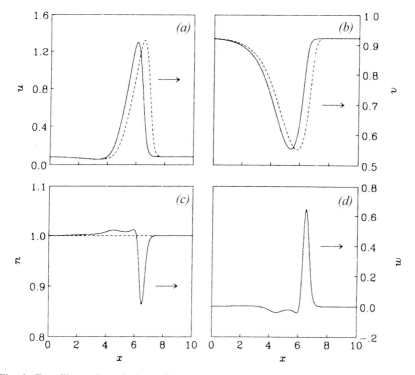

Fig. 3. Travelling pulse solutions of equations (1,3) in one space dimension (solid line), compared to wave solution of (1) with cell density n fixed at the value 1 (dashed line). (*a*) cAMP concentration, (*b*) fraction of active receptors, (*c*) cell density, (*d*) cell velocity. Parameter values and details of numerical solution are given in Höfer *et al.* (1995*b*). The numerically determined wave speed in the full model is about 4% lower than for the clamped cell density; the u and v profiles are approximately the same in both cases.

et al. (1995*a*) was formulated is amenable to many standard techniques of mathematical analysis that cannot be applied to the discrete formulation of Dallon and Othmer. This analysis was presented by Höfer *et al.* (1995*b*), and enables the separation of the cell streams during *D. discoideum* aggregation to be predicted in terms of model parameters. Most significantly, this predicts that stream separation increases with the chemotactic parameter χ, with streams not appearing if χ is less than a critical value. Höfer *et al.* (1995*b*) derived a simple formula for the dependence of this initial value on other parameters, suggesting a range of possible experimental tests.

DISCUSSION

The formation of cell streams is arguably the most visible outcome of *D. discoideum* aggregation. Both the Höfer *et al.* and Dallon–Othmer

(a)

(b)

Fig. 4. Spatiotemporal evolution of (*a*) cell density and (*b*) cAMP concentration in a numerical simulation of (3,1). Solutions are plotted every 12 minutes. The initial conditions were chosen to be a plane wavefront with a free end at the centre of the domain and homogeneous cell density, with random perturbations (±7.5%) throughout. Boundary conditions are zero flux. Parameter values and details of numerical solution are given in Höfer *et al.* (1995*b*).

mathematical models predict that this behaviour is the combined results of cAMP signalling and chemotaxis. Intuitively, the streams arise because cells both produce cAMP and move up cAMP gradients; therefore, a region of high cell density will produce cAMP at high levels, thus inducing surrounding cells to move towards the region of higher density. This is an autocatalytic mechanism that leads to stream formation. The models differ in their prediction of the extent of initial aggregation required to produce this pattern. In the Höfer *et al.* model, very slight degrees of streaming are rapidly reinforced, while Dallon and Othmer predict that a more pronounced initial pattern is required. This is an experimentally testable difference, since in the latter case, a fairly uniform initial distribution of cells would aggregate without stream formation.

A number of other mathematical models have been proposed for the aggregation phase of *D. discoideum* modelling. The continuum models of Vasiev *et al.* (1994) and van Oss *et al.* (1996) are particularly relevant to our considerations, since they propose a rather different mechanism for stream formation, in which the dependence of the speed of cAMP waves on cell density is the key phenomenon. This would imply that the directed movement of cells is less significant than their effect on cAMP wave speed. Recently, Höfer and Maini (1997) have attempted to investigate this alternative via a mathematically simpler 'caricature' model, concluding that while this density-dependent speed may contribute to the streaming phenomenon, it is not consistent as an underlying explanation. A very different mathematical model has been proposed by Savill and Hogeweg (1997), focussing on the role of direct cell–cell adhesion in *D. discoideum* morphogenesis, a phenomenon neglected in the models we have discussed. They show that this provides a quite distinct potential explanation for the formation of cell streams, although precise predictions are difficult because of an absence of appropriate data on which to base parameter values. Were such data available, their model could potentially be combined with that of Dallon and Othmer (1997), since it is also based on a discrete representation of the amoebae.

Aggregation in *D. discoideum* is an elegant example of complex behaviour coordinated by microbial signalling. At the heart of the process are the 'excitability' of extracellular cAMP kinetics and the chemotactic response of *D. discoideum* amoebae to cAMP gradients. Many experimental observations are direct consequences of these two processes. An instructive example of this is the affect of reducing the extent to which cAMP activates its own secretion, which decreases the degree of cAMP excitability. Experimentally, this can be achieved by the addition of caffeine, and causes cellular aggregation to occur around a central hole, rather than as a solid mound (Siegert & Weijer, 1989). The majority of mathematical models simulate this phenomenon (Höfer *et al.*, 1995*a*; van Oss, 1996; Dallon & Othmer, 1997), confirming that it is a simple consequence of their basic common ingredients, namely excitability and chemotaxis. Similarly, the positive correlation between wave speed and spatial wavelength, which is observed experimentally (Gross *et al.*, 1974), is a prediction shared by most models, including those of Dallon and Othmer and Höfer *et al.* In contrast, predicted explanations for cell streaming show more variation, suggesting that it represents a more delicate balance of the interaction between biochemical signalling and cell mechanics. Mathematical models provide an excellent vehicle for investigation of such interactions, with a strong track record within *D. discoideum* biology. The combination of this, and the high volume of experimental work on *D. discoideum*, suggests that the next few years will yield exciting developments in understanding *D. discoideum* morphogenesis.

ACKNOWLEDGEMENTS

JCD thanks Hans Othmer for his help and support. JAS, JCD and PKM were supported in part by grant GR/K71394 from EPSRC, and by a grant from the London Mathematical Society (scheme 3). TH acknowledges support from the Boehringer Ingelheim Fonds.

REFERENCES

Alcantara, F. & Monk, M. (1974). Signal propagation during aggregation in the slime mold *Dictyostelium discoideum*. *Journal of General Microbiology*, **85**, 321–34.

Bonner, J. T. (1982). Comparative biology of cellular slime molds. In *The Development of* Dictyostelium discoideum, ed. W. F. Loomis, pp. 1–33. Academic Press.

Collier, J. R., Monk, N. A. M., Maini, P. K. & Lewis, J. H. (1996). Pattern formation by lateral inhibition with feedback: a mathematical model of delta–notch intercellular signalling. *Journal of Theoretical Biology*, **183**, 429–46.

Dallon, J. C. & Othmer, H. G. (1997). A discrete cell model with adaptive signalling for aggregation of *Dictyostelium discoideum*. *Philosophical Transactions of the Royal Society of London, Series B*, **352**(1357), 391–417.

Devreotes, P. N. (1982). Chemotaxis. In *The Development of* Dictyostelium discoideum, ed. W. F. Loomis, pp. 117–68. Academic Press.

Goldbeter, A. & Segel, L. A. (1977). Unified mechanism for relay and oscillations of cyclic AMP in *Dictyostelium discoideum*. *Differentiation*, **17**, 127–35.

Grindrod, P. (1991). *Patterns and Waves*. Oxford University Press.

Gross, J. D., Peacey, M. J. & Trevan, D. J. (1974). Signal emission and relay propagation during early aggregation in *Dictyostelium discoideum*. *Journal of Cell Science*, **22**, 645–56.

Hagan, P. S. & Cohen, M. S. (1981). Diffusion induced morphogenesis in the development of *Dictyostelium*. *Journal of Theoretical Biology*, **93**, 881–908.

Höfer, T. & Maini, P. K. (1997). Streaming instability of slime mold amoebae: an analytical model. *Physical Review E*, **56**, 2074–80.

Höfer, T., Maini, P. K., Sherratt, J. A., Chaplain, M. A. J., Chauvet, P., Metevier, D., Montes, P. C. & Murray, J. D. (1994). A resolution of the chemotactic wave paradox. *Applied Mathematics Letters*, **7**, 1–5.

Höfer, T., Sherratt, J. A. & Maini, P. K. (1995*a*). *Dictyostelium discoideum*: cellular self-organization in an excitable biological medium. *Proceedings of the Royal Society of London, Series B*, **259**, 249–57.

Höfer, T., Sherratt, J. A. & Maini, P. K. (1995*b*). Cellular pattern formation during *Dictyostelium* aggregation. *Physica D*, **85**, 425–44.

Levine, H., Tsimring, L. & Kessler, D. (1997). Computational modeling of mound development in *Dictyostelium*. *Physica D*, **106**, 375–88.

Martiel, J. L. & Goldbeter, A. (1984). Oscillations and relay of cAMP signals in *Dictyostelium discoideum*: Analysis of a model based on the modification of the cAMP receptors. *Comptes Rendu des Seances de Academie des Sciences de Paris*, **298, Series III**, 549–52.

Meinhardt, M. (1983). A model for the prestalk/prespore patterning in the slug of the slime mold *Dictyostelium discoideum*. *Differentiation*, **24**, 191–202.

Murray, J. D. (1989). *Mathematical Biology*. Springer-Verlag.

Odell, G. M. & Bonner, J. T. (1986). How the *Dictyostelium discoideum* grex crawls. *Philosophical Transactions of the Royal Society of London, Series B*, **312**, 487–525.

Raman, R. K., Hashimoto, Y., Cohen, M. H. & Robertson, A. (1976). Differentiation for aggregation in the cellular slime molds: the emergence of autonomously signalling cells in *Dictyostelium discoideum*. *Journal of Cell Science*, **21**, 243–59.

Ross, F. M. & Newell, P. C. (1981). Streamers: chemotactic mutants of *Dictyostelium discoideum* with altered cyclic GMP metabolism. *Journal of General Microbiology*, **127**, 339–50.

Savill, N. J. & Hogeweg, P. (1997). Modelling morphogenesis: from single cells to crawling slugs. *Journal of Theoretical Biology*, **184**, 229–35.

Siegert, F. & Weijer, C. J. (1989). Digital image processing of optical density wave propagation in *Dictyostelium discoideum* and analysis of the effects of caffeine and ammonia. *Journal of Cell Science*, **93**, 325–35.

Soll, D. R., Wessels, D. & Sylwester, A. (1993). The motile behavior of amoebae in the aggregation wave in *Dictyostelium discoideum*. In *Experimental and Theoretical Advances in Biological Pattern Formation*, ed. H. G. Othmer, P. K. Maini & J. D. Murray. London: Plenum.

Tang, Y. H. & Othmer, H. G. (1994). A G-protein-based model of adaptation in *Dictyostelium discoideum*. *Mathematical Biosciences*, **120**, 25–76.

Tang, Y. H. & Othmer, H. G. (1995). Excitation, oscillations and wave propagation in a G-protein based model of signal transduction in *Dictyostelium discoideum*. *Philosophical Transactions of the Royal Society of London, Series B*, **349**, 179–95.

Tomchick, K. & Devreotes, P. N. (1981). Adenosine 3′,5′-monophosphate waves in *Dictyostelium discoideum*: a demonstration by isotope dilution fluorography. *Science*, **212**, 443–6.

Tyson, J. J., Alexander, K. A., Manoranjan, V. S. & Murray, J. D. (1989). Spiral waves of cyclic AMP in a model of slime mold aggregation. *Physica D*, **32**, 327–61.

van Oss, C., Panfilov, A. V., Hogeweg, P., Siegert, F. & Weijer, C. J. (1996). Spatial pattern formation during aggregation of the slime mould *Dictyostelium discoideum*. *Journal of Theoretical Biology*, **181**, 203–13.

Varnum, B., Edwards, K. B. & Soll, D. R. (1985). *Dictyostelium* amoebae alter motility differently in response to increasing versus decreasing temporal gradients of cAMP. *Journal of Cell Biology*, **101**, 1–5.

Vasiev, B., Siegert, F. & Weijer, C. J. (1997). A hydrodynamic model for *Dictyostelium discoideum* mound formation. *Journal of Theoretical Biology*, **184**, 441.

Vasiev, B. N., Hogeweg, P. & Panfilov, A. V. (1994). Simulation of *Dictyostelium discoideum* aggregation via reaction-diffusion model. *Physical Review Letters,* **73(23)** (December), 3173–6.

Weliky, M., Minsuk, S., Keller, R. & Oster, G. (1991). Notochord development in *Xenopus laevis*: simulation of cell behaviour underlying tissue convergence and extension. *Development,* **113,** 1231–44.

PHEROMONE COMMUNICATION IN THE FISSION YEAST *SCHIZOSACCHAROMYCES POMBE*

JOHN DAVEY

Department of Biological Sciences, University of Warwick, Coventry CV4 7AL, UK

INTRODUCTION

The conjugation of two haploid yeast cells is controlled by the reciprocal exchange of diffusible mating pheromones. Cells of each mating type release peptides, which induce changes in the target cell, leading to conjugation, meiosis and sporulation. Many of the events associated with the pheromone pathway are similar to those found in higher eukaryotes, and the experimental tractability of the yeast has made them an attractive system for studying the production and action of peptide hormones. Here current understanding of the communication process in the fission yeast *Schizosaccharomyces pombe* is reviewed. The review begins with the production of the pheromones, describes how they bring about changes at the target cell, and ends with a discussion of how the cell recovers from the effects of stimulation and returns to the resting state.

MATING TYPE

Mating type in *Sz. pombe* is determined by which of two DNA segments is carried at the *mat1* locus (Kelly *et al.*, 1988). Cells with the *mat1-P* segment, which encodes the *mat1-Pc* and *mat1-Pm* genes, are Plus (P), while those with the *mat1-M* segment, which encodes *mat1-Mc* and *mat1-Mm*, are Minus (M) (*mat1-Pm* and *mat1-Mm* are sometimes referred to as *mat1-Pi* and *mat1-Mi*). The two early subfunctions, *mat1-Pc* and *mat1-Mc*, are responsible for establishing the pheromone communication system, but all four genes are required for meiosis (Willer *et al.*, 1995). There are two further mating loci, *mat2* and *mat3*, where the *P* and *M* information is stored but not expressed. In wild-type homothallic strains, this latent information is frequently transferred to the active *mat1* locus, and cells switch mating type approximately once every three generations. Cultures of such strains are therefore a mixture of both mating types and this can complicate mating-related studies. Heterothallic strains with a stable mating type can be constructed, and these are often more suitable for analysis of pheromone communication.

THE MATING PHEROMONES

The first evidence for mating pheromones in *Sz. pombe* came from experiments where medium from one set of cells induced mating-related changes in cells of the opposite mating type (Friedmann & Egel, 1978). Later studies exploited cell challenge assays in which cells of opposite mating type were placed close to each other on an agar base and the mating-specific elongation of the isolated cells was compelling evidence for diffusible pheromones (Fukui *et al.*, 1986; Leupold, 1987). Both pheromones have since been purified and characterized. M-factor (the pheromone released by M-cells) is a peptide of nine residues in which the C-terminal cysteine residue is carboxymethylated and S-farnesylated (Davey, 1992). P-factor is an unmodified peptide of 23 residues (Imai & Yamamoto, 1994).

M-factor

M-factor was purified on the basis of its hydrophobicity (Davey, 1991) and contains a nine amino acid backbone where the C-terminal cysteine is both carboxymethylated and farnesylated on the thiol group (Davey, 1992). Such post-translational modifications are found in several fungal mating pheromones (for review, see Caldwell *et al.*, 1995; Davey, 1996), including **a**-factor from the budding yeast *Saccharomyces cerevisiae* (Anderegg *et al.*, 1988), rhodotorucine A from *Rhodosporidium toruloides* (Kamiya *et al.*, 1979), tremerogens A-10 and a-13 from *Tremella mesenterica* (Sakagami *et al.*, 1979), tremerogen A-9291-I from *Tremella brasiliensis* (Ishibashi *et al.*, 1984), the a1 and a2 pheromones from *Ustilago maydis* (Spellig *et al.*, 1994), and the MATα pheromone from *Cryptococcus neoformans* (Moore & Edman, 1993). As in all cases where these modifications have been investigated, both the methylation and farnesylation of M-factor is essential for full biological activity (Davey, 1992; Wang *et al.*, 1994).

M-factor is encoded by three genes (*mfm1*, *mfm2*, and *mfm3*), which contain the same mature pheromone sequence within slightly different precursors (Davey, 1992; Kjaerulff *et al.*, 1994). All three genes are normally used by wild-type cells, although each makes sufficient M-factor to support efficient mating (Kjaerulff *et al.*, 1994). Such redundancy may simply ensure the production of sufficient pheromone at the appropriate time, but it might also protect against mutation to sterility and could facilitate diversification of the pheromone structure. Transcription of the *mfm* genes is regulated at several levels (Kjaerulff *et al.*, 1994). The first level of control is mating type-dependent and the genes are only expressed in M-cells. There is also nutritional control such that expression is increased in nitrogen-limiting medium, conditions which initiate sexual development in *Sz. pombe*. Expression is further increased by exposing M-cells to P-factor and this may boost M-factor production in response to feedback from a stimulated P-cell.

The initial polypeptides encoded by *mfm1*, *mfm2*, and *mfm3* contain 39, 41, and 38 amino acids, respectively, and each contains a single copy of the mature M-factor sequence. The author's current model for the processing of these precursors is outlined in Fig. 1. Each precursor has an N-terminal extension but it is not hydrophobic and does not appear to act as a signal sequence for targeting the precursor to the conventional secretory pathway. Rather, all processing events seem to occur in the cytosol. The three precursors end with the sequence –Cys–Val–Ile–Ala– COOH and this CAAX motif (where C is Cys, A is an aliphatic residue, and X is any residue) serves as the signal for prenylation and carboxy-methylation (for review, see Schafer & Rine, 1992). The most likely sequence of events at the C-terminus is farnesylation via a thioether linkage, proteolysis of the C-terminal tripeptide and methyl esterification of the exposed carboxy group. Proteolytic processing of the N-terminal extension, which may occur by a two-step reaction (Davey, 1992), would generate the mature pheromone. M-factor is transported across the plasma membrane by an ATP-dependent peptide transporter that belongs to the ABC superfamily of proteins. Enzymes responsible for many of these modifications have been identified in several organisms, and progress has been made towards identifying those that process M-factor. The methyl-transferase (Mam4; Imai *et al.*, 1997) and the ABC transporter (Mam1; Christensen *et al.*, 1997) have been identified, and there are preliminary reports of enzymes that may turn out to be the AAX protease (Boy-artchuk *et al.*, 1997; Fujimura-Kamada *et al.*, 1997) and one of the N-terminal proteases (Hughes & Davey, 1997).

P-factor

P-factor is an unmodified peptide that contains 23 amino acids. It is encoded by the *map2* gene, and is initially synthesized as a precursor that contains multiple copies of the mature pheromone. These are liberated in a series of proteolytic reactions as it is transported through the secretory pathway (Fig. 2). The precursor contains four non-identical copies of the mature P-factor (Imai & Yamamoto, 1994), although the relatively conservative differences are unlikely to significantly affect activity. An N-terminal signal sequence directs the precursor to the endoplasmic reticulum and is removed during translocation across the membrane. Further processing appears to occur in the Golgi complex, where the prepheromone is split into its individual subunits by an endopeptidase that cleaves on the C-terminal side of a pair of basic residues (Lys–Arg in each case). N-terminal and C-terminal trimming of these subunits produces the mature pheromone for release into the medium.

The dibasic endopeptidase responsible for cleaving the P-factor precursor is encoded by *krp1* and is a member of the kexin family of processing enzymes

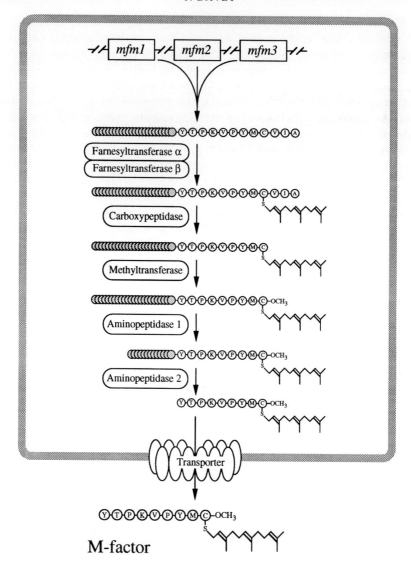

Fig. 1. Biogenesis of M-factor. Schematic illustration of the events leading to the production of M-factor. The lower part of the figure shows the sequences of the three precursor molecules. Spaces introduced to maximize the similarity are indicated by (.) and the residues in the mature pheromone are boxed.

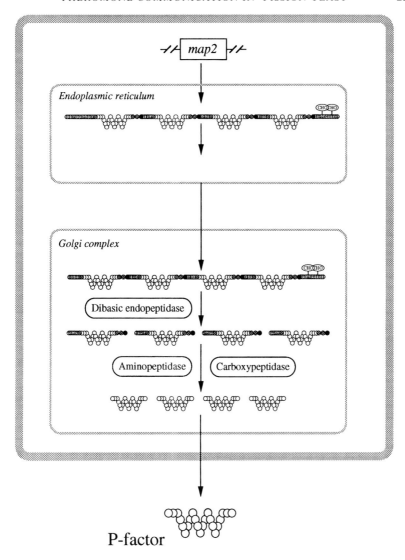

P-factor

```
  1                   MKITAVIALLFSLAAASPIPVAD
 24    PGVVSVSK SYADFLRVYQSWNTFANPDRPNL KKR
 58    EFEAAPAK TYADFLRAYQSWNTFVNPDRPNL KKR
 92    EFEAAPEK SYADFLRAYHSWNTFVNPDRPNL KKR
126    EFEAAPAK TYADFLRAYQSWNTFVNPDRPNL KKR
160    TEEDEENEEEDEEYYRFLQFYIMTVPENSTITDVNITAKFES
```

Fig. 2. Biogenesis of P-factor. Schematic illustration of the events leading to the production of P-factor. The heavily shaded circles represent the Lys–Arg motifs recognized by the dibasic endopeptidase (Krp1) in the Golgi complex. The lower part of the figure shows the sequence of the Map2 product. The sequence has been arranged to illustrate the similarity between each of the four pheromone repeats (boxed) and to highlight the location of the Lys–Arg residues (KR).

(Davey *et al.*, 1994). Kexins have been identified in many species and process a variety of proproteins during their transport through the secretory pathway (for review, see Steiner *et al.*, 1992; Seidah & Chrétien, 1997). Krp1 is a type I membrane protein that is, itself, made as an inactive precursor, and recent studies have helped define the activation process (Powner & Davey, 1998). It is initially synthesized as a preproprotein with an N-terminal signal sequence that is removed during segregation into the endoplasmic reticulum. The prosequence is thought to play a role in the correct folding of the catalytic domain, and is also removed in the endoplasmic reticulum in a reaction that is autocatalytic and probably intramolecular. Removal of the prosequence is necessary for Krp1 to become active, but it is not sufficient for activation and additional steps are required before it is able to cleave substrates presented in *trans*. It appears that the initially cleaved prosequence remains non-covalently associated with the catalytic domain of Krp1 and acts as an auto-inhibitor of the enzyme. Only when the prosequence is subsequently inactivated by autoproteolytic cleavage at an internal Lys–Arg motif is the enzyme able to become fully active. The events that trigger this final activation of the enzyme are still to be determined but, by analogy to other kexins, could be associated with the transport of the enzyme from the endoplasmic reticulum to the Golgi complex (Anderson *et al.*, 1997).

THE SIGNALLING MACHINERY

Binding of the pheromones to receptors on the surface of the target cell activates signalling machinery that consists of a heterotrimeric G protein, a kinase cascade, and a transcription factor. Stimulation induces expression of the genes required to bring about the mating-related changes in cell behaviour. These changes include an arrest of the cell cycle, shmoo formation, enhanced cell agglutination, and cell fusion. The resulting zygote then undergoes meiosis and sporulation with germination of the haploid spores completing the mating cycle (Fig. 3).

Receptors

M-cells respond to P-factor, since they express the P-factor receptor (Mam2; Kitamura & Shimoda, 1991), while P-cells possess the M-factor receptor (Map3; Tanaka *et al.*, 1993) and respond to M-factor. The receptors are the only cell-specific parts of the signalling machinery, and all other components are the same in both mating types. Consequently, ectopic expression of the 'wrong' receptor leads to an autocrine pheromone response (Tanaka *et al.*, 1993; Kitamura *et al.*, 1996). Both receptors contain seven putative trans-membrane domains and couple to a heterotrimeric guanine nucleotide-binding protein (G protein). Such G protein-coupled receptors (GPCRs) exist in all eukaryotes and detect a wide variety of extracellular signals.

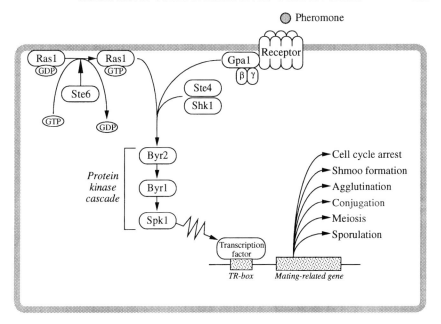

Fig. 3. Pheromone stimulation at the target cell. Pheromone stimulation activates a signalling pathway, which leads to expression of the genes required to control mating and sporulation.

Mammalian GPCRs, for example, are involved in processes that include olfaction, phototransduction, hormonal regulation, muscle contraction, and neurotransmission.

The pheromone receptors play a pivotal role in the communication process as they translate the extracellular signal into an intracellular response. By analogy to other GPCRs (Fahmy *et al.*, 1995; Büküsoglu & Jenness, 1996), binding of the pheromone is thought to induce a change in the conformation of the receptor that is then transmitted to the G protein associated with the cytoplasmic face of the membrane (Bourne, 1997).

G protein

Although precise details of the allosteric transitions remain to be determined, the changes induced by the binding of the pheromone are sufficient to bring about dissociation of the G protein. The GDP bound to the Gα subunit (Gpa1; Obara *et al.*, 1991) is released and, as a consequence of the relative concentrations of GTP and GDP in the cytosol, is replaced by GTP. The Gα-GTP dissociates from the G$\beta\gamma$ subunits which remain non-covalently coupled (the gene for the Gγ subunit has not yet been identified but the Gβ subunit is encoded by *gpb1*, Kim *et al.*, 1996). Gα-GTP then activates the kinase cascade (Obara *et al.*, 1991). This is similar to the mechanism used by the majority of receptor-coupled G proteins, but is notably different to the pheromone response in *S. cerevisiae* where the Gβγ subunits are responsible for propagating the intracellular response (Nakayama *et al.*, 1988; Blinder *et al.*, 1989; Whiteway *et al.*, 1989).

MAP kinase cascade

The mitogen-activated protein kinase (MAPK) module is a highly conserved eukaryotic signalling unit (for review, see Cano & Mahadevan, 1995; Herskowitz, 1995; Robinson & Cobb, 1997). It is composed of three protein kinases that act sequentially; a MAPK kinase kinase which phosphorylates and activates a MAPK kinase, which in turn phosphorylates and activates a MAPK. The activated MAPK then phosphorylates a variety of target proteins and it is the altered activities of these target proteins that are responsible for changing cell behaviour. This basic unit is found in signalling pathways that range from sexual differentiation in yeast to cell proliferation and differentiation in mammalian cells. The pheromone response pathway in *Sz. pombe* uses a module that comprises Byr2 (MAPKKK), Byr1 (MAPKK) and Spk1 (MAPK) (Gotoh *et al.*, 1993; Neiman *et al.*, 1993).

The key event in the stimulation of the kinase cascade is likely to be the activation of Byr2 and clues are beginning to be obtained as to how this might be achieved. Byr2 is thought to be maintained in an inactive conformation through an interaction between its C-terminal catalytic domain and an N-terminal regulatory domain (Tu *et al.*, 1997) and this is believed to be disrupted following pheromone stimulation. Genetic analyses implicate Gpa1, Ras1 (Fukui *et al.*, 1986*b*; Nielsen *et al.*, 1992), Shk1 (Ottilie *et al.*, 1995), and Ste4 (Barr *et al.*, 1996) in the activation process.

Ras1 is a small monomeric G protein that is structurally related to the mammalian ras oncoprotein. Two hybrid analysis (Van Aelst *et al.*, 1993) and *in vitro* studies (Masuda *et al.*, 1995) show that Ras1 binds to Byr2, and activated Ras1 is thought to recruit Byr2 to the plasma membrane, where it can then interact with other components of the signalling machinery. The yeast process thus appears to resemble certain mammalian signalling pathways where Raf (the functional equivalent of Byr2) is activated following its association with the membrane-bound ras protein (Leevers *et al.*, 1994). Activation of Ras1 is similar to that of the heterotrimeric G protein with Ras1-GDP being the inactive conformation and Ras1-GTP being the active form able to interact with target proteins such as Byr2. Switching between the two conformations is regulated by Ste6 (Hughes *et al.*, 1990) and Gap1 (also called Sar1) (Imai *et al.*, 1991; Wang *et al.*, 1991). Ste6 is a guanine–nucleotide exchange factor (GEF) which promotes the release of GDP and allows the formation of Ras1-GDP. In contrast, Gap1 is a GTPase-activating protein (GAP) that stimulates the intrinsic GTP hydrolysing activity of Ras1 and promotes its conversion to the inactive Ras1-GDP conformation.

Once Byr2 has been localized to the plasma membrane, it can be activated by interacting with other signalling components. Shk1, the *Sz. pombe* homologue of the STE20 kinase from *S. cerevisiae*, is believed to be instrumental in the next step of this activation process. It is thought to disrupt the interaction of the Byr2 regulatory and catalytic domains (possibly by direct phosphorylation of Byr2) and thereby encourage a more open, and thus more active, conformation of the MAPKKK (Tu *et al.*, 1997). Activated Byr2 may then be stabilized, or further activated, by the binding of its exposed regulatory domains to appropriate sites in Ste4 (Barr *et al.*, 1996). Ste4 also contains a leucine zipper motif that promotes the formation of homodimers which might facilitate Byr2 dimerization and allow auto-phosphorylation and further activation of the kinase. Byr2 could remain associated with Ras1 during this stage as Ras1 and Ste4 interact with Byr2 at partially independently sites (Barr *et al.*, 1996).

This is a reasonable and consistent model for the activation of Byr2, but there are several aspects that remain to be confirmed. One of the most significant omissions is the trigger that initiates the activation process. There may be more than one trigger as the mating response in *Sz. pombe* requires both nutritional and pheromonal signals. Nutritional control is most probably achieved through the regulated production of proteins required for the activation of Byr2 and, while most of the components are produced under all conditions, both *ste6* (Hughes *et al.*, 1994) and *ste4* (Okazaki *et al.*, 1991) are expressed only upon starvation. Either, or both, of these proteins could provide the nutritional trigger. Nutritional regulation is also applied indirectly through the regulated expression of the pheromones (Davey, 1992; Kjaerulff *et al.*, 1994; Imai & Yamamoto, 1994) and their receptors (Kitamura & Shimoda, 1991; Tanaka *et al.*, 1993). The pheromone signal is transmitted through Gpa1, but there is no evidence to suggest that this interacts directly with Byr2 and its influence is probably exerted through a mediator. In principle, this could be any of the components involved in the activation of Byr2, but Shk1 appears to be the most likely candidate as loss of Gpa1 can be suppressed by Shk1 mutants that disrupt the interaction between the regulatory and catalytic domains of Byr2 (Tu *et al.*, 1997). Similar roles have been proposed for Shk1 homologues in both *S. cerevisiae* (Leberer *et al.*, 1992; Akada *et al.*, 1996) and mammals (Knaus *et al.*, 1995; Zhang *et al.*, 1995).

Once activated, Byr2 phosphorylates Byr1 (Styrkársdóttir *et al.*, 1992; Neiman *et al.*, 1993) which then activates Spk1 (Gotoh *et al.*, 1993). Each activation step appears quite specific, and there is no significant crosstalk between the pheromone cascade and the kinase cascades involved in the osmotic and stress responses and in regulating cell wall integrity. Such specificity is probably a feature of the activity of the individual enzymes as there is no evidence for their incorporation into a multikinase complex as in *S. cerevisiae* (Elion, 1995).

Targets for Spk1

Spk1 is essential for propagating the intracellular signal (Gotoh *et al.*, 1993; Neiman *et al.*, 1993) and is presumably responsible for phosphorylating one or more target proteins. These targets have not been identified, but the transcriptional changes observed following pheromone stimulation suggest that Spk1 either directly or indirectly affects the activity of at least one transcription factor. Localization of Spk1 to the cell nucleus is consistent with this suggestion (Toda *et al.*, 1991). The regulation of transcription factors through phosphorylation is a common consequence of MAP kinase activation in mammalian cells (for review, see Karin & Hunter, 1995), and there is at least one example in *Sz. pombe*. Atf1 is a transcription factor that lies downstream of the Sty1 (also called Spc1) MAP kinase in the osmoregulation and stress response pathway in *Sz. pombe* (Degols *et al.*, 1996; Shiozaki & Russell, 1996; Wilkinson *et al.*, 1996). Exposing cells to the appropriate stimulus induces Sty1-dependent phosphorylation of Atf1 (Shiozaki & Russell, 1996; Wilkinson *et al.*, 1996).

The best candidate for a pheromone-responsive transcription factor is Ste11 (Sugimoto *et al.*, 1991), a DNA binding protein of the HMG-box family. A Ste11-binding site (the TR-box, TTTCTTTGTT) is found upstream of each of the pheromone-controlled genes so far characterized, and expression of many of these genes is reduced in a *ste11⁻* mutant (Sugimoto *et al.*, 1991). Furthermore, the TR-box has been directly implicated in pheromone-dependent transcription of *mat1-Pm* (Aono *et al.*, 1994), *fus1* (Petersen *et al.*, 1995), and *mfm1* (Kjaerulff *et al.*, 1997). Spk1-dependent phosphorylation of Ste11 has not been demonstrated but the transcription factor contains two sites (PTSP) and (PKTP), which could serve as acceptor phosphorylation sites for Spk1. It is also possible that Ste11 is not the direct target for Spk1, but is activated following Spk1-dependent phosphorylation of an associated inhibitor protein. During the pheromone response in *S. cerevisiae*, for example, the STE12 transcription factor is activated downstream of the FUS3 MAP kinase and, although activation of STE12 correlates with its phosphorylation by FUS3 (Song *et al.*, 1991; Elion *et al.*, 1993), the FUS3-dependent phosphorylation of the inhibitory RST1 and RST2 proteins (also called DIG1 and DIG2, Cook *et al.*, 1996) appears to play a more significant role in the activation of STE12 (Tedford *et al.*, 1997). A similar mechanism is thought to operate in the sevenless signalling pathway that controls Drosophila eye development (Rebay & Rubin, 1995).

PHEROMONE-DEPENDENT CHANGES

Changes in transcription

Investigation of pheromone-dependent transcription is complicated by the fact that Ste11 also mediates transcription in response to nitrogen starvation

(Sugimoto *et al.*, 1991; Li & McLeod, 1996) and certain genes are controlled by both nutritional and pheromonal signals. Expression can also be regulated by the mating type of the cell. Many of the genes necessary for sexual differentiation are required in both mating types, but there are several which are only required in one or other cell type. These factors influence various genes to different extents. At one extreme are genes such as *map1*. This is expressed in both mating types under all growth conditions, and expression is increased about threefold following nitrogen starvation but is not influenced by pheromone (Nielsen *et al.*, 1996; Yabana & Yamamoto, 1996). Many genes are not expressed during mitotic growth, but are induced by nitrogen starvation and further enhanced by pheromone stimulation. These include genes encoding the pheromones (Kjaerulff *et al.*, 1994; Imai & Yamamoto, 1994) and their receptors (Kitamura & Shimoda, 1991; Tanaka *et al.*, 1993), and such regulation may serve as a feedback mechanism to enhance pheromone communication between potential partners. There are also genes whose expression is almost completely dependent upon nitrogen starvation and pheromone stimulation. These include genes needed for cell fusion (*fus1*) (Petersen *et al.*, 1995), adaptation (*sxa2*) (Imai & Yamamoto, 1994; Ladds *et al.*, 1996) and entry into meiosis (*mat1-Pm* and *mat1-Mm*) (Nielsen *et al.*, 1992; Willer *et al.*, 1995). The simplest model is one in which the activity of Ste11 is regulated by nitrogen starvation, pheromone stimulation, and mating type. Each could either influence Ste11 directly or regulate its activity indirectly through its association with various co-factors.

There is clearly some overlap between the nutritional and pheromonal regulation of Ste11 since activation by nitrogen starvation requires an intact pheromone response pathway (Tanaka *et al.*, 1993; Aono *et al.*, 1994; Kjaerulff *et al.*, 1994; Xu *et al.*, 1994). The most likely explanation is that the signalling pathway operates at a low level even in the absence of pheromone and that this helps to mediate the nutritional response. The nutritional and pheromonal signals could activate Ste11 via the same mechanism with their different effects on transcription simply reflecting the different extents to which Ste11 is activated. It is possible, for example, that both nitrogen starvation and pheromone stimulation lead to Spk1-dependent phosphorylation of Ste11 (or an associated regulator) and that it is the extent of phosphorylation that determines which genes are expressed under the different conditions. Alternatively, the differences could be due to the different ways that the two signals activate the MAP kinase cascade (Marshall, 1995). In PC12 cells, for example, treatment with nerve growth factor leads to sustained activation of the MAP kinase and cell differentiation, while treatment with epidermal growth factor leads to transient activation of the same MAP kinase and fails to induce differentiation (York *et al.*, 1998).

Regulating the activity of Ste11 with mating type-specific factors would explain why some genes are expressed only in one of the two cell types. Such regulation is best characterized for M-cells where Mat1-Mc plays a central role in controlling the expression of M-specific genes (Kelly *et al.*, 1988; Kjaerulff *et al.*, 1994). It transpires that genes expressed in both mating types contain a complete TR-box (TTTCTTTGTT), whereas M-specific genes contain a truncated motif (TCTTTGTT) that only binds Ste11 efficiently in the presence of Mat1-Mc (Kjaerulff *et al.*, 1997). Mat1-Mc contains a HMG-box which binds to both the complete and truncated TR-boxes (Dooijes *et al.*, 1993; Kjaerulff *et al.*, 1997) but binds even more strongly to a so-called M-box (ACAATG) that is adjacent to the truncated TR-boxes in M-specific genes (Kjaerulff *et al.*, 1997). Binding of Mat1-Mc to the M-box probably induces localized bending of the DNA (Dooijes *et al.*, 1993) which promotes binding of Ste11 to the truncated TR-box (Bustin & Reeves, 1996). The M-box also appears to be important in discriminating between the M-specific genes that require a pheromone signal (such as *mfm3* and *sxa2*) and those that are highly expressed in the absence of pheromone (such as *mam1*, *mam2*, *mfm1*, and *mfm2*). Whereas pheromone-independent promoters contain complete M-boxes, the genes that require a pheromone signal have truncated M-boxes (AACAAT) that bind Mat1-Mc with a much lower affinity (Kjaerulff *et al.*, 1997). This weaker interaction is presumably unable to promote expression in the absence of a pheromone signal but, again, the molecular explanation for this regulation remains to be determined.

Regulated expression in P-cells is less well understood. The M-specific genes are not transcribed since there is no Mat1-Mc to promote the binding of Ste11 to the truncated TR-boxes, but it is not yet clear whether there is an analogous mechanism for P-specific genes. Expression of these genes requires both Mat1-Pc (Kelly *et al.*, 1988; Nielsen *et al.*, 1996) and Map1 (Nielsen *et al.*, 1996; Yabana & Yamamoto, 1996) and two hybrid analysis suggests that the two proteins act as a heterodimer (Yabana & Yamamoto, 1996). This would have DNA-binding capabilities via the MADS-box present in the Map1 protein (for review, see Treisman & Ammerer, 1992) but no target sequences have been defined. Map1 is also required for expression of M-specific genes (Yabana & Yamamoto, 1996).

Cell cycle arrest

Pheromone stimulation induces an arrest of cell division at the G1 stage of the cell cycle (Davey & Nielsen, 1994; Imai & Yamamoto, 1994). This presumably ensures that both cells are at the correct stage of the cycle at the time of conjugation. Demonstrating the arrest is complicated by the nutritional signals required to allow a pheromone response, for the traditional approach of starving cells of nitrogen is sufficient to cause a G1 arrest (Egel & Egel-Mitani, 1974; Costello *et al.*, 1986). These difficulties are overcome by using mutants that are derepressed for sexual activity during mitotic growth

(Davey & Nielsen, 1994; Imai & Yamamoto, 1994). Mutants defective in *cyr1*, the gene encoding adenylate cyclase (Yamawaki-Kataoka *et al.*, 1989; Young *et al.*, 1989), have no detectable cAMP and undergo sexual differentiation in rich medium (Maeda *et al.*, 1990; Kawamukai *et al.*, 1991; Sugimoto *et al.*, 1991). Alternatively, strains possessing the temperature sensitive *pat1–114* allele will respond to pheromone during mitotic growth at 23°C (Davey & Nielsen, 1994). Although the arrest is most clearly demonstrated in such derepressed mutants, it does occur in wild-type cells (Egel & Egel-Mitani, 1974; Nielsen & Davey, 1995).

The G1 to S transition in eukaryotic cells requires stimulation of the appropriate cyclin-dependent kinase (CDK) and activation of the transcription factor that controls expression of the genes required for S phase. Inhibiting either of these events will induce a G1 arrest of the cycle. The pheromone-induced arrest in *Sz. pombe* is mediated through inhibition of the G1 CDK (Cdc2) (Stern & Nurse, 1997), and there is no apparent change in the activity of the G1-specific transcription factor (Cdc10-Res1) (Stern & Nurse, 1997). Inhibition centres on the interaction of Cdc2 with the B-type cyclins. Cdc2 is normally activated by association with Cig2 (Martin *et al.*, 1996; Mondesert *et al.*, 1996) (although Cdc13 and Cig1 can also activate the kinase; Fisher & Nurse, 1996) and this association is inhibited by pheromone stimulation. Details are still to be resolved, but the inhibition process does not affect the production of the cyclins and can be overcome by overexpressing either Cig2 or a stable form of Cdc13 (Stern & Nurse, 1997). Degradation or sequestration of the cyclins are attractive possibilities, although a more indirect effect mediated through ancillary proteins cannot be excluded. Rum1, for example, is a potent inhibitor of the G1 Cdc2 (Moreno & Nurse, 1994; Correa-Bordes & Nurse, 1995) and its activity could be regulated by pheromone stimulation.

Mating pheromone also affects cells in G2, where it reduces growth and advances mitosis (Stern & Nurse, 1997). These effects do not appear to be mediated through the mitotic CDK (Stern & Nurse, 1997) and the physiological consequences of the resulting reduction in cell size (Davey & Nielsen, 1994; Stern & Nurse, 1997) is unclear.

Arrested cells continue to grow, and there is a considerable increase in cell size before division resumes (Davey & Nielsen, 1994). This appears to help trigger re-entry into the cell cycle. Cdc2 activity increases as the cell elongates and pheromone-induced arrest is compromised in enlarged mutant cells (Stern & Nurse, 1997). It is not yet clear, however, whether this mechanism is specific for overcoming pheromone-induced arrest or whether it is more generally used to overcome any treatment that causes a transient arrest of the cycle and a corresponding increase in cell size.

Pheromone-induced G1 arrest in budding yeast is also through inhibition of the CDK. Activation of the FUS3 MAP kinase leads to phosphorylation of FAR1 (Chang & Herskowitz, 1990, 1992) which then binds to, and inactivates, the G1 form of CDC28 (CLN-CDC28) (Elion *et al.*, 1993; Peter *et al.*, 1993; Tyers & Futcher, 1993; Peter & Herskowitz, 1994). Unlike *Sz. pombe*, however, the *S. cerevisiae* CDK is also required to activate G1-specific transcription factors (Cross & Tinkelenberg, 1991; Nasmyth & Dirick, 1991) and there is a two-stage inhibition of the cell cycle. Ubiquitin-mediated degradation of the phosphorylated FAR1 releases the G1 arrest and allows cells to proceed into S phase (Henchoz *et al.*, 1997).

Morphological response

Cell growth continues during the pheromone-induced G1 arrest and there is an increase in cell size (Davey, 1991; Davey & Nielsen, 1994). Growth still occurs from the cell tip, but it is no longer limited to being parallel to the long axis, and the cell elongates towards the source of the pheromone (Fukui *et al.*, 1986a; Leupold, 1987). This chemotropic elongation leads to the formation of a shmoo. Binding and subsequent fusion of mating partners occurs at the tips of the shmoos. Little is known about the fusion process except that it requires Fus1 which, appropriately, becomes localized to the tip of the elongated cell (Petersen *et al.*, 1995).

The direction of growth during the mitotic cycle is tightly regulated (for review, see Mata & Nurse, 1998). Growth occurs first at one end of the cell, switches to both ends at NETO (new end take-off), and is then targeted to the centre to form the septum for cytokinesis (Mitchison & Nurse, 1985). It is controlled by machinery that first marks (or tags) the site of growth and then redirects the cytoskeleton to the chosen site on the cortex. The tag is provided by Tea1, and targeting to the correct location is achieved through an association with microtubules (Mata & Nurse, 1997). Loss of either Tea1 or the microtubules disrupts polarized growth and leads to bending and branching of the growing cell (Mata & Nurse, 1997).

Elongation of shmoos towards the pheromone source implies that stimulation generates new signals for controlling the direction of cell growth. Normal mitotic tags are either removed or ignored and replaced by signals generated by the pheromone gradient. The down-regulation of Tea1 in pheromone-stimulated cells (Mata & Nurse, 1997) would help facilitate this change. It is possible presently only to speculate about the nature of the new signal, but it is likely to involve the pheromone signalling machinery. The distribution of pheromone-coupled receptors, for example, will reflect the external pheromone gradient, while the cytoplasmic tails of these receptors could serve to tag the appropriate site on the cortex. Any of the components that assemble at the activated receptor would also be able to act as the tag.

Once a growth site has been selected, its position is relayed to the actin cytoskeleton. This requires a functional Ras1 protein and *ras1* mutants are shorter and plumper than wild-type cells (Fukui *et al.*, 1986b). The mutants also fail to elongate when exposed to mating pheromone and are sterile

(Fukui *et al.*, 1986*a*). Screening for mutants with phenotypes similar to those with defects in *ras1* led to the isolation of other genes required for shmoo formation (Fukui & Yamamoto, 1988; Chang *et al.*, 1994) and two hybrid analyses (Fawell *et al.*, 1992; Miller & Johnson, 1994) revealed pairwise interactions between Ras1 and Ral1 (Scd1), Ral1 and Ral3 (Scd2), and Ral3 and Cdc42. Ral1 is a guanine nucleotide exchange factor (GEF) for the Rho-like Cdc42 GTPase while the two SH3 domains within Ral3 may help to stabilize the complex. Rho-like GTPases control the actin cytoskeleton in mammalian cells and mediate many of the morphological changes associated with stimulation by growth factors (Ridley, 1995). Activation of the associated GEF leads to nucleotide exchange and the GTP-bound GTPase binds to, and activates, a PAK protein kinase (Manser *et al.*, 1994). Phosphorylation of PAK targets is then thought to control actin reorganization. PCR-based screening identified a PAK homologue in *Sz. pombe* (Ottilie *et al.*, 1995). Pak1 preferentially binds to GTP-Cdc42, and mutational analysis suggests it is involved in controlling cell morphology during mitotic growth and following pheromone stimulation (Ottilie *et al.*, 1995). No Pak1 targets have yet been identified.

Pheromone-dependent changes in the shape of budding yeast seem to occur by a mechanism that is similar to that in *Sz. pombe* (for review, see Roemer *et al.*, 1996). Tags that identify the mitotic bud site are depleted following pheromone stimulation (Sanders & Herskowitz, 1996; Roemer *et al.*, 1996*b*), possibly by a mechanism that involves FAR1 (Dorer *et al.*, 1995; Valtz *et al.*, 1995), and are replaced by signals that require the receptor and its associated G$\beta\gamma$ subunits but does not involve components from the MAP kinase cascade (Schrick *et al.*, 1997). Homologues of Ral1 (CDC24), Ral3 (BEM1), Cdc42 (CDC42) and Pak1 (STE20) appear to play similar roles in reorganizing the actin cytoskeleton. The two yeast are not identical, however, and the *S. cerevisiae* pheromone response does not involve a Ras1 homologue, its proposed role in activating the Rho-GEF apparently being replaced by the G$\beta\gamma$ subunits (Nern & Arkowitz, 1998).

Induction of meiosis

Conjugation of the haploid cells produces a diploid zygote. This can be maintained as a diploid if immediately transferred to rich medium, but it normally enters meiosis directly. The cells initiate premeiotic DNA synthesis and proceed through meiosis to form four haploid spores which subsequently germinate to complete the life cycle. Entry into meiosis is directly controlled by mating pheromones (Willer *et al.*, 1995) and involves inhibition of the Pat1 protein kinase (Beach *et al.*, 1985; Iino & Yamamoto, 1985*a*; McLeod & Beach, 1986; Nielsen & Egel, 1990) and dephosphorylation of Mei2 (Beach *et al.*, 1985; Iino & Yamamoto, 1985*b*; Watanabe *et al.*, 1988). Pat1 (also called Ran1) is a negative regulator of sexual differentiation and *pat1* mutants initiate meiosis regardless of nutritional signals, pheromonal signals, or cell ploidy. Mei2 is an RNA-binding protein that associates with meiRNA (the product of the *sme2* gene) to promote entry into meiosis I and with another, as yet unidentified RNA, to promote premeiotic DNA synthesis (Watanabe & Yamamoto, 1994). Mei2 is inhibited during mitotic growth by phosphorylation with Pat1 (Watanabe *et al.*, 1997)

As a consequence of pheromone stimulation, the prezygotic P-cell contains Mat1-Pm and the M-cell contains Mat1-Mm. Conjugation creates a zygote containing both of the pheromone-induced products and this leads to expression of the *mei3* gene (McLeod *et al.*, 1987). Mei3 then uses a pseudosubstrate mechanism to inhibit Pat1 (Li & McLeod, 1996), and this allows dephosphorylation and activation of Mei2 (Watanabe *et al.*, 1997). This relatively simple model for the control of meiosis is sufficient to explain our observations, but it may not be the full story. Pat1 also phosphorylates Ste11 (at least *in vitro*) (Li & McLeod, 1996) and this could provide additional control for the expression of meiotic genes, including *mei2* (Sugimoto *et al.*, 1991).

A similar regulation mechanism exists in diploid cells where meiosis is triggered by starvation (Willer *et al.*, 1995). Transferring a heterozygous diploid to nitrogen-free conditions induces production of the two pheromones and both receptors. As in the haploid cells, this leads to expression of *mat1-Pm* and *mat1-Mm* which triggers production of Mei3 and inhibition of Pat1.

ADAPTATION

Cells that fail to mate recover from the effects of pheromone stimulation and resume mitotic growth (Davey & Nielsen, 1994; Imai & Yamamoto, 1994). Recovery occurs in the presence of extracellular pheromone and confirms that the cells become desensitized to continued stimulation. Such desensitization, or adaptation, also appears to be important in the mating process as mutations that reduce the ability to recover from stimulation also lead to reduced fertility (Egel, 1992; Imai & Yamamoto, 1992; Rusu, 1992). There are likely to be many mechanisms that contribute to the adaptation process, and most of the components activated by pheromone stimulation will need to be deactivated during the recovery process.

Degradation of the pheromone

Removing the extracellular signal would aid the recovery process. It would not only prevent further stimulation but also provide an opportunity for the cell to turn off the components activated during the initial response. A general dilution in the growth medium helps to reduce the local concentration of most ligands, but many cells have also developed mechanisms that either remove or degrade the extracellular signal (for review, see Ladds *et al.*, 1998). The surface of many mammalian cells, for example, has a battery of endopeptidases, aminopeptidases, and dipeptidyl peptidases that act in concert to modulate and terminate the action of various regulatory peptides (for review, see Turner & Tanzawa, 1997).

Cells of both mating types secrete a variety of proteases. These have relatively low substrate specificities, and could make a general contribution to the degradation of the pheromones. M-cells also produce a specific enzyme that provides an elegant feedback mechanism for regulating stimulation by P-factor. Exposing M-cells to P-factor results in the secretion of Sxa2 (Imai & Yamamoto, 1992), a serine carboxypeptidase that inactivates P-factor by removing the C-terminal leucine residue (Ladds *et al.*, 1996). Sxa2 is not produced by P-cells or unstimulated M-cells and appears to act on a very limited range of substrates (Ladds *et al.*, 1996).

M-factor is farnesylated and carboxymethylated. Both modifications are essential for full biological activity, and inactivation of M-factor could be achieved not only by peptidases but also by enzymes that remove or alter these modifications. Esterase activity might, for example, hydrolyse the C-terminal methyl ester. However, no enzyme has yet been identified that specifically inactivates M-factor. There has only been one, rather inconclusive, report concerning the degradation of any of the fungal prenylated pheromones (Marcus *et al.*, 1991), and it is possible that these lipophilic molecules are inactivated by a different mechanism. For example, M-factor is susceptible to chemical inactivation and the peptide backbone, the methyl group, and the farnesyl moiety are all potential targets for modification (Davey, 1992). Alternatively, the pheromone may not be inactivated but could be sequestered into a location where it can no longer function. M-factor is extremely hydrophobic (Davey, 1992) and readily binds to many surfaces so that the amount available to the receptors on the surface of the P-cells might quickly drop below the concentration required to maintain the response.

Internalization of the receptors

Endocytosis of the pheromone receptors is likely to make several contributions to the recovery process. It could, for example, help to dissociate the receptor from its G protein, which probably remains at the cell surface (Hirschman *et al.*, 1997), while the degradation of the receptors in the vacuole would make a more long-term contribution to desensitization. Internalization of the pheromone bound to the receptor would also help reduce the concentration of extracellular signal.

A detailed description of the internalization process is not yet available in *Sz. pombe* but it is likely to be similar to that found in other systems (for review, see Koenig & Edwardson, 1997). In *S. cerevisiae*, for example, pheromone binding changes the conformation of the receptor and the subsequent phosphorylation and ubiquitination of the cytoplasmic tails triggers a series of events that ends with the degradation of the receptors in the vacuole (Riezman, 1998). Defects in any of these steps reduces the ability of the cells to recover from stimulation (Reneke *et al.*, 1988; Raths *et al.*, 1992, Rohrer *et al.*, 1993; Hicke & Riezman, 1996).

Modification of the receptors

Receptor phosphorylation is often the most rapid means of attenuating GPCR responsiveness (for review, see Ferguson *et al.*, 1996). It occurs within seconds or minutes of agonist stimulation and causes the receptor to dissociate from its G protein. Phosphorylation is usually mediated by either second messenger-dependent kinases (such as protein kinase A) or G protein-coupled receptor kinases (GRKs). GRKs specifically phosphorylate agonist-activated receptors, and therefore only affect recovery to the stimulating agonist (homologous desensitization). In contrast, PKA and related kinases phosphorylate and desensitize unoccupied receptors and even target receptors for other agonists (heterologous desensitization). Phosphorylation by GRKs also promotes the binding of arresting proteins (arrestins) to the receptor tail and these further inhibit the signalling process.

The yeast pheromone receptors are phosphorylated following stimulation and this contributes to the recovery process (Reneke *et al.*, 1988; Zanolari *et al.*, 1992; Chen & Konopka, 1996). It is not yet clear, however, if this affects the association between the receptor and the G protein or whether it simply reflects the role of phosphorylation in promoting endocytosis. Searches of the genomic databases and PCR-based screens (see, for example, Watson & Davey, 1998) have failed to identify a yeast member of this highly conserved family of kinases. Perhaps the fact that yeast seems to contain only one GPCR negates the need for homologous desensitization.

Regulating the G protein

The G proteins are also subject to desensitization (for review, see Dohlman & Thorner, 1997; Dohlman *et al.*, 1998). A recently identified family of RGS proteins (regulators of G protein signalling) contributes to the recovery process by enhancing the intrinsic GTPase activity of the Gα subunits. Decreasing the lifetime of the active Gα-GTP promotes its reassociation with the Gβγ subunits and dampens the response. The first RGS protein (SST2) was discovered during studies of the pheromone response in *S. cerevisiae* (Chan & Otte, 1982; Dohlman *et al.*, 1995, 1996). Strains lacking SST2 have a severely reduced ability to recover from pheromone-induced G1 arrest, while overproduction of SST2 makes cell resistant to pheromone stimulation (Konopka, 1993; Dohlman *et al.*, 1996). Searches of the genomic database have identified a potential RGS protein in *Sz. pombe* but, as yet, there is no biochemical evidence to indicate if it affects the pheromone response.

G proteins can also be regulated by events that affect their assembly or targeting within the cell. The G protein can only be activated by the receptor if it is targeted to the correct part of the plasma membrane. Targeting appears to involve post-translational fatty acylation of the subunits (Wedegaertner *et al.*, 1995) and regulation of this process could be an effective mechanism for controlling signalling and recovery (Song & Dohlman, 1996; Song *et al.*, 1996).

Protein phosphatases

Activation of the MAP kinase cascade involves phosphorylation of the individual enzymes, and dephosphorylation will contribute to the recovery process. Each of the enzymes in the cascade could be subject to regulation, but the MAP kinase is a particularly attractive target as it becomes phosphorylated on both threonine and tyrosine residues. Dephosphorylation of either residue will inactivate Spk1, and serine/threonine phosphatases or tyrosine-specific phosphatases could therefore be involved in the recovery process. Alternatively, both residues could be dephosphorylated by a single dual-specificity threonine/tyrosine phosphatase. First described in mammalian cells (Keyse & Emslie, 1992), MAP kinase phosphatases (MKPs) play an important role in regulating the intensity and duration of signalling through MAP kinase cascades (for review, see Keyse, 1998). In *S. cerevisiae,* for example, MSG5 dephosphorylates FUS3/KSS1 and promotes recovery from pheromone stimulation (Doi *et al.*, 1994). Pmp1 is the only MKP so far identified in *Sz. pombe* (Sugiura *et al.*, 1998). Isolated as a suppressor of the chloride sensitivity caused by loss of calcineurin, Pmp1 dephosphorylates the Pmk1 MAP kinase that regulates cell integrity and cation sensitivity (Toda *et al.*, 1996). As many MKPs are not restricted to a single substrate (Watanabe *et al.*, 1995), it will be interesting to discover whether Pmp1 has any influence on the pheromone response. MAP kinases are also regulated by tyrosine-specific phosphatases. Osmoregulation and stress response in *Sz. pombe* involves activation of the Sty1 (Spc1) MAP kinase, and this is dephosphorylated by Pyp1 and Pyp2 (Millar *et al.*, 1995; Shiozaki & Russell, 1995). Whether these also regulate Spk1 remains to be determined, but the analogous enzymes in *S. cerevisiae* (PTP2 and PTP3; Wurgler-Murphy *et al.*, 1997; Jacoby *et al.*, 1997) dephosphorylate FUS3/KSS1 as well as the HOG1 kinase from the osmolarity response pathway (Zhan *et al.*, 1997).

CONCLUSION

Work over the past 10 years has provided a reasonable understanding of the pheromone communication process in the fission yeast *Sz. pombe*. Many of the key players have been identified and most of the main events have been described in some detail. The next challenge is to discover how the machinery is regulated and to define more precisely how activation brings about the changes necessary for mating.

ACKNOWLEDGEMENTS

I thank Paul Nurse for encouraging me to begin studying the pheromones and I am grateful to all of the people who have helped in those studies. I am particularly indebted to the Cancer Research Campaign for their vision in the early days and their continued support of the work, and I also thank the Wellcome Trust and BBSRC for their support. Finally, I thank the Lister Institute of Preventive Medicine for the fellowship that allows me to concentrate on my research.

REFERENCES

Akada, R., Kallal, L., Johnson, D. I. & Kurjan, J. (1996). Genetic relationships between the G protein $\beta\gamma$ complex, Ste5p, Ste20p and Cdc42p: investigation of effector roles in the yeast pheromone response pathway. *Genetics*, **143**, 103–17.

Anderegg, R. J., Betz, R., Carr, S. A., Crabb, J. W. & Duntze, W. (1988). Structure of *Saccharomyces cerevisiae* mating hormone a-factor. Identification of S-farnesyl cysteine as a structural component. *Journal of Biological Chemistry*, **263**, 18 236–40.

Anderson, E. D., VanSlyke, J. K., Thulin, C. D., Jean, F. & Thomas, G. (1997). Activation of furin endoprotease is a multiple-step process: requirements for acidification and internal propeptide cleavage. *EMBO Journal*, **16**, 1508–18.

Aono, T., Yanai, H., Miki, F., Davey, J. & Shimoda, C. (1994). Mating pheromone-induced expression of the *mat1-Pm* gene of *Schizosaccharomyces pombe*: identification of signalling components and characterisation of upstream controlling elements. *Yeast*, **10**, 757–70.

Barr, M. M., Tu, H., Van Aelst, L. & Wigler, M. (1996). Identification of Ste4 as a potential regulator of Byr2 in the sexual response pathway of *Schizosaccharomyces pombe*. *Molecular and Cellular Biology*, **16**, 5597–603.

Beach, D., Rodgers, L. & Gould, J. (1985). RAN1 + controls the transition from mitotic division to meiosis in fission yeast. *Current Genetics*, **10**, 297–311.

Blinder, D., Bouvier, S. & Jenness, D. D. (1989). Constitutive mutants in the yeast pheromone response: ordered function of the gene products. *Cell*, **56**, 479–86.

Bourne, H. R. (1997). How receptors talk to trimeric G proteins. *Current Opinion in Cell Biology*, **9**, 134–42.

Boyartchuk, V. L., Ashby, M. N. & Rine, J. (1997). Modulation of Ras and a-factor function by carboxyl-terminal proteolysis. *Science*, **275**, 1796–800.

Büküsoglu G. & Jenness, D. D. (1996). Agonist-specific conformational changes in the yeast α-factor pheromone receptor. *Molecular and Cellular Biology*, **16**, 4818–23.

Bustin, M. & Reeves, R. (1996). High-mobility-group chromosomal proteins: architectural components that facilitate chromatin function. *Progress in Nucleic Acid Research and Molecular Biology*, **54**, 35–100.

Caldwell, G. A., Naider, F. & Becker, J. M. (1995). Fungal lipopeptide mating pheromones: a model system for the study of protein prenylation. *Microbiological Reviews*, **59**, 406–22.

Cano, E. & Mahadevan, L. (1995). Parallel signal processing among mammalian MAPKs. *Trends in Biochemical Sciences*, **20**, 117–22.

Chan, R. K. & Otte, C. A. (1982). Isolation and genetic analysis of *Saccharomyces cerevisiae* mutants supersensitive to G1 arrest by **a**-factor and α-factor pheromones. *Molecular and Cellular Biology*, **2**, 11–20.

Chang, E. C., Barr, M., Wang, Y., Jung, V., Xu, H-P. & Wigler, M. H. (1994). Cooperative interaction of *S. pombe* proteins required for mating and morphogenesis. *Cell*, **79**, 131–41.

Chang, F. & Herskowitz, I. (1990). Identification of a gene necessary for cell cycle arrest by a negative growth factor of yeast: FAR1 is an inhibitor of a G1 cyclin, CLN2. *Cell*, **63**, 999–1011.

Chang, F. & Herskowitz, I. (1992). Phosphorylation of FAR1 in response to α-factor: a possible requirement for cell-cycle arrest. *Molecular Biology of the Cell*, **3**, 445–50.

Chen, Q. J. & Konopka, J. B. (1996). Regulation of the G protein-coupled α-factor pheromone receptor by phosphorylation. *Molecular and Cellular Biology*, **16**, 247–57.

Christensen, P. U., Davey, J. & Nielsen, O. (1997). The *Schizosaccharomyces pombe mam1* gene encodes an ABC transporter mediating secretion of M-factor. *Molecular and General Genetics*, **255**, 226–36.

Cook, J. G., Bardwell, L., Kron, S. J. & Thorner, J. (1996). Two novel targets of the MAP kinase Kss1 are negative regulators of invasive growth in the yeast *Saccharomyces cerevisiae*. *Genes and Development*, **15**, 2831–48.

Correa-Bordes, J. & Nurse, P. (1995). p25^{rum1} orders S phase and mitosis by acting as an inhibitor of the p34^{cdc2} mitotic kinase. *Cell*, **83**, 1001–9.

Costello, G., Rodgers, L. & Beach, D. (1986). Fission yeast enters the stationary phase G0 state from either mitotic G1 or G2. *Current Genetics*, **11**, 119–25.

Cross, F. R. & Tinkelenberg, A. H. (1991). A potential positive feedback loop controlling CLN1 and CLN2 gene expression at the start of the yeast cell cycle. *Cell*, **65**, 875–83.

Davey, J. (1991). Isolation and quantitation of M-factor, a diffusible mating factor from the fission yeast *Schizosaccharomyces pombe*. *Yeast*, **7**, 357–66.

Davey, J. (1992). Mating pheromones of the fission yeast *Schizosaccharomyces pombe*: purification and structural characterisation of M-factor and isolation and structural characterisation of two genes encoding the pheromone. *EMBO Journal*, **11**, 951–60.

Davey, J. (1996). M-factor, a farnesylated mating factor from the fission yeast *Schizosaccharomyces pombe*. *Biochemical Society Transactions*, **24**, 718–23.

Davey, J. & Nielsen, O. (1994). Mutations in *cyr1* and *pat1* reveal pheromone-induced G1 arrest in the fission yeast *Schizosaccharomyces pombe*. *Current Genetics*, **26**, 105–12.

Davey, J., Davis, K., Imai, Y., Yamamoto, M. & Matthews, G. (1994). Isolation and characterisation of krp, a dibasic endopeptidase required for cell viability in the fission yeast *Schizosaccharomyces pombe*. *EMBO Journal*, **13**, 5910–21.

Degols, G., Shiozaki, K. & Russell, P. (1996). Activation and regulation of the Spc1 stress-activated protein kinase in *Schizosaccharomyces pombe*. *Molecular and Cellular Biology*, **16**, 2870–7.

Dohlman, H. G. & Thorner, J. (1997). RGS proteins and signaling by heterotrimeric G proteins. *Journal of Biological Chemistry*, **272**, 3871–74.

Dohlman, H. G., Apaniesk, D., Chen, Y., Song, J. & Nusskern, D. (1995). Inhibition of G protein signalling by dominant gain-of-function mutations in Sst2p, a pheromone desensitisation factor in *Saccharomyces cerevisiae*. *Molecular and Cellular Biology*, **15**, 3635–43.

Dohlman, H. G., Song, J. P., Apanovitch, D. M., DiBello, P. R. & Gillen, K. M. (1998). Regulation of G protein signaling in yeast. *Seminars in Cell and Developmental Biology*, **9**, 135–41.

Dohlman, H. G., Song, J. P., Ma, D. R., Courchesne, W. E. & Thorner, J. (1996). SST2, a negative regulator of pheromone signalling in the yeast *Saccharomyces cerevisiae*: expression, localisation and genetic interaction and physical association with GPA1 (the G protein α-subunit). *Molecular and Cellular Biology*, **16**, 5194–209.

Doi, K., Gartner, A., Ammerer, G., Errede, B., Shinkawa, H., Sugimoto, K. & Matsumoto, K. (1994). MSG5, a novel protein phosphatase promotes adaptation to pheromone response in *S. cerevisiae*. *EMBO Journal*, **13**, 61–70.

Dooijes, D., Wetering, M. V. D., Knippels, L. & Clevers, H. (1993). The *Schizosaccharomyces pombe* mating-type gene *mat1-Mc* encodes a sequence-specific DNA-binding high mobility group box protein. *Journal of Biological Chemistry*, **268**, 24 813–17.

Dorer, R., Pryciak, P. M. & Hartwell, L. H. (1995). *Saccharomyces cerevisiae* cells execute a default pathway to select a mate in the absence of pheromone gradients. *Journal of Cell Biology*, **131**, 845–61.

Egel, R. (1992). Pheromone detection and pheromone hypersensitive mutants in the fission yeast *Schizosaccharomyces pombe*. *Yeast*, **8**, S363.

Egel, R. & Egel-Mitani, M. (1974). Premeiotic DNA synthesis in fission yeast. *Experimental Cell Research*, **88**, 127–34.

Elion, E. A. (1995). Ste5: a meeting place for MAP kinases and their associates. *Trends in Cell Biology*, **5**, 322–7.

Elion, E. A., Satterberg, B. & Kranz, J. (1993). FUS3 phosphorylates multiple components of the mating signal transduction cascade; evidence for STE12 and FAR1. *Molecular Biology of the Cell*, **4**, 495–510.

Fahmy, K., Siebert, F. & Sakmar, T. P. (1995). Photoactivated state of rhodopsin and how it can form. *Biophysical Chemistry*, **56**, 171–81.

Fawell, E., Bowden, S. & Armstrong, J. (1992). A homologue of the ras-related *CDC42* gene from *Schizosaccharomyces pombe*. *Gene*, **114**, 153–4.

Ferguson, S. S. G., Barak, L. S., Zhang, J. & Caron, M. G. (1996). G protein-coupled receptor regulation: role of G protein-coupled receptor kinases and arrestins. *Canadian Journal of Physiology and Pharmacology*, **74**, 1095–110.

Fisher, D. L. & Nurse, P. (1996). A single fission yeast mitotic cyclin B-p34^{cdc2} kinase promotes both S-phase and mitosis in the absence of G1 cyclins. *EMBO Journal*, **15**, 850–60.

Friedmann, K. L. & Egel, R. (1978). Protein patterns during sporulation in fission yeast. *Zeitschrift Naturforschung*, **33c**, 84–91.

Fujimura-Kamada, K., Nouvet, F. J. & Michaelis, S. (1997). A novel membrane-associated metalloprotease, Ste24p, is required for the first step of N-terminal processing of the yeast a-factor precursor. *Journal of Cell Biology*, **136**, 271–85.

Fukui, Y. & Yamamoto, M. (1988). Isolation and characterisation of *Schizosaccharomyces pombe* mutants phenotypically similar to *ras1-*. *Molecular and General Genetics*, **215**, 26–31.

Fukui, Y., Kaziro, Y. & Yamamoto, M. (1986*a*). Mating pheromone-like diffusible factor released by *Schizosaccharomyces pombe*. *EMBO Journal*, **5**, 1991–3.

Fukui, Y., Kozasa, T., Kaziro, Y., Takeda, T. & Yamamoto, M. (1986*b*). Role of a ras homolog in the life cycle of *Schizosaccharomyces pombe*. *Cell*, **44**, 329–36.

Gotoh, Y., Nishida, E., Shianuki, M., Toda, T., Imai, Y. & Yamamoto, M. (1993).

Schizosaccharomyces pombe Spk1 is a tyrosine-phosphorylated protein functionally related to *Xenopus* mitogen-activated protein kinase. *Molecular and Cellular Biology*, **13**, 6427–34.

Henchoz, S., Chi, Y., Catarin, B., Herskowitz, I., Deshaies, R. J. & Peter, M. (1997). Phosphorylation and ubiquitin-dependent degradation of the cyclin-dependent kinase inhibitor Far1p in budding yeast. *Genes and Development*, **11**, 3046–60.

Herskowitz, I. (1995). MAP kinase pathways in yeast: for mating and more. *Cell*, **80**, 187–97.

Hicke, L. & Riezman, H. (1996). Ubiquitination of a yeast plasma membrane receptor signals its ligand-stimulated endocytosis. *Cell*, **84**, 277–87.

Hirschman, J. E., De Zutter, G. S., Simonds, W. F. & Jenness, D. D. (1997). The G$\beta\gamma$ complex of the yeast pheromone response pathway: subcellular fractionation and protein-protein interactions. *Journal of Biological Chemistry*, **272**, 240–8.

Hughes, M. & Davey, J. (1997). Proteases involved in the maturation of the M-factor mating pheromone in fission yeast. *Biochemical Society Transactions*, **25**, 447.

Hughes, D. A., Fukui, Y. & Yamamoto, M. (1990). Homologous activators of ras in fission and budding yeast. *Nature*, **344**, 355–7.

Hughes, D. A., Yabana, N. & Yamamoto, M. (1994). Transcriptional regulation of a Ras nucleotide-exchange factor gene by extracellular signals in fission yeast. *Journal of Cell Science*, **107**, 3635–42.

Iino, Y. & Yamamoto, M. (1985*a*). Mutants of *Schizosaccharomyces pombe* which sporulate in the haploid state. *Molecular and General Genetics*, **198**, 416–21.

Iino, Y. & Yamamoto, M. (1985*b*). Negative control for the initiation of meiosis in *Schizosaccharomyces pombe*. *Proceedings of the National Academy of Sciences, USA*, **82**, 2447–51.

Imai, Y. & Yamamoto, M. (1992). *Schizosaccharomyces pombe sxa1*$^+$ and *sxa2*$^+$ encode putative proteases involved in the mating response. *Molecular and Cellular Biology*, **12**, 1827–34.

Imai, Y. & Yamamoto, M. (1994). The fission yeast mating pheromone P-factor: its molecular structure, gene structure and physiological activities to induce gene expression and G1 arrest in the mating partner. *Genes and Development*, **8**, 328–38.

Imai, Y., Davey, J., Kawagishi-Kobayashi, M. & Yamamoto, M. (1997). Genes encoding farnesyl cysteine carboxyl methyltransferase in *Schizosaccharomyces pombe* and *Xenopus laevis*. *Molecular and Cellular Biology*, **17**, 1543–51.

Imai, Y., Miyake, S., Hughes, D. A. & Yamamoto, M. (1991). Identification of a GTPase-activating protein homolog in *Schizosaccharomyces pombe*. *Molecular and Cellular Biology*, **11**, 3088–94.

Ishibashi, Y., Sakagami, Y., Isogai, A. & Suzuki, A. (1984). Structures of Tremerogens A-9291-I and A-9291-VIII: peptidyl sex hormones of *Tremella brasiliensis*. *Biochemistry*, **23**, 1399–404.

Jacoby, T., Flanagan, H., Faykin, A., Seto, A. G., Mattison, C. & Ota, I. (1997). Two protein-tyrosine phosphatases inactivate the osmotic stress response pathway in yeast by targeting the mitogen-activated protein kinase, HOG1. *Journal of Biological Chemistry*, **272**, 17 749–55.

Kamiya, Y., Sakurai, A., Tamura, S., Takahashi, N., Tsuchiya, T., Abe, K. & Fukui, S. (1979). Structure of rhodotorucine A, a peptidyl factor inducing mating tube formation in *Rhodosporidium toruloides*. *Agricultural and Biological Chemistry*, **43**, 363–9.

Karin, M. & Hunter, T. (1995). Transcriptional control by protein phosphorylation: signal transmission from the cell surface to the nucleus. *Current Biology*, **5**, 747–57.

Kawamukai, M., Ferguson, M., Wigler, M. & Young, D. (1991). Genetic and biochemical analysis of the adenylyl cyclase of *Schizosaccharomyces pombe*. *Cell Regulation*, **2**, 155–64.

Kelly, M., Burke, J., Smith, M., Klar, A. & Beach, D. (1988). Four mating-type genes control sexual differentiation in the fission yeast. *EMBO Journal*, **7**, 1537–47.

Keyse, S. M. (1998). Protein phosphatases and the regulation of MAP kinase activity. *Seminars in Cell and Developmental Biology*, **9**, 143–52.

Keyse, S. M. & Emslie, E. A. (1992). Oxidative stress and heat shock induce a human gene encoding a protein-tyrosine phosphatase. *Nature*, **359**, 644–7.

Kim, D. U., Park, S. K., Chung, K. S., Choi, M. U. & Yoo, H. S. (1996). The G protein β-subunit GPB1 of *Schizosaccharomyces pombe* is a negative regulator of sexual development. *Molecular and General Genetics*, **252**, 20–32.

Kitamura, K. & Shimoda, C. (1991). The *Schizosaccharomyces pombe mam2* gene encodes a putative pheromone receptor which has significant homology with the *Saccharomyces cerevisiae* Ste2 protein. *EMBO Journal*, **10**, 3743–51.

Kitamura, K., Nakamura, T., Miki, F. & Shimoda, C. (1996). Autocrine response of *Schizosaccharomyces pombe* haploid cells to mating pheromones. *FEMS Microbiology Letters*, **143**, 41–5.

Kjaerulff, S., Davey, J. & Nielsen, O. (1994). Analysis of the structural genes encoding the M-factor in the fission yeast *Schizosaccharomyces pombe*: identification of a third gene *mfm3*. *Molecular and Cellular Biology*, **14**, 3895–905.

Kjaerulff, S., Dooijes, D., Clevers, H. & Nielsen, O. (1997). Cell differentiation by interaction of two HMG-box proteins: Mat1-Mc activates M cell-specific genes in *S. pombe* by recruiting the ubiquitous transcription factor Ste11 to weak binding sites. *EMBO Journal*, **16**, 4021–33.

Knaus, U. G., Morris, S., Dong, H. J., Chernoff, J. & Bokoch, G. M. (1995). Regulation of human leukocyte p21-activated kinases through G protein-coupled receptors. *Science*, **269**, 221–3.

Koenig, J. A. & Edwardson, J. M. (1997). Endocytosis and recycling of G protein-coupled receptors. *Trends in Pharmacological Sciences*, **18**, 276–86.

Konopka, J. B. (1993). AFR1 acts in conjunction with the α-factor receptor to promote morphogenesis and adaptation. *Molecular and Cellular Biology*, **13**, 6876–88.

Ladds, G., Hughes, M. & Davey, J. (1998). Extracellular degradation of agonists as an adaptive mechanism. *Seminars in Cell and Developmental Biology*, **9**, 111–18.

Ladds, G., Rasmussen, E. M., Young, T., Nielsen, O. & Davey, J. (1996). The *sxa2*-dependent inactivation of the P-factor mating pheromone in the fission yeast *Schizosaccharomyces pombe*. *Molecular Microbiology*, **20**, 35–42.

Leberer, E., Dignard, D., Harcus, D., Thomas, D. Y. & Whiteway, M. (1992). The protein kinase homologue Ste20p is required to link the yeast pheromone response G-protein $\beta\gamma$ subunits to downstream signalling components. *EMBO Journal*, **11**, 4815–24.

Leevers, S. J., Paterson, H. F. & Marshall, C. J. (1994). Requirement for Ras in Raf activation is overcome by targeting Raf to the plasma membrane. *Nature*, **369**, 411–14.

Leupold, U. (1987). Sex appeal in fission yeast. *Current Genetics*, **12**, 543–5.

Li, P. & McLeod, M. (1996). Molecular mimicry in development: identification of ste11[+] as a substrate and mei3[+] as a pseudosubstrate inhibitor of ran1[+] kinase. *Cell*, **87**, 869–80.

McLeod, M. & Beach, D. (1986). Homology between the *ran1[+]* gene of fission yeast and protein kinases. *EMBO Journal*, **5**, 3665–71.

McLeod, M., Stein, M. & Beach, D. (1987). The product of the *mei3*[+] gene, expressed under control of the mating-type locus, induces meiosis and sporulation in fission yeast. *EMBO Journal*, **6**, 729–736.

Maeda, T., Mochizuki, N. & Yamamoto, M. (1990). Adenylyl cyclase is dispensable for vegetative cell growth in the fission yeast *Schizosaccharomyces pombe*. *Proceedings of the National Academy of Sciences, USA*, **87**, 7814–18.

Manser, E., Leung, T., Salifuddin, H., Zhao, Z-S. & Lim, L. (1994). A brain serine threonine kinase activated by Cdc42 and Rac1. *Nature*, **367**, 40–6.

Marcus, S., Xue, C-B., Naider, F. & Becker, J. M. (1991). Degradation of **a**-factor by a *Saccharomyces cerevisiae* α-mating-type-specific endopeptidase: evidence for a role in recovery of cells from G1 arrest. *Molecular and Cellular Biology*, **11**, 1030–9.

Marshall, C. J. (1995). Specificity of receptor tyrosine kinase signaling: transient versus sustained extracellular signal-regulated kinase activation. *Cell*, **80**, 179–85.

Martin, C., Labib, K. & Moreno, S. (1996). B-type cyclins regulate G1 progression in fission yeast in opposition to the p25[rum1] CDK inhibitor. *EMBO Journal*, **15**, 839–49.

Masuda, T., Kariya, K., Shinkai, M., Okada, T. & Kataoka, T. (1995). Protein kinase Byr2 is a target of Ras1 in the fission yeast *Schizosaccharomyces pombe*. *Journal of Biological Chemistry*, **270**, 1979–82.

Mata, J. & Nurse, P. (1997). Tea1 and the microtubular cytoskeleton are important for generating global spatial order within the fission yeast cell. *Cell*, **89**, 939–49.

Mata, J. & Nurse, P. (1998). Discovering the poles in yeast. *Trends in Cell Biology*, **8**, 163–7.

Millar, J. B. A., Buck, V. & Wilkinson, M. G. (1995). Pyp1 and Pyp2 PTPases dephosphorylate an osmosensing MAP kinase controlling cell size at division in fission yeast. *Genes and Development*, **9**, 2117–30.

Miller, P. J. & Johnson, D. I. (1994). Cdc42p GTPase is involved in controlling polarised cell growth in *Schizosaccharomyces pombe*. *Molecular and Cellular Biology*, **14**, 1075–83.

Mitchison, J. M. & Nurse, P. (1985). Growth in cell length in the fission yeast *Schizosaccharomyces pombe*. *Journal of Cell Science*, **75**, 357–76.

Moore, T. D. E. & Edman, J. C. (1993). The α-mating type locus of *Cryptococcus neoformans* contains a peptide pheromone gene. *Molecular and Cellular Biology*, **13**, 1962–70.

Mondesert, O., McGowan, C. & Russell, P. (1996). Cig2, a B-type cyclin, promotes the onset of S in *Schizosaccharomyces pombe*. *Molecular and Cellular Biology*, **16**, 1527–33.

Moreno, S. & Nurse, P. (1994). Regulation of progression through the G1 phase of the cell cycle by the *rum1*[+] gene. *Nature*, **367**, 236–42.

Nakayama, N., Kaziro, Y., Arai, K-I. & Matsumoto, K. (1988). Role of *STE* genes in the mating factor signaling pathway mediated by GPA1 in *Saccharomyces cerevisiae*. *Molecular and Cellular Biology*, **8**, 3777–83.

Nasmyth, K. & Dirick, L. (1991). The role of SWI4 and SWI6 in the activity of G1 cyclins in yeast. *Cell*, **66**, 995–1013.

Neiman, A. M., Stevenson, B. J., Xu, H-P., Sprague, G. F., Herskowitz, I., Wigler, M. & Marcus, S. (1993). Functional homology of protein kinases required for sexual differentiation in *Schizosaccharomyces pombe* and *Saccharomyces cerevisiae* suggests a conserved signal transduction module in eukaryotic organisms. *Molecular Biology of the Cell*, **4**, 107–20.

Nern, A. & Arkowitz, R. A. (1998). A GTP-exchange factor required for cell orientation. *Nature*, **391**, 195–98.

Nielsen, O. & Davey, J. (1995). Pheromone communication in the fission yeast *Schizosaccharomyces pombe*. *Seminars in Cell Biology*, **6**, 95–104.

Nielsen, O. & Egel, R. (1990). The pat1 protein kinase controls transcription of the mating-type genes in fission yeast. *EMBO Journal*, **9**, 1401–6.

Nielsen, O., Davey, J. & Egel, R. (1992). The ras1 function of *Schizosaccharomyces pombe* mediates pheromone-induced transcription. *EMBO Journal*, **11**, 1391–5.

Nielsen, O., Friis, T. & Kjaerulff, S. (1996). The *Schizosaccharomyces pombe map1* gene encodes an SRF/MCM1-related protein required for P-cell specific gene expression. *Molecular and General Genetics*, **253**, 387–92.

Obara, T., Nakafuku, M., Yamamoto, M. & Kaziro, Y. (1991). Isolation and characterisation of a gene encoding a G-protein α-subunit from *Schizosaccharomyces pombe*: involvement in mating and sporulation pathways. *Proceedings of the National Academy of Sciences, USA*, **88**, 5877–81.

Okazaki, N., Okazaki, K., Tanaka, K. & Okayoma, H. (1991). The *ste4*[+] gene, essential for sexual differentiation of *Schizosaccharomyces pombe*, encodes a protein with a leucine zipper motif. *Nucleic Acids Research*, **19**, 7043–7.

Ottilie, S., Miller, P. J., Johnson, D. I., Creasy, C. L., Sells, M. A., Bagrodia, S., Forsburg, S. L. & Chernoff, J. (1995). Fission yeast *pak1*[+] encodes a protein kinase that interacts with Cdc42p and is involved in the control of cell polarity and mating. *EMBO Journal*, **14**, 5908–19.

Peter, M. & Herskowitz, I. (1994). Direct inhibition of the yeast cyclin-dependent kinase CDC28-CLN by FAR1. *Science*, **265**, 1228–31.

Peter, M., Gartner, A., Horecka, J., Ammerer, G. & Herskowitz, I. (1993). FAR1 links the signal transduction pathway to the cell cycle machinery in yeast. *Cell*, **73**, 747–60.

Petersen, J., Wilguny, D., Egel, R. & Nielsen, O. (1995). Characterisation of *fus1* of *Schizosaccharomyces pombe*: a developmentally controlled function needed for conjugation. *Molecular and Cellular Biology*, **15**, 3697–707.

Powner, D. & Davey, J. (1998). Activation of the kexin from *Schizosaccharomyces pombe* requires internal cleavage of its initially cleaved prosequence. *Molecular and Cellular Biology*, **18**, 400–8.

Raths, S., Rohrer, J., Crausaz, F. & Riezman, H. (1992). *end3* and *end4*: two mutants defective in receptor-mediated and fluid-phase endocytosis in *Saccharomyces cerevisiae*. *Journal of Cell Biology*, **120**, 55–65.

Rebay, I. & Rubin, G. M. (1995). Yan functions as a general inhibitor of differentiation and is negatively regulated by activation of the Ras1/MAPK pathway. *Cell*, **81**, 857–66.

Reneke, J. E., Blumer, K. J., Courchesne, W. E. & Thorner, J. (1988). The carboxy-terminal segment of the yeast α-factor receptor is a regulatory domain. *Cell*, **55**, 221–34.

Ridley, A. J. (1995). Rho-related proteins: actin cytoskeleton and cell cycle. *Current Opinion in Genetics and Development*, **5**, 24–30.

Riezman, H. (1998). Down-regulation of yeast G protein-coupled receptors. *Seminars in Cell and Developmental Biology*, **9**, 129–34.

Robinson, M. J. & Cobb, M. H. (1997). Mitogen-activated protein kinase pathways. *Current Opinion in Cell Biology*, **9**, 180–6.

Roemer, T., Madden, K., Chang, J. T. & Snyder, M. (1996*a*). Selection of axial growth sites in yeast requires Axl2p, a novel plasma membrane glycoprotein. *Genes and Development*, **10**, 777–93.

Roemer, T., Vallier, L. G. & Snyder, M. (1996*b*). Selection of polarised growth sites in yeast. *Trends in Cell Biology*, **6**, 434–41.

Rohrer, J., Benedetti, H., Zanolari, B. & Riezman, H. (1993). Identification of a novel sequence mediating regulated endocytosis of the G protein-coupled α-pheromone receptor in yeast. *Molecular Biology of the Cell*, **4**, 511–21.

Rusu, M. (1992). A mating deficient and temperature-sensitive lethal mutant of *Schizosaccharomyces pombe* defines a new fertility locus. *Current Genetics*, **21**, 17–22.

Sakagami, Y., Isogal, A., Suzuki, A., Tamura, S., Kitada, C. & Fujino, M. (1979). Structure of tremerogen A-10, a peptidyl hormone inducing conjugation tube formation in *Tremella mesenterica*. *Agricultural and Biological Chemistry*, **43**, 2643–5.

Sanders, S. L. & Herskowitz, I. (1996). The BUD4 protein of yeast, required for axial budding, is localized to the mother/bud neck in a cell cycle-dependent manner. *Journal of Cell Biology*, **134**, 413–27.

Schafer, W. R. & Rine, J. A. (1992). Protein prenylation: genes, enzymes, targets, and functions. *Annual Reviews of Genetics*, **30**, 209–37.

Schrick, K., Garvik, B. & Hartwell, L. H. (1997). Mating in *Saccharomyces cerevisiae*: the role of the pheromone signal transduction pathway in the chemotropic response to pheromone. *Genetics*, **147**, 19–32.

Seidah, N. G. & Chrétien, M. (1997). Eukaryotic protein processing: endoproteolysis of precursor proteins. *Current Opinion in Biotechnology*, **8**, 602–7.

Shiozaki, K. & Russell, P. (1995). Counteractive roles of protein phosphatase 2C (PP2C) and a MAP kinase kinase homolog in the osmoregulation of fission yeast. *EMBO Journal*, **14**, 492–502.

Shiozaki, K. & Russell, P. (1996). Conjugation, meiosis, and the osmotic stress response are regulated by Spc1 kinase through Atf1 transcription factor in fission yeast. *Genes and Development*, **10**, 2276–88.

Song, J. & Dohlman, H. G. (1996). Partial constitutive activation of pheromone responses by a palmitoylation site mutant of a G protein α-subunit in yeast. *Biochemistry*, **35**, 14 806–17.

Song, J., Hirschman, J., Gunn, K. & Dohlman, H. G. (1996). Regulation of membrane and subunit interactions by N-myristoylation of a G protein α-subunit in yeast. *Journal of Biological Chemistry*, **271**, 20 273–83.

Song, O., Dolan, J. W., Yuan, Y. L. & Fields, S. (1991). Pheromone-dependent phosphorylation of the yeast STE12 protein correlates with transcriptional activation. *Genes and Development*, **5**, 741–50.

Spellig, T., Bölker, M., Lottspeich, F., Frank, R. W. & Kahmann, R. (1994). Pheromones trigger filamentous growth in *Ustilago maydis*. *EMBO Journal*, **13**, 1620–7.

Steiner, D. F., Smeekens, S. P., Ohagi, S. & Chan, S. J. (1992). The new enzymology of precursor processing endoproteases. *Journal of Biological Chemistry*, **267**, 23 435–38.

Stern, B. & Nurse, P. (1997). Fission yeast pheromone blocks S-phase by inhibiting the G1 cyclinB-p34^{cdc2} kinase. *EMBO Journal*, **16**, 534–44.

Styrkársdóttir, U., Egel, R. & Nielsen, O. (1992). Functional conservation between *Schizosaccharomyces pombe* ste8 and *Saccharomyces cerevisiae* STE11 protein kinases in yeast signal transduction. *Molecular and General Genetics*, **235**, 122–30.

Sugimoto, A., Iino, Y., Maeda, T., Watanabe, Y. & Yamamoto, M. (1991). *Schizosaccharomyces pombe* ste11$^+$ encodes a transcription factor with an HMG motif that is a critical regulator of sexual development. *Genes and Development*, **5**, 1990–9.

Sugiura, R., Toda, T., Shuntoh, H., Yanagida, M. & Kuno, T. (1998). pmp1$^+$, a suppressor of calcineurin deficiency, encodes a novel MAP kinase phosphatase in fission yeast. *EMBO Journal*, **17**, 140–8.

Tanaka, K., Davey, J., Imai, Y. and Yamamoto, M. (1993). *Schizosaccharomyces pombe* map3$^+$ encodes the putative M-factor receptor. *Molecular and Cellular Biology*, **13**, 80–8.

Tedford, K., Kim, S., Sa, D., Stevens, K. & Tyers, M. (1997). Regulation of the mating pheromone and invasive growth responses in yeast by two MAP kinases substrates. *Current Biology*, **7**, 228–38.

Toda, T., Dhut, S., Supertifurga, G., Gotoh, Y., Nishida, E. & Sugiura, R. (1996). The fission yeast *pmk1*⁺ gene encodes a novel mitogen-activated protein kinase homolog which regulates cell integrity and functions co-ordinately with the protein kinase C pathway. *Molecular and Cellular Biology*, **16**, 6752–64.

Toda, T., Shimanuki, M. & Yanagida, M. (1991). Fission yeast genes that confer resistance to staurosporine encode an AP-1-like transcription factor and a protein kinase related to the mammalian ERK1/MAP2 and budding yeast FUS3 and KSS1 kinases. *Genes and Development*, **5**, 60–73.

Treisman, R. & Ammerer, G. (1992). The SRF and MCM1 transcription factors. *Current Opinion in Genetics and Development*, **2**, 221–6.

Tu, H., Barr, M., Dong, D. L. & Wigler, M. (1997). Multiple regulatory domains on the Byr2 protein kinase. *Molecular and Cellular Biology*, **17**, 5876–87.

Turner, A. J. & Tanzawa, K. (1997). Mammalian membrane metallopeptidases: NEP, ECE, KELL and PEX. *FASEB Journal*, **11**, 355–64.

Tyers, M. & Futcher, B. (1993). Far1 and Fus3 link the mating pheromone signal transduction pathway to three G1-phase Cdc28 kinase complexes. *Molecular and Cellular Biology*, **13**, 5659–69.

Valtz, N., Peter, M. & Herskowitz, I. (1995). FAR1 is required for oriented polarization of yeast cells in response to mating pheromones. *Journal of Cell Biology*, **131**, 863–73.

Van Aelst, L., Barr, M., Marcus, S., Polverino, A. & Wigler, M. (1993). Complex formation between Ras and Raf and other protein kinases. *Proceedings of the National Academy of Sciences, USA*, **90**, 6213–17.

Wang, S-H., Xue, C-B., Nielsen, O., Davey, J. & Naider, F. (1994). Chemical synthesis of the M-factor mating pheromone from *Schizosaccharomyces pombe*. *Yeast*, **10**, 595–601.

Wang, Y., Xu, H-P., Riggs, M., Rodgers, L. & Wigler, M. (1991). *byr2*, a Schizosaccharomyces pombe gene encoding a protein kinase capable of partial suppression of the *ras1* mutant phenotype. *Molecular and Cellular Biology*, **11**, 3554–63.

Watanabe, Y. & Yamamoto, M. (1994). *S. pombe mei2*⁺ encodes an RNA-binding protein essential for premeiotic DNA synthesis and meiosis I, which co-operates with a novel RNA species meiRNA. *Cell*, **78**, 487–98.

Watanabe, Y., Iino, Y., Furuhata, K., Shimoda, C. & Yamamoto, M. (1988). The *S. pombe mei2* gene encoding a crucial molecule for commitment to meiosis is under the regulation of cAMP. *EMBO Journal*, **7**, 761–7.

Watanabe, Y., Irie, K. & Matsumoto, K. (1995). Yeast *RLM1* encodes a serum response factor-like protein that may function downstream of the Mpk1 (Slt2) mitogen-activated protein kinase pathway. *Molecular and Cellular Biology*, **15**, 5740–9.

Watanabe, Y., Shinozaki-Yabana, S., Chikashige, Y., Hiraoka, Y. & Yamamoto, M. (1997). Phosphorylation of RNA-binding protein controls cell cycle switch from mitotic to meiotic in fission yeast. *Nature*, **386**, 187–90.

Watson, P. & Davey, J. (1998). Characterization of the Prk1 protein kinase from *Schizosaccharomyces pombe*. *Yeast*, **14**, 485–92.

Wedegaertner, P. B., Wilson, P. T. & Bourne, H. R. (1995). Lipid modifications of trimeric G proteins. *Journal of Biological Chemistry*, **270**, 503–6.

Whiteway, M., Hougan, L., Dignard, D., Thomas, D. Y., Bell, L., Saari, G. C., Grant, F. J., O'Hara, P. & MacKay, V. L. (1989). The *STE4* and *STE18* genes of yeast encode potential β and γ subunits of the mating factor receptor-coupled G protein. *Cell*, **56**, 467–77.

Wilkinson, M. G., Samuels, M., Takeda, T., Toone, W. M., Shieh, J-C., Toda, T., Millar, J. B. A. & Jones, N. (1996). The Atf1 transcription factor is a target for the Sty1 stress-activated MAP kinase pathway in fission yeast. *Genes and Development*, **10**, 2289–301.

Willer, M., Hoffmann, L., Styrkársdóttir, U., Egel, R., Davey, J. & Nielsen, O. (1995). Two-step activation of meiosis by the *mat1* locus in *Schizosaccharomyces pombe*. *Molecular and Cellular Biology*, **15**, 4964–70.

Wurgler-Murphy, S. M., Maeda, T., Witten, E. A. & Saito, H. (1997). Regulation of the *Saccharomyces cerevisiae* HOG1 mitogen-activated protein kinase by the PTP2 and PTP3 protein tyrosine phosphatases. *Molecular and Cellular Biology*, **17**, 1289–97.

Xu, H-P., White, M., Marcus, S. & Wigler, M. (1994). Concerted action of RAS and G proteins in the sexual response pathways of *Schizosaccharomyces pombe*. *Molecular and Cellular Biology*, **14**, 50–8.

Yabana, N. & Yamamoto, M. (1996). *Schizosaccharomyces pombe map1*$^+$ encodes a MADS-box family protein required for cell type-specific gene expression. *Molecular and Cellular Biology*, **16**, 3420–8.

Yamawaki-Kataoka, Y., Tamaoki, T., Choe, H-R., Tanaka, H. & Kataoka, T. (1989). Adenylate cyclases in yeast: a comparison of the genes from *Schizosaccharomyces pombe* and *Saccharomyces cerevisiae*. *Proceedings of the National Academy of Sciences, USA*, **86**, 5693–7.

York, R. D., Yao, H., Dillon, T., Ellig, C. L., Eckert, S. P., McCleskey, E. W. & Stork, P. J. S. (1998). Rap1 mediates sustained MAP kinase activation induced by nerve growth factor. *Nature*, **392**, 622–6.

Young, D., Riggs, M., Field, J., Vojtek, A., Broek, D. & Wigler, M. (1989). The adenylyl cyclase gene from *Schizosaccharomyces pombe*. *Proceedings of the National Academy of Sciences, USA*, **86**, 7989–93.

Zanolari, B., Raths, S., Singer-Krüger, B. & Riezman, H. (1992). Yeast pheromone receptor endocytosis and hyperphosphorylation are independent of G protein-mediated signal transduction. *Cell*, **71**, 755–63.

Zhan, X. L., Deschenes, R. J. & Guan, K. L. (1997). Differential regulation of FUS3 MAP kinase by tyrosine-specific phosphatases PTP2/PTP3 and dual-specificity phosphatase MSG5 in *Saccharomyces cerevisiae*. *Genes and Development*, **11**, 1690–702.

Zhang, S. J., Han, J. H., Sells, M. A., Chernoff, J., Knaus, U. G., Ulevitch, R. J. & Bokoch, G. M. (1995). Rho-family GTPases regulate p38 mitogen activated protein kinase through the downstream mediator Pak1. *Journal of Biological Chemistry*, **270**, 23 934–6.

SIGNALS AND INTERACTIONS BETWEEN PHYTOPATHOGENIC ZOOSPORES AND PLANT ROOTS

N. A. R. GOW, T. A. CAMPBELL, B. M. MORRIS, M. C. OSBORNE, B. REID, S. J. SHEPHERD AND P. VAN WEST

Department of Molecular and Cell Biology, Institute of Medical Sciences, University of Aberdeen, Aberdeen AB25 2ZD, UK

INTRODUCTION

Members of the genera *Phytophthora* and *Pythium* include some of the most important plant pathogens of commercial crops and other plants (for a comprehensive review, see Erwin & Ribeiro, 1996). These organisms are not true fungi in the phylogenetic sense, but are two families in the order Pythiaceae of the Oomycota, which are relatives of a group that include the golden algae. In all practical aspects they resemble true fungi, growing by elaboration of filamentous branching hyphae and the formation of asexual and sexual spores and are usually treated as fungi by mycologists (Money, 1998). One hallmark of these organisms is their ability to form biflagellated, swimming zoospores that are important for dispersal of the organism through films of water within wet soils at times of mild or warm humid weather. For plant pathogenic species, the aquatic zoospore plays a key role in the ability to find a new host to parasitize. Recent research has demonstrated that they have evolved a sophisticated battery of tactic and tropic responses, as well as mechanisms for encysting selectively on target root surfaces to facilitate their ability to locate and colonize potential plant hosts. The signalling systems involve the sensing of both chemical and electrical signals generated by plants that are used first to guide zoospores to the plant surface and then to bring about their immobilization, encystment and invasion of the host. In addition, the encysted zoospores form germ tubes that also exhibit tropic responses in relation to plant exudates and signals generated by groups of encysted zoospores. This chapter aims to review and evaluate these various zoospore–root and zoospore–zoospore interactions to try to establish the main cues that facilitate efficient plant targeting by zoospores of oomycete pathogens.

STRATEGY OF THE ZOOSPORE

Although some species of *Phytophthora* and *Pythium* can spread by the direct germination of air-borne sporangia, most pathogenic varieties can also locate new host plants by targeted homing responses of aquatic zoospores (Deacon & Donaldson, 1993). In the life of a zoospore everything is done in haste. They are produced in a hurry when the weather is wet and warm and then swim frantically through rain-drenched soils to attempt to find a host plant to reproduce on before dry conditions close the window of opportunity for further spread. For example, under ideal conditions it takes only 20–30 minutes for zoospores of *Phytophthora palmivora* to be formed via cytoplasmic cleavage then released via the sporangial papilla (S. Shepherd & N. A. R. Gow, unpublished data). Because zoospores are liberated from a sporangium within an hour, diseases caused by these fungi can be multicyclic, resulting in severe epidemics that can decimate whole crops within a single season. Most notable of these was the outbreak of potato blight in Ireland in the 1840s leading to widespread starvation and the emigration of local communities.

The role of the zoospore of a plant pathogen is to locate and encyst on the surface of a potential host. Failure to do so means death for that propagule. The discharged zoospore is not thought to assimilate nutrients, but rather it oxidizes stored lipids during its swimming phase (Bimpong, 1975). Hence the duration of activity of zoospores is constrained by its own limited energy supply and by environmental conditions (Erwin & Ribeiro, 1996). The duration of swimming is reduced at sub- or supra-optimal temperatures or by acidic pH values, and for most *Phytophthora* species would typically not exceed 24–48 h, even under ideal conditions (Erwin & Ribeiro, 1996). In most cases the active period would be very much shorter than this.

The frequency of collisions between zoospores and solid objects also reduces swimming time by increasing the likelihood of the zoospore encysting prematurely (Hickman & Ho, 1966, 1967; Bimpong & Clerk, 1970; Benjamin & Newhook, 1982). Overall, the range of the zoospore is in the region of millimetres to a few centimetres (Kulman, 1964) depending on soil water content (matric potential), porosity and presence of encystment agents. The journey time and range of an individual zoospore is such that it is not likely to encounter many roots of alternative hosts.

It is perhaps not surprising that the strategy of most zoospores is therefore to try to settle on, and infect, any plant it encounters in the limited time available, rather than to detect and selectively colonize only host species. As will be seen, many of the tactic and tropic mechanisms displayed by zoospores and early germ tubes lack any marked host specificity. Therefore, the host range of these organisms is not determined at the zoosporic phase (Gow, 1993). Gene-for-gene signalling (de Wit, 1995), which ultimately determines whether colonization will proceed as far as infection of the plant is determined later as the germ tube forms an appressorium and attempts to penetrate the host. Discussion here is limited to the critical early events that allow the fungus to build up its inoculum potential at the surface of a plant and position itself in the right place to attempt to overcome the host resistance mechanisms.

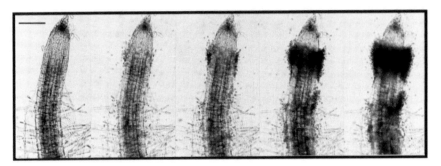

Fig. 1. Zoospores of *Phytophthora palmivora* accumulating around a rye grass root (*Lolium perenne*). Photographs were taken at time 0 min (left) and then at 15 min intervals after the addition of the zoospores. The scale bar is 200 μm.

ZOOSPORE DEVELOPMENTAL CYCLE

A striking demonstration of the speed at which zoospores can locate and encyst on a plant root can be provided by simply observing the zoospore–root interactions with a microscope (Figs. 1, 2). Within a minute, a swarm of zoospores can be seen in the vicinity of the root surface, often around the apex. A few minutes later, encysted zoospores will be piling up on one another and, within an hour, certain regions of the root will be entirely coated with a thick layer of cysts, many of which will have germinated (Figs. 1, 2). The speed of the process reflects the mechanics of the swimming process, the array of mechanisms that exist to bring the zoospore to the plant surface and the nature of the encystment mechanism.

Zoospore swimming and docking

The zoospore of oomycetes such as *Pythium* and *Phytophthora* species is a kidney-shaped cell with a flexible cortex and two flagella emanating from a groove on the ventral side (Cho & Fuller, 1989; Carlile, 1986). These two flagella bring about a helical swimming path with velocities up to 180 μm s^{-1} (Allen & Newhook, 1973; Fig. 3) punctuated by periodic turns in direction (Morris & Gow, 1993; Morris *et al.*, 1995). The pattern of swimming of zoospores of *Pythium* species has been shown to be affected by additions of the divalent cation chelator EGTA, and by various Ca^{2+}-modulating drugs, which modify the normal trajectory to one that is straight, circular, irregular or jerky (Donaldson & Deacon, 1993a). Ca^{2+} has been shown to play a critical role not only in swimming but also in the signals involved in encystment and germination stages of the zoospore life cycle (Hemmes & Da Silva, 1980; Byrt *et al.*, 1982a; Irving *et al.*, 1984; Iser *et al.*, 1989; Gübler *et al.*, 1990; Donaldson & Deacon, 1992; von Broembsen & Deacon, 1996,

(a)

(b)

(c)

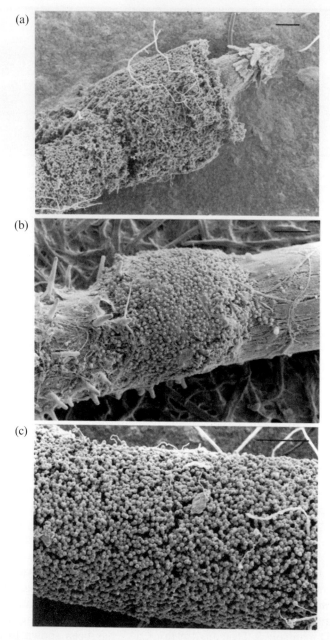

Fig. 2. Scanning electron micrographs of accumulated cysts of *Ph. palmivora* (a), (c) and *Py.*
aphanidermatum (b) on intact (a), (c) and wounded (b) rye grass roots. In (b) the wound has been
filled in completely by *Py. aphanidermatum* cysts. The scale bars are all 100 μm.

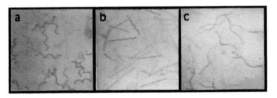

Fig. 3. Computer-generated traces of the swimming patterns of (a) *Py. aphanidermatum*, (b) *Ph. palmivora* and (c) *Py. catenulatum* recorded over 3 second intervals.

1997). The frequency of turning increases as the zoospore moves up a gradient of attractant (Carlile, 1983) or in the presence of an electrical field (Morris & Gow, 1993), suggesting a klinokinetic model for tactic swimming, as found in many bacteria. The anterior flagellum has been proposed to provide 90% of the propulsive power for swimming (Carlile, 1983; Holwill, 1985), while the longer posterior flagellum has been suggested to be involved in steering (Carlile, 1983). However, recent observations of zoospores of *Ph. palmivora* showed that, during a turn the zoospore soma turned prior to the reorientation of the posterior flagellum, which always assumed a trailing position (Morris *et al.*, 1995). This suggests that the anterior flagellum may also play a role in determining the direction of swimming and that the posterior flagellum may act to stabilize the zoospore, much like the tail of a kite (Morris *et al.*, 1995).

Zoospores swim for a greater or lesser period until mechanical collision or the presence of various chemical agents brings about encystment. Immediately prior to encystment, the zoospore may become sluggish in their movements before they stop, round up and shed or retract their two flagella (Hemmes, 1983). Prior to encysting on a plant, zoospores of *Phytophthora cinnamomi* exhibit a scanning behaviour, gliding ventral side down, back and forwards along on the surface of a root. They may then spin intermittently before settling and encysting (Hardham & Gübler, 1990). In other cases, zoospores exhibit an excited behaviour with frequent turns prior to encystment (Jones *et al.*, 1991; Reid *et al.*, 1995).

Encystment

The docking and encystment phase has been characterized in most detail by Hardham's group for zoospores of *Phytophthora cinnamomi* (Hardham & Gübler, 1990; Hardham *et al.*, 1991; Chambers *et al.*, 1995; Dearnaley *et al.*, 1996). These zoospores contain at least three types of vesicles in the peripheral cytoplasm: large peripheral vesicles that are located predominantly around the dorsal face of the zoospore (Gübler & Hardham, 1990) and two smaller vesicle populations distributed dorsally and ventrally. Glycoprotein within the two smaller vesicle populations is exocytosed within 2 min of the initiation of encystment (Hardham & Gübler, 1990).

The small dorsal vesicles secrete glycoproteins of a molecular weight exceeding 300 kDa which forms the cyst primary coat. The small ventral vesicles contain a 220 kDa proteinaceous Ca^{2+}-dependent adhesive which is exocytozed to form an adhesive pad which attaches the ventral face of the cyst to the substrate (Gübler & Hardham, 1988, 1990; Gübler *et al.*, 1989). The function of the high molecular weight proteins within the large peripheral proteins is less clear, since they become randomly distributed during encystment and do not form a structural part of the cyst wall. Therefore, adhesive material and the primary cyst coat are preformed within the zoospore and can be discharged rapidly to the surface to convert the zoospore into the attached early cyst. Finally, peripheral cisternae in the zoospore vesiculate and form the cyst wall, which is strengthened by *de novo* synthesis of various β-glucans. Zoospores are therefore equipped with preformed adhesins and wall components that enable them to anchor themselves rapidly to their targets and lay down a protective shield prior to mounting their ultimate assault on the plant. Cysts germinate rapidly, often forming a germ tube within 20–30 minutes. The site of germ tube formation of *Pythium* (Mitchell & Deacon, 1986) and *Phytophthora* (Paktitis *et al.*, 1986) has been shown to be fixed, so that evagination occurs on the former ventral face of the zoospore. Because zoospores encyst with their ventral surface towards the root (Gübler & Hardham, 1990), this positions the evaginating germ tube in an ideal orientation, tip towards the root, for penetration.

CHEMICAL SIGNALS IN THE RHIZOSPHERE

Chemotaxis

The rhizosphere is a densely populated niche. One major reason for this is that as much as one-third of the photoassimilate produced in the green tissues is eventually leached out of the plant at the root tips around the meristematic and cell elongation regions. These exudates clearly stimulate microbial growth. Hence, it makes intuitive sense that microbes, such as zoospores, would have evolved chemotactic systems that would naturally lead them to roots which they could subsequently parasitize. In this discussion of chemotaxis, two general points will be underlined. First, there is little evidence for specific chemotactic responses of zoospores to the roots of their host plants and secondly many reports of chemotaxis do not distinguish between *bona fide* chemotaxis, taken to mean directional swimming in a chemical gradient, and apparent chemotaxis due to chemical induction of encystment. Other aspects of chemotaxis are dealt with in detail elsewhere (Wynne, 1981; Carlile, 1983, 1986; Deacon & Donaldson, 1993).

Despite considerable effort, reflected in an extensive literature, there is little evidence that specific zoospore species exhibit preferential chemotaxis towards the specific components of the exudates from host plants (Cunningham & Hagedorn, 1961; Royle & Hickman, 1964; Hickman & Ho, 1966; Ho & Hickman, 1967; Deacon & Donaldson, 1993; Gow, 1993). Zoospores of *Ph. cinnamomi* have been shown to accumulate preferentially on avocado roots over non-host roots (Zentmyer, 1961) although this fungus has a broad host range and selectivity for other hosts has not been demonstrated. *Phytophthora sojae* zoospores have been reported to respond to three soybean isoflavones at nanomolar concentrations and zoospores of other non-soybean parasites failed to exhibit taxis to these compounds (Morris & Ward, 1992). A few other reports exist of differential accumulation of zoospores, but in no cases has it been possible to relate the findings unequivocally to a host-specific attractant. *In vitro* experiments also provide little evidence of specificity of chemotaxis, although amino acids have been found in general to be more powerful attractants than sugars or alcohols (Khew & Zentmyer, 1973; Carlile, 1983; Donaldson & Deacon, 1993*b*). Only in the case of taxis to isovaleraldehyde by zoospores of *Ph. palmivora* has any differential competition for binding to a putative chemoreceptor been demonstrated (Cameron & Carlile, 1981). In many cases, compounds that are reported to cause chemotaxis also accelerate encystment and this complicates the interpretation of most chemotaxis assays. The application of exudates of plant roots has also been reported to shorten the time of zoospore swimming by inducing encystment (Ho & Hickman, 1967). The immobilization of zoospores in a region of high local concentration of an encystment-inducing agent may give the appearance that chemotaxis has taken place, yet this accumulation may not involve any directional swimming response. A further complication in assessing the role of chemotaxis results from what is now known about autotaxis and autoaggregation and is discussed later.

Although it is clear that zoospores are highly attracted towards wound sites while others are not (see Figs. 2, 7), wound-exudates are likely to contain most or all of the same low molecular weight solutes exuded from root apices. Given that chemotaxis is relatively non-specific, and most zoospores do not even distinguish between roots of host and non-host plants, the basis for the differential accumulation of zoospores at the apex or at wound sites is very difficult to explain by what is known about zoospore chemotaxis.

Induced encystment

Although there is little evidence for specificity in chemotaxis of zoospores towards specific roots, some oomycetes have been found to encyst preferentially on host rather than on non-host roots (Mitchell & Deacon, 1986*a*). Some low molecular weight components of plant exudates can, and do, induce encystment. Other encystment triggers exist that may also play important roles in the rhizosphere. A likely signal leading to encystment results from the physical association of the flagella of the zoospore with specific components of the root, in particular with certain saccharide components of complex carbohydrates in the root mucilage (Donaldson & Deacon, 1993*b*). Cellulose and chitin caused selective encystment of a variety of zoosporic fungi, but encystment on cellulose was not limited to plant pathogenic varieties (Mitchell & Deacon, 1986*b*). A variety of experiments using *in vitro* application of extracts of components from root mucilage, oxidative treatment of the root and the addition of blocking agents such as lectins (Hinch & Clark, 1980; Longman & Callow, 1987) and dyes that complexed with the root surface (Mitchell & Deacon, 1986*a*) all lend support to the view that components of the plant wall can cause species or isolate-specific encystment. In particular, certain fucosyl-containing compounds (Hinch & Clark, 1980; Longman & Callow, 1987) and polyuronates, which are major components of root mucilage (Byrt *et al.*, 1982*b*; Zhang *et al.*, 1990; Donaldson & Deacon, 1993*c*) are implicated. In addition, monoclonal antibodies that recognize specifically flagellar epitopes trigger encystment (Hardham & Suzaki, 1986; Estrada-Garcia *et al.*, 1990). Therefore, some sort of receptor–ligand interaction is suggested between the zoospore flagella and the root mucilage. However, experiments in which a thick coat of calcium-alginate was applied to a root have been used to prevent physical interaction between zoospore flagella and the plant root surface and the resulting findings that cannot be explained by such a mechanism (Jones *et al.*, 1991; Donaldson & Deacon, 1993*c*). The presence of these gel coatings did not affect the localized encystment patterns on roots, indicating that direct contact of the root and zoospores was not necessary for normal encystment. The gels would not represent a barrier to root exudate diffusion or resistance to electrical current. Taken together, these studies show that a variety of signals and behavioural responses are involved in bringing about the immobilisation of zoospores at the plant root surface.

AUTOAGGREGATION

In the absence of a target root or any chemical encystment agent, zoospores of certain species have been shown to form clumps of encysted cells called 'autoaggregates'. Such aggregates have been described for zoospores of *Phytophthora dreschsleri* (Porter & Shaw, 1978), *Ph. palmivora* (Ko & Chase, 1973; Reid *et al.*, 1995), *Py. dissotocum* (Reid *et al.*, 1995) and various *Achlya* species (Thomas & Peterson, 1990). Some species do not form such aggregates. Direct examination of the pattern of motility of zoospores of *Ph. palmivora* adjacent to an autoaggregate showed that the zoospores exhibited taxis towards the aggregate centre (Reid *et al.*, 1995). This is presumed to represent a chemotactic mechanism, although no chemotactic compound for autotaxis has thus far been identified. In addition, germ tubes emanating from cysts adjacent to an aggregate exhibited strong tropic orientation towards the aggregate. However, follow-

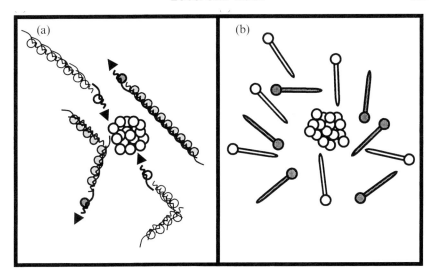

Fig. 4. Genus and species-specific taxis and tropism of zoospores and germ tubes respectively of *Phytophthora* and *Pythium* species. In (a) the shaded zoospores do not exhibit taxis towards the aggregate of cysts formed by non-shaded zoospores. However, swimming zoospores of the same type as those in the aggregate do exhibit taxis towards, and encystment on, the aggregate. In (b) the aggregate elicits tropic orientation from germinated cysts of the same species, but not of a different species.

ing germination, germ tubes grew away from the mature aggregate, suggesting that a secondary tropic mechanism was overriding the autotropic mechanism, or that the chemoattractant produced by the aggregation centre was reduced or saturated (Reid *et al.*, 1995). The aggregation centre was shown using microelectrodes to be a source of calcium ion secretion, and $CaCO_3$ crystals acted as nuclei for aggregate formation, while the calcium ion chelator EGTA prevented autoaggregation.

Aggregates of *Ph. palmivora* failed to attract zoospores of *Py. catenulatum* and *vice versa*, aggregates of *Py. catenulatum* were not attractive to *Ph. palmivora* (Reid *et al.*, 1995). It has also been shown recently that *Phytophthora parastica* zoospores and germ tubes do not respond to *Ph. palmivora* aggregates (T. Campbell & N. A. R. Gow, unpublished data). The tactic and tropic responses of zoospores and germ tubes of *Pythium* and *Phytophthora* species to aggregates are therefore both genus and species-specific (Fig. 4). This conclusion suggests that some specific pheromone or chemotactic signalling system is required for autoaggregation. Calcium ions are unlikely to act in this capacity, since it is difficult to see how species-specificity could be imparted by a ubiquitous divalent cation.

There are a number of important implications and deductions that can be made based on these observations. First, although the experiments described above were done in the absence of a host plant, it is very likely that autoaggregation plays a key role in the recruitment of zoospores to plant

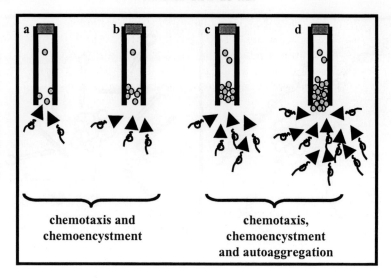

Fig. 5. Interpretation of the factors that play a role in swim-in tests used to measure chemotaxis. Initially, zoospores are attracted towards and then encyst at the high part of the nutrient gradient established within and adjacent to a capillary tube containing a test nutrient (a), (b). As the number of encysted zoospores in the capillary increases the effect of chemotactic factors released from the aggregates of cysts has a greater influence on the overall accumulation that is observed (c), (d).

surfaces. Plant-generated chemical or electrical signals may be responsible for the initial attraction phase and for the induction of encystment of the first zoospores to encyst on the surface. However, for zoospores that exhibit aggregation phenomena, the continued build-up of zoospores on the root surface is very likely to also involve autosignalling. The net effect is that the strength of the signal for zoospore targeting to a root will increase as a function of the number of zoospores that arrive at the surface. Since this autorecruitment signal is species-specific such a build-up in inoculum potential would occur without causing the attraction of competitive species, and may therefore function in excluding competitors at the infection court.

Autoaggregation is also likely to complicate the interpretation of commonly used chemotaxis or electrotaxis assays. For example, 'swim-in tests' are used to assay potential chemoattractants. The large cyst balls that accumulate at the mouths of the pipettes may accentuate the overall tactic response (Fig. 5). The greater the number of cysts, the larger will be the influence of autotaxis relative to the imposed primary chemical gradient.

Finally, it is possible that autoaggregation *per se* is adaptive and may function as a survival mechanism, particularly under conditions of low internal energy and when no external source of nutrients has been detected (Reid *et al.*, 1995). Zoospores do not take up exogenous nutrients, which are evidently only accessible at the stage when mature hyphae are first formed

(Penington *et al.*, 1989; Madsen *et al.*, 1995). Within an autoaggregate, nutrients may be recycled and used by germinating cysts to increase survival and promote germ tube growth. In any case, it is clear that zoosporic fungi may make use of both endogenous and exogenous signals in root targeting via attraction to the root itself and to aggregates of cysts that have already docked successfully with a root.

ELECTRICAL SIGNALS

It has long been assumed that secreted exudates from the meristematic region, zone of cell elongation and wounds represent the key chemical signals that are sensed by zoospores searching for potential host plants. It is now known that plant roots also generate electrical signals within the rhizosphere. The need to maintain ionic balance within root cortical cells results in the generation of ionic circulations by electrogenic ion transporters in the cortical cell membrane. Asymmetries in the spatial distribution of these ion pumps and channels along the root generate electrical currents in the rhizosphere. These currents can be detected by ultrasensitive vibrating microelectrodes (Jaffe & Nuccitelli, 1974; Weisenseel *et al.*, 1979; Behrens *et al.*, 1982; Miller *et al.*, 1991), and the associated electrical field can be calculated from Ohms law and a knowledge of the resistivity of the liquid in which the root is growing. This type of high-resolution voltage-sensitive microelectrode can be used to map the sites at which there is a net inward (cathodic) or outward (anodic) flux of positive electrical current. Vibrating electrodes do not invade the cell and therefore can be used to map the currents of growing cells or tissues without damaging them. To date, such ionic currents have been used to detect endogenous electrical currents of a huge variety of eukaryotic cell types from fungi to chick embryos (Jaffe, 1981; Gow, 1989; Nuccitelli, 1988). In plant roots, the pattern of these electrical currents has been correlated with root growth, polarity, gravitropism and wound healing activities (see, for example, Rathore *et al.*, 1990; Weisenseel *et al.*, 1979; Collings *et al.*, 1992; Hush *et al.*, 1992). In this review the context of such electrical currents is, however, ecological rather than physiological.

Because these currents are carried by ions, standing gradients of pH, Ca^{2+} or other ions that are components of the ionic circulation are also generated along the axis of the electrical circuit. For example, root currents are carried predominantly by protons (e.g. Miller & Gow, 1989*b*; Collings *et al.*, 1992; Hush *et al.*, 1992). Consequently, a pH gradient is generated within the rhizosphere between sites of proton efflux and influx (Weisenseel *et al.*, 1979; Miller *et al.*, 1991) and is locally acidic at regions of proton efflux and alkaline at regions of proton uptake. Therefore, the rhizosphere can be regarded as an electrochemically dynamic environment with distinct anodic, cathodic, acidic and alkaline domains. In this section, evidence is summarized that suggests that zoospores take advantage of these electrochemical signals to guide their movements towards the root surface.

Specifications of the root battery

A number of parameters affect the magnitude and direction of current flow – plant species, orientation of the root, salt content of the medium and hence its resistivity, pH, the source of combined nitrogen and the presence of plant growth regulators, such as indole acetic acid (IAA) or fusicoccin (for a review, see Gow *et al.*, 1992). The current densities adjacent to the surface of healthy, growing roots are generally between 0.1 to 10 μA cm^{-2}, which give rise to electrical fields of 0.5 to 50 mV cm^{-1} for roots growing in soil water of a moderate resistivity of around 5000 Ω cm (Gow *et al.*, 1992; Gow & Morris, 1995). The electrical current density and associated electrical field adjacent to a wound can be several-fold higher than this (Miller *et al.*, 1988; Hush *et al.*, 1992). The magnitude of electrical fields decreases as the square of the distance from the current-generating root surface. Therefore, the zone of influence on zoospore movement only extends over a distance of several hundred microns from the root. Current at the apical region can be inward (cathodic) or outward (anodic), depending on the plant root species (Gow *et al.*, 1992) and can be influenced by local concentrations of growth regulators (Miller *et al.*, 1989*b*; Rathore *et al.*, 1990). The direction of flow of apical current may also be different at the root cap and meristematic and cell elongation regions. The mature region further behind the root apex is associated with current flow in the opposite direction to the apical current so that net charge flow in and out of the root is conserved over the root as a whole. As a consequence, the electrical field varies both quantitatively and qualitatively along the root.

One consistent finding is that wound currents are always sites of inward flow and hence are cathodic (Miller *et al.*, 1988; Hush *et al.*, 1992). This is the case, irrespective of whether a wound is made in a region of the root that was previously anodic or cathodic with respect to the endogenous ionic circulation. Wound currents and electrical fields are generally larger than the currents and fields around intact roots, and are confined spatially to the immediate vicinity of the wound.

Zoospore responses to applied and endogenous electrical fields

Given that plant roots generate electrical fields in the rhizosphere, it is of interest to establish whether electrical fields influence zoospore swimming. Several early reports exist of zoospore electrotaxis *in vitro* (Troutman & Wills, 1964; Katsura *et al.*, 1966; Ho & Hickman, 1967; Khew & Zentmeyer, 1974). Most of these investigations employed bare electrical wires to apply local electrical fields and did not consider possible artefacts that inevitably result from the formation of local products of hydrolysis next to the electrodes. In addition, these studies preceded the advent of the vibrating probe, and hence the means to measure accurately the size of physiological electrical fields. Mostly, they employed fields that were one or two orders of magnitude larger than those recorded around roots. Although these reports are certain to be undermined to some extent by experimental artefacts, they do suggest that electrical fields can simulate either positive or negative electrotaxis or can cause the immobilization of zoospores by induced encystment (Gow $et\ al.$, 1992).

More recently, the possible role of electrotaxis in zoospore targeting to the root was re-evaluated. Using a chamber that protected zoospores from electrode products it was shown that zoospores of $Ph.\ palmivora$ exhibited anodotaxis in electrical fields $\geqslant 5\ mV\ cm^{-1}$, and that the response saturated at around $100\ mV\ cm^{-1}$ (Morris $et\ al.$, 1992). The sensitivity of the electrotactic response is therefore well within the range of endogenous electrical fields found around roots. Control experiments showed that electrotaxis was not due to net electrophoresis or electroosmosis of the zoospore (Morris $et\ al.$, 1992). In a further study, it was demonstrated that zoospores of $Py.$ $aphanidermatum$ were cathodotactic in physiological electrical fields, and that those of $Pythium\ catenulatum$ and $Pythium\ dissotocum$ were only anodotactic in strong fields that were above those normally found in the rhizosphere (Morris & Gow, 1993).

The mechanism of electrotaxis was studied by observing the effects of physiological electrical fields on swimming zoospores and on zoospores that had been paralysed with sodium azide. Swimming zoospores were not affected in the velocity of movement to the anode or cathode, but the turning frequency increased three to five times (Morris & Gow, 1993). Paralysed zoospores were allowed to settle on the base of a chamber and observed under dark field microscopy to enable visualization of the anterior and posterior flagella. In an electrical field, the posterior flagella of anodotactic $Ph.\ palmivora$ zoospores swung to point to the cathode, while the posterior flagellum of cathodotactic $Py.\ aphanidermatum$ zoospores swung to point to the anode (Fig. 6; Morris & Gow, 1993). This suggested that the zoospore and its flagella act as a charged dipole and were orientated electrophoretically in an electrical field so that the anterior points to one or other electrical pole. The mechanism of electrotaxis is therefore due to a combination of factors including physical alignment in the field and modulation of turning frequency.

These in vitro experiments suggest that electrotaxis may also operate within soils in the rhizosphere. The currents around roots of host and non-host plants were therefore mapped and then the zones of accumulation of anodotactic and cathodotactic zoospores observed (Fig. 7; B. M. Morris, B. Reid & N. A. R. Gow, unpublished data). Cathodotactic zoospores of $Py.$ $aphanidermatum$ did not encyst at the anodic apical region of rye grass roots,

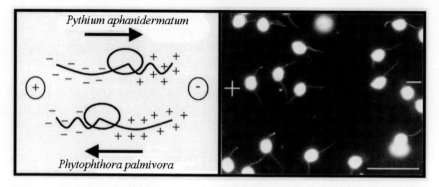

Fig. 6. Physical alignment of zoospores in an electrical field due to a charge dipole across the anterior and posterior flagella. The cartoon on the left shows the inferred charge distribution on the two flagella of anodotactic zoospores of *Ph. palmivora* and cathodotactic zoospores of *Py. aphanidermatum*. On the right is a dark field micrograph showing the alignment of flagella of paralysed zoospores of *Ph. palmivora* in an electrical field of 0.5 V/cm (from Morris & Gow, 1993). Note the orientation of the longer posterior flagella towards the cathode. The scale bar is 30 µm.

but accumulated rapidly at sites of local wounds and at the distal cathodic region including the root hair zone (Figs. 1, 2). When a root that had first been exposed to *Py. aphanidermatum* zoospores was then wiped gently clean of cysts and placed in a suspension of anode-seeking zoospores of *Ph. palmivora*, these accumulated at mutually exclusive regions of the roots. Zoospores of *Ph. palmivora* were attracted to the anodic apical region and repelled from cathodic wound sites (Figs. 2, 7). However, *Ph. palmivora* zoospores also accumulated and encysted at the distal root hair zone, which was cathodic, suggesting that more than one guidance cue was being employed in this case. In general the *in vitro* electrotactic behaviour predicted the zones where zoospores accumulated on roots, suggesting that the electrical field had an important role in root-targeting by zoospores. There is very little evidence for root exudates causing specific chemotactic responses of different zoospore types (see above); therefore, these observations of differential accumulations of zoopores of *Ph. palmivora* and *Py. aphanider-matum* are difficult to explain by a chemical-based chemotactic signalling system. To do so it would be necessary to invoke the presence of wound and root-apex specific attractants and/or repellents that differentially affected the swimming behaviour of *Ph. palmivora* and *Py. aphanidermatum* zoospores.

In addition to having a role in attraction to the root, the endogenous electrical field may also influence zoospore targeting by stimulating encystment. It has been observed that point sources of electrical currents passed through micropipettes not only attract zoospores of the appropriate electrotactic behaviour but also induce their encystment (B. M. Morris, B. Reid & N. A. R. Gow, unpublished data). These micropipette-generated focal electrical fields could also be used to bring zoospores to the surface of a

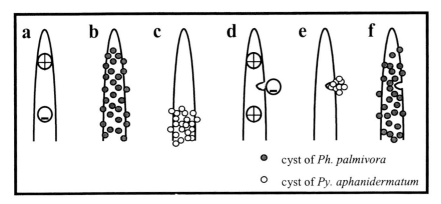

Fig. 7. Cartoon summarizing an experiment in which the spatial distribution of anodotactic zoospores of *Ph. palmivora* and cathodotactic zoospores of *Py. aphanidermatum* are described in relation to the endogenous anodic and cathodic regions of a root of rye grass as determined with a vibrating microelectrode. In the experiments the roots are exposed first to zoospores of *Ph. palmivora* then wiped clean and exposed to zoospores of *Py. aphanidermatum* or vice versa. The pattern for a non-wounded root (a)–(c) and a wounded root (d)–(f) are shown.

plant root at a location from which it would normally be excluded. For example, *Py. aphanidermatum* zoospores could be recruited and then made to encyst at the anodic apex of a rye grass root by local application of negative current adjacent to the surface of the root tip. Normally, these zoospores would avoid this region at the root. Therefore, the endogenous electrical field of a root may play roles both in the initial tactic response to the surface and in the immobilization of the zoospore by encystment.

pH gradients and pH taxis

Protons are the main ion accounting for the transcellular electrical current of plant roots (Weisenseel *et al.*, 1979; Miller & Gow, 1989*b*). These currents therefore also create pH gradients within the rhizosphere that may result in the local pH being almost three pH units different between sites of proton influx and efflux (Miller *et al.*, 1991; Morris *et al.*, 1995). pH-taxis of zoospores has been reported *in vivo* several times (Allen & Harvey, 1974; Cameron & Carlile, 1980; Morris *et al.*, 1995). The pH tactic behaviour of zoospores of *Ph. palmivora* towards acidic buffers is consistent with their attraction to local zones of acidity measured with a pH microelectrode placed adjacent to the apex of host cocoa roots (Morris *et al.*, 1995). Similarly, these zoospores were not attracted to wounds, which were found to be transiently alkaline. However, the pattern of accumulation around wounded and non-wounded roots was not affected by the presence of buffers that could completely abolish the pH gradients (Morris *et al.*, 1995). Therefore, pH-taxis is unlikely to play a major role in zoospore targeting of plant roots.

CONCLUSIONS AND FUTURE DIRECTIONS

It was once assumed that chemotaxis was sufficient to explain the marked accumulations of zoospores around plant roots; however, things now appear to be more complex. A multitude of chemical, electrical, endogenous and exogenous factors have been implicated in having more or less significant roles to play in zoospore-targeting of plants (Table 1). Recent studies of electrotaxis go a long way to providing an explanation for the conundrum of why some zoospores ignore, or are repelled from, a nutrient rich wound site, while others are highly attracted by a wound and do not encyst on adjacent regions of a root. However, it is clear that chemotaxis and induced encyst-ment mechanisms also play important roles. The fact that so many signals, cues and behavioural responses can be employed by zoospores highlights their very specific and urgent role homing in on potential hosts during limited periods in which the soil is wet enough to swim. Decaying plant material and other inert sources of nutrients in soil could act as decoys that would deflect zoospores from their task of finding a viable host that would allow reproduction and survival of the fungus. Perhaps such a combination of signalling systems allows the zoospore not only to find a source of nutrients but also to avoid accidental encystment on dead materials.

Much of the work described above deals with the phenomenology of zoospore–root and zoospore–zoospore interactions. *Phytophthora* genomics projects are now well under way, and the molecular analysis of such phenomena is coming rapidly of age. Gene-silencing technologies have also been developed that will permit functional analysis of genes implicated in the interaction of *Phytophthora* and eventually other plant pathogenic zoosporic fungi, with the their hosts (Judelson *et al.*, 1991, 1992, 1993; Kamoun *et al.*, 1998). The ability to make strains that are silenced at different developmental stages in the infection cycle and in specific host-fungus recognition processes

Table 1. *Factors that may be involved in root-targeting by phytopathogenic zoospores*

Signal or behavioural response	Specificity of response
Zoospore chemotaxis	Not specific
Chemically induced zoospore encystment	Variable
Germ tube chemotropism	Not specific
Autoaggregation (autotaxis)	Specific
Auto-encystment	Not known
Germ tube autotropism	Specific
Electrotaxis to either anodic or cathodic regions of roots	Not specific
Electrically induced encystment	Not specific
Zoospore pH taxis	Not specific

will help clarify and disentangle the true significance of each of these processes.

ACKNOWLEDGEMENTS

Work in the authors' laboratory was supported by grants from BBSRC and NERC. We thank Debbie Marshall for help with the SEM in Fig. 2.

REFERENCES

Allen, R. N. & Harvey, J. D. (1974). Negative chemotaxis of zoospores of *Phytophthora cinnamomi*. *Journal of General Microbiology*, **84**, 28–38.

Allen, R. N. & Newhook, F. J. (1973). Chemotaxis of zoospores of *Phytophthora cinnamomi* to ethanol in capillaries of soil pore dimensions. *Transactions of the British Mycological Society*, **61**, 287–302.

Behrens, H. M., Weisenseel, M. H. & Sievers, A. (1982). Rapid changes in the pattern of electric current around the root tip of *Lepidium sativum* L. following gravistimulation. *Plant Physiology*, **70**, 1079–83.

Benjamin, M. & Newhook, F. J. (1982). Effect of glass microbeads on *Phytophthora* zoospore motility. *Transactions of the British Mycological Society*, **78**, 43–6.

Bimpong, C. E. (1975). Changes in metabolic reserves and enzyme activities during zoospore motility and cyst germination *in Phytophthora palmivora*. *Canadian Journal of Botany*, **53**, 1411–16.

Bimpong, C. E. & Clerk, G. C. (1970). Motility and chemotaxis in zoospores of *Phytophthora palmivora* (Butl.) Butl. *Annals of Botany*, **34**, 617–24.

Byrt, P. N., Irving, H. R. & Grant, B. R. (1982*a*). The effect of cations on zoospores of the fungus *Phytophthora cinnamomi*. *Journal of General Microbiology*, **128**, 1189–98.

Byrt, P. N., Irving, H. R. & Grant, B. R. (1982*b*). The effect of organic compounds on the encystment, viability and germination of zoospores of *Phytophthora cinnamomi*. *Journal of General Microbiology*, **128**, 2343–51.

Cameron, J. N. & Carlile, M. J. (1981). Binding of isovaleraldehyde, an attractant, to zoospores of the fungus *Phytophthora palmivora* in relation to zoospore chemotaxis. *Journal of Cell Science*, **49**, 273–81.

Cameron, J. N. & Carlile, M. J. (1978). Fatty acids, aldehydes and alcohols as attractants for zoospores of *Phytophthora palmivora*. *Nature* (London), **271**, 448–9.

Cameron, J. N. & Carlile, M. J. (1980). Negative chemotaxis of zoospores of the fungus *Phytophthora palmivora*. *Journal of General Microbiology*, **120**, 347–53.

Carlile, M. J. (1983). Motility, taxis and tropism in *Phytophthora*. In *Phytophthora: Its Biology, Taxonomy, Ecology, and Pathology*, ed. D. C. Erwin, S. Bartnicki-Garcia & P. H. Tsao, pp. 95–107. St Paul: The American Phytopathological Society.

Carlile, M. J. (1986). The zoospore and its problems. In *BMS Symposium 11, Water, Fungi and Plants*, ed. P. G. Ayres & L. Boddy, pp. 105–18. Cambridge: Cambridge University Press.

Chambers, S. M., Hardham, A. R. & Scott, E. S. (1995). *In planta* immunolabelling of three types of peripheral vesicles in cells of *Phytophthora cinnamomi* infecting chestnut roots. *Mycological Research*, **99**, 1281–8.

Cho, C. W. & Fuller, M. S. (1989). Ultrastructural organisation of freeze-substituted zoospores of *Phytophthora palmivora*. *Canadian Journal of Botany*, **67**, 1493–9.

Collings, D. A., White, R. G. & Overall, R. L. (1992). Ionic current changes associated with the gravity-induced bending response in roots of *Zea mays* L. *Plant Physiology*, **100**, 1417–26.

Deacon, J. W. & Donaldson, S. P. (1993). Molecular recognition in the homing responses of zoosporic fungi, with special reference to *Pythium* and *Phytophthora*. *Mycological Research*, **97**, 1153–71.

Dearnaley, J. D. W., Maleszka, J. M. & Hardham, A. R. (1996). Synthesis of zoospore peripheral vesicles during sporulation of *Phytophthora cinnamomi*. *Mycological Research*, **100**, 39–48.

De Wit, P. J. G. M. (1995). Fungal avirulence genes and plant resistance genes: unravelling the molecular basis of gene-for-gene interactions. In *Advances in Botanical Research*, ed. J. H. Andrews & I. C. Tommerup, Vol. 21, pp. 147–84. London: Academic Press.

Donaldson, S. P. & Deacon, J. W. (1992). Role of calcium in adhesion and germination of zoospore cysts of *Pythium*: a model to explain infection of host plants. *Journal of General Microbiology*, **138**, 2051–9.

Donaldson, S. P. & Deacon, J. W. (1993a). Changes in motility of *Pythium* zoospores induced by calcium and calcium-modulating drugs. *Mycological Research*, **97**, 877–83.

Donaldson, S. P. & Deacon, J. W. (1993b). Effects of amino acids and sugars on zoospore taxis, encystment, and cyst germination in *Pythium aphanidermatum* (Edson) Fitzp., *P. catenulatum* Mathews and *P. dissotocum* Drechs. *New Phytologist*, **123**, 289–95.

Donaldson, S. P. & Deacon, J. W. (1993c). Differential encystment of zoospores of *Pythium* species by saccharides in relation to establishment on roots. *Physiological and Molecular Plant Pathology*, **42**, 177–84.

Erwin D. C. & Ribiero, O. K. (1996). *Phytophthora Diseases Worldwide* St Paul: APS Press.

Estrada-Garcia, T., Ray, T. C., Green, J. R., Callow, J. A. & Kennedy, J. F. (1990). Encystment of *Pythium aphanidermatum* zoospores is induced by root mucilage polysaccharides, pectin and a monoclonal antibody to a surface antigen. *Journal of Experimental Botany*, **41**, 693–9.

Gow, N. A. R. (1989). Ciculating ionic currents in micro-organisms. *Advances in Microbial Physiology*, **30**, 89–123.

Gow, N. A. R. (1993). Non-chemical signals used for host location and invasion by fungal pathogens. *Trends in Microbiology*, **1**, 45–50.

Gow, N. A. R. & Morris, B. M. (1995). The electric fungus. *Botanical Journal of Scotland*, **47**, 263–77.

Gow, N. A. R., Morris, B. M. & Reid, B. (1992). The electrophysiology of root–zoospore interactions. In *Perspectives in Plant Cell Recognition*, Society for Experimental Biology Seminar Series, ed. J. A. Callow & J. R. Green, Vol. 48, pp. 173–92. Cambridge: Cambridge University Press.

Grant, B. R., Irving, H. R. & Radda, M. (1985). The effect of pectin and related compounds on encystment and germination of *Phytophthora palmivora* zoospores. *Journal of General Microbiology*, **131**, 669–76.

Gübler F. & Hardham, A. R. (1988). Secretion of adhesive material during encystment of *Phytophthora cinnamomi* zoospores, characterised by immunogold labelling with monoclonal antibodies to components of peripheral vesicles. *Journal of Cell Science*, **90**, 225–35.

Gübler F. & Hardham, A. R. (1990). Protein storage in large peripheral vesicles in *Phytophthora* zoospores and its breakdown after cyst germination. *Experimental Mycology*, **14**, 393–404.

Gübler, F., Hardham, A. R. & Duniec, J. (1989). Characterising adhesiveness of *Phytophthora cinnamomi* zoospores during encystment. *Protoplasma*, **149**, 24–30.

Gübler, F., Jablonsky, P. P., Jadwiga, D. & Hardham, A. R. (1990). Localisation of calmodulin in flagella of zoospores of *Phytophthora cinnamomi*. *Protoplasma*, **155**, 233–8.

Hardham, A. R. & Gubler, F. (1990). Polarity of attachment of zoospores of a root pathogen and pre-alignment of the emerging germ tube. *Cell Biology International Report*, **14**, 947–56.

Hardham, A. R. & Suzaki, E. (1986). Encystment of zoospores of the fungus, *Phytophthora cinnamomi*, is induced by specific lectin and monoclonal antibody binding to the cell surface. *Protoplasma*, **133**, 165–73.

Hardham, A. R., Gübler, F., Duniec, J. & Eliott, J. (1991). A review of methods for the production and use of monoclonal antibodies to study zoosporic plant pathogens. *Journal of Microscopy*, **162**, 305–18.

Hemmes, D. E. (1983). Cytology of *Phytophthora*. In *Phytophthora: Its Biology, Taxonomy, Ecology and Pathology*, ed. D. C. Erwin, S. Bartnicki-Garcia & P. H. Tsao, pp. 9–40. St. Paul: American Phytopathological Society.

Hemmes, D. E. & Pinto Da Silva, P. (1980). Localisation of secretion-related, calcium-binding substrates in encysting zoospores of *Phytophthora palmivora*. *Biologique Cellulaire*, **37**, 235–40.

Hickman, C. J. & Ho, H. H. (1966). Behaviour of zoospores in plant-pathogenic phycomycetes. *Annual Reviews of Phytopathology*, **4**, 195–220.

Hinch, J. M. & Clarke, A. E. (1980). Adhesion of fungal zoospores to root surfaces is mediated by carbohydrate determinants of the root slime. *Physiological Plant Pathology*, **16**, 303–7.

Ho, H. H. & Hickman, C. J. (1967). Factors governing zoospore responses of *Phytophthora megasperma* (var. *Sojae*) to plant roots. *Canadian Journal of Botany*, **45**, 1983–94.

Holwill, M. E. J. (1985). Dynamics of eukaryotic flagellar movement. In *Symposia of the Society for Experimental Biology XXXV: Prokaryotic and Eukaryotic Flagella*, ed. W. B. Amos & J. G. Ducket, pp. 289–312. Cambridge: Cambridge University Press.

Hush, J. H., Newman, I. A. & Overall, R. L. (1992). Utilization of the vibrating probe and ion-selective microelectrode techniques to investigate electrophysiological responses to wounding in pea roots. *Journal of Experimental Botany*, **43**, 1251–7.

Irving, H. R. & Grant, B. R. (1984). The effect of calcium on zoospore differentiation in *Phytophthora cinnamomi*. *Journal of General Microbiology*, **130**, 1569–76.

Irving, H. R., Griffith, J. M. & Grant, B. R. (1984). Calcium efflux associated with encystment of *Phytophthora palmivora* zoospores. *Cell Calcium*, **5**, 487–500.

Iser, J. R., Griffith, J. M., Balson, A. & Grant, B. R. (1989). Accelerated ion fluxes during differentiation in zoospores of *Phytophthora palmivora*. *Cell Differentiation and Development*, **26**, 29–38.

Jaffe, L. F. (1981). The role of ionic currents in establishing developmental pattern. *Philosophical Transactions of the Royal Society of London*, **B295**, 553–66.

Jaffe, L. F. & Nuccitelli, R. (1974). An ultrasensitive vibrating probe for measuring steady extracellular currents. *Journal of Cell Biology*, **63**, 614–28.

Jones, S. W., Donaldson, S. P. & Deacon, J. W. (1991). Behaviour of zoospores and zoospore cysts in relation to root infection by *Pythium aphanidermatum*. *New Phytologist*, **117**, 289–301.

Judelson, H. S., Tyler, B. M. & Michelmore, R. W. (1991). Stable transformation of the oomycete pathogen *Phytophthora infestans*. *Molecular Plant–Microbe Interactions*, **4**, 602–7.

Judelson, H. S., Tyler, B. M. & Michelmore, R. W. (1992). Regulatory sequences for expressing genes in oomycete fungi. *Molecular and General Genetics*, **234**, 138–46.

Judelson, H. S., Dudler, R., Pieterse, C. M., Unkles, S. E. & Michelmore, R. W.

(1993). Expression and antisense inhibition of transgenes in *Phytophthora infestans* is modulated by choice of promoter and position effects. *Gene*, **133**, 63–9.

Kamoun, S., van West, P., Vleeshouwers, V. G. A. A., de Groot, K. E. & Govers, F. (1998). Resistance of *Nicotiana benthamiana* to *Phytophthora infestans* is mediated by the recognition of the elicitors protein INF1. *The Plant Cell*, **10**, 1413–25.

Katsura, K., Masago, H. & Miyata, Y. (1966). Movements of zoospores of of *Phytophthora capsici*. I. Electrotaxis in some organic solutions (abstract). *Annals of the Phytopathological Society of Japan*, **32**, 215–20.

Khew, K. L. & Zentmeyer, G. A. (1973). Chemotactic response of five species of Phytophthora. *Phytopathology*, **63**, 1511–17.

Khew, K. L. & Zentmeyer, G. A. (1974). Electrotactic response of zoospores of seven species of *Phytophthora*. *Phytopathology*, **64**, 500–7.

Ko, W. H. & Chase, L. L. (1973). Aggregation of zoospores of *Phytophthora palmivora*. *Journal of General Microbiology*, **78**, 79–82.

Kuhlman, E. G. (1964). Survival and pathogenicity of *Phytophthora cinnamomi* in several western Oregon soils. *Forestry Science*, **10**, 151–8.

Longman, D. & Callow, J. A. (1987). Specific saccharide residues are involved in the recognition of plant root surfaces by zoospores of *Pythium aphanidermatum*. *Physiological and Molecular Plant Pathology*, **30**, 139–50.

Madsen, A. M., Robinson, H. R. & Deacon, J. W. (1995). Behaviour of zoospore cysts of the mycoparasite *Pythium oligandrum* in relation to their potential for biocontrol of plant pathogens. *Mycological Research*, **99**, 1417–24.

Miller, A. L. & Gow, N. A. R. (1989*a*). Correlation between profile of ion-current circulation and root development. *Physiological Plantarum*, **75**, 102–8.

Miller, A. L. & Gow, N. A. R. (1989*b*). Correlation between root-generated ionic currents, pH, fusicoccin, indole acetic acid and growth of the primary root of *Zea mays*. *Plant Physiology*, **89**, 1198–206.

Miller, A. L., Shand, E. & Gow, N. A. R. (1988). Ion currents associated with root tips, emerging laterals and induced wound sites in *Nicotiana tabacum*: spatial relationship proposed between resulting electrical fields and phytophthoran zoospore infection. *Plant Cell and Environment*, **11**, 21–5.

Miller, A. L., Smith, G. N., Raven, J. A. & Gow, N. A. R. (1991). Ion currents and the nitrogen status of roots of *Hordeum vulgare* and non-nodulated *Trifolium repens*. *Plant Cell and Environment*, **14**, 559–67.

Mitchell, R. T. & Deacon, J. W. (1986*a*). Differential (host specific) accumulation of zoospores of *Pythium* on roots of graminaceous and non-graminaceous plants. *New Phytologist*, **102**, 113–22.

Mitchell, R. T. & Deacon, J. W. (1986*b*). Selective accumulation of zoospores of chytridiomycetes and oomycetes on cellulose and chitin. *Transactions of the British Mycological Society*, **86**, 219–23.

Money, N. P. (1998). Why oomycetes have not stopped being fungi. *Mycological Research*, **102**, 767–8.

Morris, B. M. & Gow, N. A. R. (1993). Mechanism of electrotaxis of zoospores of phytopathogenic fungi. *Phytopathology*, **83**, 877–82.

Morris, B. M., Reid, B. & Gow, N. A. R. (1992). Electrotaxis of zoospores of *Phytophthora palmivora* at physiologically relevant field strengths. *Plant Cell and Environment*, **15**, 345–53.

Morris, B. M., Reid, B. & Gow, N. A. R. (1995). Tactic responses of zoospores of the fungus *Phytophthora palmivora* to solutions of different pH in relation to plant infection. *Microbiology*, **141**, 1231–7.

Morris, P. F. & Ward, E. W. B. (1992). Chemoattraction of zoospores of the soybean pathogen *Phytophthora sojae*, by isoflavones. *Physiological and Molecular Plant Pathology*, **40**, 17–22.

Nuccitelli, R. (1988). Physiological electric fields can influence cell motility, growth and polarity. *Advances in Cell Biology*, **2**, 213–33.

Paktitis, S., Grant, B. & Lawrie, A. (1986). Surface changes in *Phytophthora palmivora* zoospores during induced differentiation. *Protoplasma*, **135**, 119–29.

Pennington, C. J., Iser, J. R., Grant, B. R. & Gayler, K. R. (1989). Role of RNA and protein synthesis in stimulated germination of zoospores of the phytopathogenic fungus *Phytophthora palmivora*. *Experimental Mycology*, **13**, 158–68.

Porter, J. R. & Shaw, D. S. (1978). Aggregation of *Phytophthora drechsleri* zoospores; pattern analysis suggests a taxis. *Transactions of the British Mycological Society*, **71**, 515–18.

Rathore, K. S., Hotary, K. B. & Robinson, K. R. (1990). A two-dimensional vibrating probe study of currents around lateral roots of *Raphanus sativus* developing in culture. *Plant Physiology*, **92**, 543–6.

Reid, B., Morris, B. M. & Gow, N. A. R. (1995). Calcium-dependent, genus-specific, autoaggregation of zoospores of phytopathogenic fungi. *Experimental Mycology*, **19**, 202–13.

Royle, D. J. & Hickman, C. J. (1964). Analysis of factors governing *in vitro* accumulation of zoospores of *Pythium aphanidermatum* on roots. II Substances causing response. *Canadian Journal of Microbiology*, **10**, 201–19.

Thomas, D. D. & Peterson, A. P. (1990). Chemotactic autoaggregation in the water mould *Achlya*. *Journal of General Microbiology*, **136**, 847–53.

Troutman, J. L. & Wills, W. H. (1964). Electrotaxis of *Phytophthora parasitica* zoospores and its possible role in infection of tobacco by the fungus. *Phytopathology*, **54**, 225–8.

von Broembsen, S. L. & Deacon, J. W. (1996). Effects of calcium on germination and further zoospore release from zoospore cysts of *Phytophthora parasitica*. *Mycological Research*, **100**, 1498–504.

von Broembsen, S. L. & Deacon, J. W. (1997). Calcium interference with zoospore biology and infectivity of *Phytophthora parasitica* in nutrient irrigation solutions. *Phytopathology*, **87**, 522–8.

Weisensee, M. H., Dorn, A. & Jaffe, L. F. (1979). Natural H^+ currents traverse growing roots and root hairs of barley (*Hordeum vulgare* L.). *Plant Physiology*, **64**, 512–18.

Weisensee, M. H., Becker, H. F. & Ehlogotz, J. G. (1992). Growth, gravitropism and endogenous ion currents of cress roots (*Lepidium-satinum* L.). Measurements using a novel 3-dimensional recording probe. *Plant Physiology*, **100**, 16–25.

Wynn, W. K. (1981). Tropic and taxic responses of pathogens to plants. *Annual Reviews of Phytopathology*, **19**, 237–55.

Zentmyer, G. A. (1961). Chemotaxis of zoospores for root exudates. *Science*, **133**, 1595–6.

Zhang, Q., Griffith, J. M., Moore, J. G., Iser, J. R. & Grant, B. R. (1990). The effect of modified pectin, pectin fragments and cations on *Phytophthora palmivora* zoospores. *Phytochemistry*, **29**, 695–700.

HYPHAL INTERACTIONS

GRAHAM W. GOODAY

Department of Molecular and Cell Biology, University of Aberdeen,
Aberdeen AB25 2ZD, UK

INTRODUCTION

The hypha is the key characteristic of a fungus. This ramifying tubular structure enables the fungus to invade and utilize fresh nutrient sources in soil, in detritus and litter, and in living or dead plants or animals. Hyphae from different individuals will encounter each other in these environments, and will interact in a variety of fashions. They may repel each other. If they are of the same species, they may fuse with each other. This may lead to an interconnected mycelium with greater resources, or to mating with eventual formation of sexual spores; or it may lead to death of parts of the hyphae of one or both individuals. One individual may parasitize the other. These interactions entail responses by hyphae to signals from each other. The major senses of hyphae are chemical, i.e. taste and smell, responding to soluble and volatile chemicals, which require a battery of specific chemoreceptors. These senses will be the most important in encounters between hyphae. Other senses that may be involved in hypha–hypha interactions are: contact, i.e. touch, which may lead to thigmotropism; electrical fields, as hyphae both generate and respond to electrical currents; and gravity, e.g. within fruit bodies. This account describes some aspects of encounters between hyphae, with an emphasis on chemical communication. Hyphal interactions within multihyphal structures are not dealt with here, but their involvement in tissue formation is discussed in detail by Moore (1995), Rayner (1996) and Rayner *et al.* (1985).

NEGATIVE AUTOTROPISM

Observation of hyphae growing on solid medium indicates that there must be mechanisms regulating the development of a well-ordered mycelium in which hyphae avoid each other, rarely crossing each other. This avoidance response, negative autotropism, can be observed directly by watching the behaviour of individual hyphae, which tend to turn away or grow alongside neighbouring hyphae as they encounter them. As part of a study characterizing spiral growth of hyphae, Trinci *et al.* (1979) recorded the minimum distances at which autotropism was observed for *Neurospora crassa*, *Aspergillus nidulans* and *Mucor hiemalis* as 30, 27 and 24 μm, respectively.

Hutchinson *et al.* (1980) observed similar behaviour of *M. hiemalis*, with hyphal tips growing away from each other when they came within 10–20 μm apart. This could be interpreted as growth towards fresh nutrients, but contrary to an often-held assumption, there is no evidence of chemotropic growth of hyphae of fungi *sensu stricto* (i.e. excepting those of Oomycotina) to nutrients such as sugars and amino acids (Gooday, 1975). Musgrave *et al.* (1977) also observed negative autotropism of germ tube hyphae of the Oomycete *Achlya bisexualis*, using an experimental system designed to unequivocally study concentration-dependent chemotropic responses to amino acids. Another explanation is that hyphae grow away from each other's metabolic 'staling substances'. This explanation was rejected by Stadler (1952, 1953) in an extensive series of experiments showing that germ tubes of *Rhizopus nigricans* grew away from each other in a variety of circumstances, including presence of staled growth media, and predominantly grew upstream in flowing medium. Oh *et al.* (1997), however, interpret the growth upstream of hyphae of several species in their experiments as being the result of the hyphae sensing the flow of liquid.

Although still lacking irrefutable evidence, the most likely mechanism for the universal negative autotropic response is positive aerotropism, i.e. tropism towards oxygen. This was suggested by Robinson (1973*a*,*b*,*c*), to explain the patterns of germ tube formation and germ tube orientations of spores of *Geotrichum candidum*. The negative autotropic germination of pairs of spores was very marked, with germ tubes always emerging at opposite ends of the two spores. Germ tubes also grew towards the edge when spores had been inoculated under a coverslip, and towards the pore when they were separated from uninoculated medium by a perforated plate, a technique pioneered for investigating hyphal chemotropisms by Miyoshi (1894).

Palková *et al.* (1997), however, present evidence that, under certain growth conditions, neighbouring colonies of a variety of species of yeasts communicate between each other with pulses of ammonia, resulting in inhibition of growth of their nearer ridges and so favouring growth of the colonies away from each other. Amino acids in the medium were required for this behaviour, which Palková *et al.* interpret as the colonies orienting their growth so as to minimize competition for nutrients. Some effects that have been observed between colonies of filamentous colonies may involve a similar system.

POSITIVE VEGETATIVE AUTOTROPISM

Anastomoses between vegetative hyphae of the same species are common among members of the Ascomycetes, Basidiomycetes and Mitosporic fungi, but also occur rarely among members of the Oomycetes and Zygomycetes, notably in *Endogone*. These result in the formation of reticulate mycelia which allow damaged hyphae to be bypassed readily, and allow the formation of multihyphal structures such as coremia, mycelial strands, sclerotia and fruit bodies (Carlile, 1995). Their importance is emphasized by Gregory (1984), who credits Ward (1888) as the first to observe the phenomenon of vegetative hyphal fusion in 4-day-old cultures of *Botrytis*. They were extensively studied by Buller (1933), who classified four types of fusion: hypha-to-hypha fusions, between two hyphal tips that attract each other from about 15 μm apart; hypha-to-peg fusions, between a hyphal tip and a peg produced as a side branch from an older hypha in telemorphotic response to this advancing tip; peg-to-peg fusions, between two pegs produced opposite to one another from adjacent hyphae; and the hook-to-peg fusions during clamp connection formation in dikaryotic basidiomycete hyphae (discussed later). Ward (1888) and Buller (1933) observed that sometimes two pegs were formed by the one hypha in response to an approaching hypha or peg, and that, if only one peg fused successfully, the unsuccessful peg quickly ceased growing, as if the growth stimulus ceased to be produced after the two tips had fused. Watkinson (1978) observed a different type of fusion, tip-to-side, with hyphal tips in very young mycelium of *Penicillium claviformae* fusing to lateral hyphal walls, with no discernible peg formation, either visually or autoradiographically. Aylmore and Todd (1984) also observed tip-to-side fusions between hyphae of different isolates of the Basidiomycete *Coriolus versicolor*. Fusions did not always occur; often hyphae grew across or along one another, frequently with the hyphal tip becoming flattened and bifurcating. When fusion did occur, a single enlarging pore was formed between the two hyphae. A variety of events followed fusion, sometimes with nuclear degeneration and replacement. Ainsworth and Rayner (1986) describe the processes involved in self- and non-self fusions of hyphae of *Phanerochaete velutina*. Hyphal tips responded at distances of up to 250 μm by curving growth toward specific sites on nearby hyphae. Hyphal apices repelled each other if they grew near to one another, and receptive sites generally were in compartments close to the apical compartment. Formation of H-bridges between adjacent hyphae was frequent. In several cases, the Spitzenkörper in the apex of the approaching hypha became assymetrically displaced just before contact so that it became re-aligned with the receptive site of fusion. In self-fusions, the fusion pore usually enlarged to occupy nearly all of the contact area, whereas with non-self-fusions it never expanded fully. The fine structures of these self- and non-self-fusions of *P. velutina* are described by Aylmore and Todd (1986a,b). In both cases, at the point of contact, a central pore was formed, which enlarged by highly localized lysis. In self-fusions, a new septum was formed within 2 hours. In contrast, a somatic incompatibility reaction occurred in hyphae involved in non-self-fusions. These hyphae rapidly showed increased vacuolation and the development of autophagic bodies, followed by widespread degeneration. The spread of this incompatibility response was restricted by the plugging of septal pores.

Hyphal fusions between different isolates and different species of the Basidiomycete genus *Stereum* have been studied in detail by Ainsworth and Rayner (1989). Summarizing the wide range of interactions that they observed, and drawing on observations with other Basidiomycetes, they produced a flow diagram of recognition events and options prior to, during, and after hyphal fusions. Thus hyphal proximity can lead to repulsion, chemotropic homing, telemorphotic tip induction or growth arrest; hyphal contact can lead to continued extension, fusion or formation of penetration pegs and sheaths or coils; fusion can lead to self-recognition with nuclear division and septation, or non-self-recognition with acceptance and nuclear migration or rejection with lysis. The pathway through this that any pair of hyphae will take will depend primarily on their genotypes with respect to mating-type loci and incompatibility loci. Ainsworth and Rayner (1990) analysed a similar range of hyphal interactions in the basidiomycete *Coniophora puteana*, some of which are illustrated in Fig. 1.

Homing to oidia and basidiospores

Hyphae of Hymenomycetes, as well as fusing between themselves, also show the phenomenon of homing; directed growth towards oidia. Thus hyphae of *Schizophyllum commune* grow towards viable oidia from a distance of 15 µm, but do not respond to non-viable spores (Voorhees & Peterson, 1986). Kemp (1970) showed that homing was species-specific with four species of *Coprinus*. Fusion did occur when *Coprinus bisporus* and *Coprinus congregatus* were tested against each other, but was followed within 3 hours by vacuolation and apical cell death. In *Psathyrella stercorea*, hyphae grew towards oidia from a distance of 75 µm. Fries (1981, 1983) describes homing of vegetative hyphae of six species of *Leccinum* to their basidiospores. From about 100 µm, hyphal tips responded chemotropically, and grew towards germinating spores in a spiral fashion. This was not species specific, but interspecies fusions led to vacuolation and death of the apical cell.

Nature of the attractants

There seems to have been a reluctance to accept the involvement of chemotropism in hyphal fusions, despite the fact that action at a distance was clearly described by Ward (1888) for *Botrytis*, who concluded 'it seems to me impossible to avoid the impression that some attraction is exerted'. Jaffe (1966) and Müller and Jaffe (1965) rigorously investigated positive autotropism of germinating spores of *Botrytis cinerea*. When pairs of spores were observed, statistical analysis showed that there was a strong tendency for a germ tube to emerge towards a neighbouring spore and then grow towards it. One set of experiments involved the orientation of germ tubes of spores

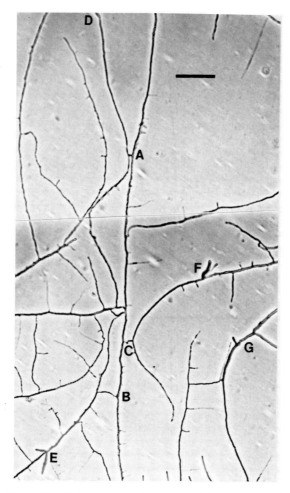

Fig. 1. Hyphal interactions between two mating-compatible homokaryons of the basidiomycete *Coniophora puteana*. Self-fusions are shown at A and B. A non-self fusion, with both partners showing growth curvature to the point of fusion, is shown at C. Emergent dikaryotic secondary hyphae, respectively from both progenitors, are shown at D, E and F, G. Scale bar represents 50 mm. From Ainsworth and Rayner (1990).

sparsely seeded onto a polythene film and then subjected to medium flowing past at different rates. Results were analysed by comparing the behaviour of the germinating spores with theoretical models predicting diffusion of factors in different conditions. Contrary to Stadler's (1952) result with high concentrations of spores of *R. nigricans*, these germ tubes preferentially grew downstream, and this tendency increased with increasing flow rates. In multichambered experiments, the presence of high concentrations of spores upstream reduced the preferential downstream orientation. After considering several possibilities, such as effects of molecular orientation, wetting and pressure, Müller and Jaffe (1965) concluded that their observations could only be explained by the production by cells of a macromolecular diffusible growth stimulator, with a half-life of about 10 seconds, that would be washed downstream from the spores, and which would have a radius of action of about 10 μm in still medium. These properties are similar to those proposed by Raper (1952) for the putative series of species-specific chemoattractants involved in vegetative hyphal fusions in Basidiomycetes, as they have to be large enough to be specific and short-lived, with a radius of action of about 10 μm (Gooday, 1975). Although as yet no knowledge of specific molecules exists, the recent elucidation of multiple complex mating pheromone/pheromone receptor systems (discussed later) demonstrates that fungi have complex signal production, reception and transduction mechanisms. It may be that some parts of the mating signalling systems could also be used during vegetative autotropisms and fusions.

Vegetative incompatibility

As described for several Basidiomycetes earlier, many intraspecific hyphal fusions lead to the phenomenon of vegetative incompatibility. This is a mechanism preventing the formation of heterokaryotic strains, the existence of genetically different nuclei within a common cytoplasm. This phenomenon has been investigated in detail in the Ascomycete *Podospora anserina* (Bégueret *et al.*, 1994).

Incompatible strains of this, and many other fungi, produce a 'barrage' where they meet, which is a line of cells that have fused and subsequently died. This reaction is regulated by a series of *het* genes. In *P. anserina* there are nine *het* loci, and, for a compatible fusion to occur, the alleles at all the *het* loci must be identical. Bégueret *et al.* (1994) suggest that complexes between products of incompatible genes are lethal to the cell. The *het-s* locus encodes polypeptides of no similarity to any known protein, which are non-essential to the cell; the *het-c* encodes a protein with similarity to a mammalian protein catalysing exchange of glycolipids between cellular membranes, and is required for control of meiosis during ascospore formation; and the *het-e* locus encodes a protein with regions of similarity to β-subunits of trimeric G proteins and to GTP-binding domains. Mutation of a single amino acid in the GTP-binding domain gives a protein that is no longer active in triggering the lethal reaction (Saupe *et al.*, 1995). Coupled with the finding that adenylate cyclase is involved in vegetative incompatibility (Loubradou *et al.*, 1996), this suggests that a G protein transduction system plays a role in regulating this system.

Neurospora crassa has at least ten incompatibility loci, and very unusually, the mating-type locus is one of these. This acts to prevent formation of mixed mating type heterokaryons. When fusion occurs between two vegetative hyphae which differ at the mating type locus but not at the other incompatibility loci, their cytoplasms become vacuolated and they die (Kronstad & Staben, 1997).

SEXUAL INTERACTIONS

Sexual reproduction is a major factor aiding adaptability and fitness in all organisms. Fungal hyphae in natural environments are faced with the problem of finding a compatible partner. Their major senses are chemical, so in its native environment each individual hypha can be imagined exuding its own specific repertoire of chemical sexual signals. These chemicals have to be at least reasonably specific to fungal species, and completely specific to mating type within that species, so that attempts at mating stand a good chance of being successful. Thus potentially there are more different chemicals than there are species. Such specific chemicals can be termed hormones, but an increasingly used synonym in the fungal literature is 'pheromone' for a chemical acting at a distance. The very small number of such compounds that have been identified to date fall into two chemical classes: isoprenoids (derived from mevalonic acid) among the 'lower fungi' (a very diverse phylogenetic group, the Oomycetes, Chytridiomycetes and Zygomycetes), and hydrophobic peptides, mostly isoprenylated, among Ascomycetes and Basidiomycetes (Gooday & Adams, 1993; Gooday, 1998). As well as these very specific effectors, there are numerous reports throughout the fungi of apparently less specific sex factors, which include growth substances regulating development by their overall concentration, and morphogens regulating localized differentiation by providing positional information (Dyer et al., 1992).

Steps leading up to successful mating can be summarized as follows, based on the very few examples where this has been investigated in detail. As mature mycelia approach each other, hyphae of one or both of the potential partners are constitutively releasing specific pheromones. The opposite partner responds because it has a specific receptor, which can trigger the appropriate signal transduction pathway. An initial response is to initiate or to greatly increase production and release of the complementary pheromone. Commonly, there is then growth arrest, if this has not already occurred through starvation, and then there is initiation of sexual differentiation, leading to the formation of the specific mating hyphae that will undergo plasmogamy. An essential feature of the process is also specific destruction of the pheromone by the recipient, to maintain the activity of the receptors. The key to the mating process is thus chemical communication between the mating hyphae.

Oomycetes, Chytridiomycetes, Zygomycetes

Among the Oomycetes, there is only detailed information on the water mould genus *Achlya*. Their vegetative hyphae are diploid, and sexual reproduction entails the production of antheridial hyphae and oogonial initials on the same mycelium or two mycelia that are growing close to each other. The antheridial hyphae grow towards the oogonial initials, and when in contact with each other, both mature by delimitation of the hyphal apices by septal formation and meiosis, giving rise to egg cells and antheridial fertilization tubes. Most strains are homothallic, but some are heterothallic, self-sterile with three types of behaviour, solely male, solely female or male–female according to the nature of the strain with which they are paired. Sexual reproduction in these fungi is regulated by complementary sterol hormones, antheridiol and oogoniol, biosynthesized by two separate pathways from the common plant sterol fucosterol (Mullins, 1994). Female hyphae constitutively produce antheridiol, and under conditions of low nutrients male hyphae respond to increasing concentrations by: (i) ceasing apical growth; (ii) forming antheridial branches; (iii) chemotropic growth of antheridial branches towards antheridiol, which is produced at increasing concentrations by the developing oogonial initials; (iv) maturation of the antheridia, which become delimited by septa, and undergo meiosis to form the male gametic nuclei (Barksdale, 1967; Gow & Gooday, 1987). Counting the number of antheridial branches is the basis of the bioassay, which can detect 10 pg ml^{-1}. Accompanying these morphological changes are a series of biochemical responses, which include: (i) synthesis and release of oogoniol; (ii) induced metabolism of antheridiol to inactive metabolites, enabling the antheridial hyphae to remain responsive to the gradient of antheridiol; (iii) increase in activity and secretion of cellulase to allow antheridial branching, by softening the lateral cellulosic wall; (iv) marked enhancement of synthesis of rRNA, mRNA and protein, and of histone acetylation; (v) up-regulation of heat-shock proteins correlated with branching and secretion of glycoproteins (Thomas & Mullins, 1969; Musgrave & Niewenhuis, 1975; McMorris, 1978; Timberlake & Orr, 1984; Silver *et al.*, 1993). Although much less investigated, responses of female hyphae to the most potent oogoniol, the minor metabolite dehydro-oogoniol, are complementary to those of male hyphae to antheridiol (McMorris *et al.*, 1993). They include the formation of oogonial initials and enhanced synthesis of antheridiol.

There is good biological evidence for equivalent hormone systems involved in hyphal interactions in other Oomycete species, but these remain to be characterized (Gooday & Adams, 1993). For example, formation of antheridia and oogonia occurs in male and female strains of *Pythium sylvaticum*, respectively, when they are grown separated by a permeable membrane (Gall & Elliott, 1985). There is also evidence for a hormonal system regulating mating in *Phytophthora* and *Pythium* species that is different from this putative sterol hormone system. This has been termed hormonal hetero-thallism, as both A1 and A2 mating types of heterothallic species form oospores when paired with the opposite mating type of the same or different species on the opposite side of a polycarbonate membrane. This shows that a heterothallic strain is induced to become self-fertile by a mating-type specific hormone: A1 strains produce a1 hormone which stimulates A2 but not A1 to produce oospores, while A2 strains produce a2 hormone with the comple-mentary specificity (Ko, 1988).

Sexual reproduction in the filamentous Chytridiomycetes can take several forms, with fusion taking place between whole thalli, rhizoids, or motile anisogametes (Beakes, 1994). The best studied system, that of *Allomyces* species, involves the fusion of motile male and female gametes mediated by complementary hormones. The bicyclic sesquiterpene sirenin is produced by female gametes and chemotropically attracts male gametes (Pommerville *et al.*, 1988), but also parisin is produced by males and attracts females (Pommerville & Olsen, 1987). It is likely that similar hormones regulate mating between hyphae in chytrid species where this occurs, but these have not been characterized.

Among the Zygomycetes, mating between (+) and (−) strains of *Mucor mucedo* and other mucoraceous fungi involves two levels of chemical signalling between the two partners, first between the two colonies, and then between individual hyphae. As two compatible cultures grow towards each other, initial contact is the exchange via diffusion of mating-type specific prohormones, chiefly methyl-4-dihydrotrisporate from (+) and trisporol from (−). These are metabolized by (−) and (+), respectively, to trisporic acids, which act on both (+) and (−) mating types, resulting in the formation of zygophores, the specialized mating hyphae. Trisporic acids B and C are apocarotenoids, C18 oxidative metabolites of β-carotene. Their biosynthesis is via a remarkable metabolic collaboration by the two mating types, whereby (+) and (−) each possesses an incomplete enzymatic pathway, accumulating intermediates that can only be further metabolized by the opposite mating type. This has been elucidated chiefly by studies with (+) and (−) strains of *Blakeslea trispora* (Bu'Lock *et al.*, 1976; Van den Ende, 1984; Gooday & Adams, 1993). Both (+) and (−) cells then respond to trisporic acids by switching from vegetative growth to sexual differentia-tion. The trisporic acids are not species specific, either in production or effect. *M. mucedo*, which is used in the bioassay of trisporic acids, is especially sensitive to them, concentrations of 10 pM being detectable (Gooday, 1978). Cultures of *Phycomyces blakesleeanus* need to be in a state of growth arrest before responding (Drinkard *et al.*, 1982). Only negligible amounts of trisporic acids are detectable in unmated cultures, but their synthesis is enormously enhanced in mated colonies by a 'cascade mechanism', as trisporic acid greatly stimulates the biosynthesis of β-carotene and the formation of its own precursors (Bu'Lock *et al.*, 1976; Van den Ende 1984).

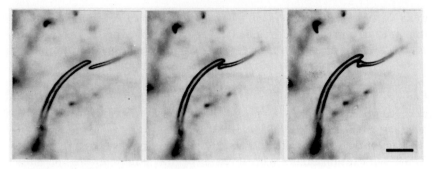

Fig. 2. Zygotropism in *Mucor mucedo*: showing oriented growth leading to fusion of tips of (+) and (−) zygophores (left and right respectively). Photographs at 6 min intervals. Scale bar represents 100 mm.

Progress continues to be made in unravelling this fascinating story. The (−)-specific methyl-4-dihydrotrisporate dehydrogenase, with NADP as co-factor, shown to be localized in (−) zygophores of *M. mucedo* and in one of the two conjugating hyphae of the homothallic species *Zygorhynchus moelleri* by Werkman (1976), has been purified and its gene cloned by Czempinski *et al.* (1996).

In *M. mucedo*, aerial zygophores are produced, which grow towards each other in mated pairs, from distances up to 2 mm (equivalent to 60 m on a human scale). When they meet, pairs of zygophores fuse, usually just behind their tips (Fig. 2), leading to karyogamy and zygospore formation (Gooday, 1973, 1978). This process of zygotropism must be in response to complementary volatile chemicals. These have not been definitively identified but are most likely to be the mating-type specific prohormones, methyl-4-dihydro-trisporate and trisporol. Mesland *et al.* (1974) demonstrated that zygophore formation could be induced when (+) and (−) mycelia of *M. mucedo* were separated by an air gap, showing that the trisporate precursors are volatile enough to have biological activity. The observation that zygophores cease to be attractive immediately they have fused with mating partners (Gooday, 1975) can be explained as the prohormones would then immediately be metabolized through to trisporate.

Ascomycetes and Basidiomycetes

There are many observations suggesting the activity of diffusible hormones during sexual interactions between hyphae, both between and within colonies of filamentous ascomycetes, but none of these compounds has been char-acterized. In contrast, pheromonal regulation of mating of hemiascomyce-tous yeasts, particularly *Saccharomyces cerevisiae* and *Schizosaccharomyces pombe*, has been defined in great detail (Gooday & Adams, 1993; Caldwell *et al.*, 1995). For each species, there are complementary pairs of pheromones and pheromone receptors. In *S. cerevisiae*, α-cells secrete α-factor and have a membrane-bound **a**-factor receptor, **a**-cells secrete **a**-factor and have a membrane-bound α-factor receptor. In *Sz. pombe*, plus cells secrete P-factor and have a membrane-bound M-factor receptor, minus cells secrete M-factor and have a membrane-bound P-factor receptor. α-Factor and P-factor are unmodified peptides; **a**-factor and M-factor are peptides with COOH-terminal cysteines modified by S-farnesylation and methyl esterification. The receptors for these latter two lipopeptides are encoded by the genes *STE3* and *map3*[+], respectively. In each case, the yeast cells respond to the appropriate pheromone by producing a conjugation tube which grows towards a cell of opposite mating type.

It is fairly certain that filamentous Ascomycetes have analogous phero-mone systems, but direct evidence is still lacking. In *Neurospora crassa*, the mating-type loci, **A** and **a**, encode complimentary regulatory genes (Kron-stad & Staben, 1997). Presumably these control expression of genes encoding pheromones and pheromone receptors. There is good evidence for such hormones in *N. crassa*, in particular controlling chemotropic growth of the female filament (trichogyne) of each mating type (*A* and *a*) to the male fertilizing conidium (Bistis, 1983). Other pheromone systems in filamentous ascomycetes ripe for reinvestigation are those in the genera *Glomerella*, *Bombardia*, *Ascobolus* and *Nectria* (Gooday & Adams, 1993). A different type of chemical communication has been elucidated by Champe and El-Zayat (1989), who describe diffusible psi (precocious sexual induction) factors from the homothallic ascomycete *Aspergillus nidulans*, which induce premature development of sexual structures. Psi activity was assayed by observing the response of mycelium growing from a uniform inoculum of conidia over an agar plate to filter discs impregnated with test samples. The activity resulted in formation of cleistothecia instead of conidiophores, and in the release of yellow pigment into the medium. A family of psi factors have been characterized as derivatives of linoleic acid, with minor components as the corresponding oleic acid derivatives (Champe *et al.*, 1994). They appear to be species-specific. A further type of chemical control of sexual morpho-genesis has been suggested by Siddiq *et al.* (1989), who have isolated a class of sexual morphogens (SF, sexual factors) involved in sexual reproduction in the heterothallic plant pathogen *Pyrenopeziza brassicae*. When lipid extracts from mated cultures were added to unmated cultures, they suppressed asexual sporulation and induced the formation of immature, sterile apothe-cia. When added to mated cultures, they greatly increased the speed of production and number of fertile apothecia. These activities were not species-specific, as effects on growth and sporulation were observed when the extracts were added to a wide range of fungi. Extracts from unmated cultures of *P. brassicae* had none of these effects. It is likely that this wide range of effects is due to several components acting as morphogens and growth substances. Enhancement of fertile perithecial development of *Nectria haematococca* by the addition of lipid extracts from mated cultures of this fungus has been described by Dyer *et al.* (1993). Linoleic acid had the same effect, and was detected as a metabolite of the mated cultures. Dyer *et al.* (1993) suggest that this and related fatty acids are produced during mating as sex factors enhancing the development of perithecia. There is evidence for linoleic acid being involved in sexual reproduction in several other fungi as an endogenous sexual growth substance (Dyer *et al.* 1993).

The complexity of mating systems in the Basidomycetes has long fascinated mycologists, but as is the case with the Ascomycetes, the yeast genera have proved to be most amenable to experimentation. The best characterized sexual signalling systems among the Basidiomycetes are those involving the lipopeptide sex pheromones produced by yeast cells of heterobasidiomycetous yeasts, smuts and jelly fungi (for review see Gooday & Adams, 1993; Caldwell *et al.*, 1995; Vaillancourt & Raper, 1996). These pheromones are encoded by the mating type loci, along with genes for the complementary pheromone receptors. Mating in species of *Tremella*, *Rhodosporidium*, *Ustilago* and *Cryptococcus* occurs between compatible pairs of haploid yeast cells, which respond to cells of opposite mating type by developing conjugation hyphae which grow chemotropically to the sources of the pheromones.

Farnesylated peptide pheromones have been characterized from culture filtrates and/or their identity has been inferred from sequences of mating-type loci in these fungi. The characteristic amino acid sequence of the propheromones is a COOH-terminal 'CaaX box'; i.e. -cysteine–aliphatic amino acid-aliphatic amino acid–any other amino acid, and an amino-terminal methionine. By analogy with the *S. cerevisiae* mating pheromone **a**-factor, which has been studied in detail (Caldwell *et al.*, 1995), each of these pheromone precursors is probably processed by initial farnesylation of the cysteine sulphur atom accompanied by proteolytic cleavage of the COOH-terminal 'aaX' tripeptide, and most likely by methylation of the new terminal carboxyl group. The resultant membrane-associated lipopeptide would then undergo one or more proteolytic cleavages at its NH_2-terminus, and the mature pheromone would be exported directly across the plasma membrane by an ATP-driven peptide pump. The formation of *Ustilago maydis* mating hyphae is triggered by a complementary pair of such pheromones, a1 (13 aa) and a2 (9 aa) (Spellig *et al.*, 1994). The resultant mating hyphae then grow chemotactically towards the sources of the pheromone (Snetselaar *et al.*, 1996).

Mating among the Homobasidiomycetes is characterized by the extraordinary lengths to which some of them go in order to maximize their chances of outbreeding. The tetrapolar mating systems of *Schizophyllum commune* and *Coprinus cinereus* give rise to more than 20 000 and 12 000 mating types, respectively, giving outbreeding potentials better than 98%

(Raper, 1966; Vaillancourt & Raper, 1996; Kothe, 1997; Casselton & Olesnicky, 1998). In *S. commune*, mating type is determined collectively by two tightly linked multi-allelic loci, α and β, for A; and two, α and β, for B (Kothe, 1997; Wessels *et al.*, 1998). Compatible mating requires a difference at either of the two A loci and either of the two B loci. The complex A loci code for interacting homeodomain transcription factors, and the complex B loci code for complementary pheromones and their receptors. Two complex B mating type loci, Bα1 and Bβ1, have been cloned (Vaillancourt & Raper, 1996; Vaillancourt *et al.*, 1997). Each contains a single pheromone receptor gene and three putative pheromone genes. The pheromone receptor proteins encoded by these genes, Bar1 and Bbr1, and a further one, Bar2, have homology with the pheromone receptors *S. cerevisiae*, *U. maydis* and *Sz. pombe* (Vaillancourt & Raper, 1996; Vaillancourt *et al.*, 1997). These pheromone receptor proteins are part of the large family of receptor proteins with seven transmembrane domains, involved in signal transduction through G-proteins. The six putative pheromone precursors encoded by Bα1 and Bβ1 are recognized by having the 'CaaX' box. There are no obvious differences between the α and β series, and each pheromone gene is unique in sequence.

DNA sequencing of the *B6* locus of *C. cinereus* shows the presence of six genes for putative pheromone precursors (O'Shea *et al.*, 1998). These genes code for peptides with 53–72 amino acids with the 'CaaX' box for farnesylation at their COOH-termini, but no conservation at their NH$_2$-termini. With proteolytic cleavage of the C-terminal 'aaX', this predicts that there are six mature peptide pheromones of 11–13 amino acids. Three other genes in the *B6* locus encode putative pheromone receptors with seven transmembrane domains, again with homology to receptors of *S. cerevisiae*, and *U. maydis*.

The discovery that the **B** complexes of *S. commune* and *C. cinereus* encode pheromones and pheromone receptors was a surprise, as initial fusion between two hyphae of these species is independent of mating type. The **B** mating type genes control two distinct processes in mating. First, they regulate the remarkable phenomenon of nuclear migration. Following fusion of two monokaryons with different **B** genes, there is extensive bidirectional invasion of the compatible nuclei throughout the mycelia. This process involves enzymatic dissolution of the complex dolipore septa, and nuclear migration occurs at great speed, estimated at up to 3 mm/hour, to give rise to very rapid establishment of dikaryotic mycelium. Northern blotting shows that pheromone genes are up-regulated during this process (Vaillancourt *et al.*, 1997). This suggests that the pheromones have a direct role in nuclear migration. The pheromones may diffuse extracellularly (perhaps along the cell walls) ahead of the nuclei, preparing the hyphae for nuclear migration by triggering the activation of lytic enzymes involved in septal dissolution, and by regulating organization of cytoskeletal motor protein functions (Kothe, 1997; Vaillancourt & Raper, 1996; Vaillancourt *et* *al.*, 1997). The **B** mating type genes also control the hook–cell fusion during clamp connection formation (Raper, 1966) and this process gives the appearance of involving specific chemotropic attractants. Thus Buller (1933) describes the turning and re-fusion of the hook cell with the subapical wall: '... the hook of a clamp-connexion does not fuse directly with the main hypha as hitherto has been supposed, but with a blunt process or peg sent out by the main hypha in response to a stimulus given by the apex of the hook'. The process of hook–cell fusion has two elements in common with initial dikaryon formation; nuclear movement and localized cell wall dissolution, so it seems reasonable that both processes could share the same pheromone/pheromone receptor systems. Wessels *et al.* (1998) and Schuurs *et al.* (1998) suggest that the pheromones could be secreted locally into the wall adjacent to each nucleus, where they would regulate the spatial distribution of the nuclei, by diffusing laterally to interact with complementary receptors adjacent to the complementary nuclei. In this model, however, the pheromones would be retained in the wall, and would have no extracellular role.

These findings still do not clarify the relationship discussed by Raper (1952) and Gooday (1975) between chemotropic vegetative fusions of basidiomycete hyphae of the same species and of monokaryotic hyphae and oidia, which are both independent of mating type, and chemotropism of the hook cell. However, the finding of multiple pheromones and pheromone receptors in *S. commune* and *C. cinereus* demonstrates that these fungi possess complex specific recognition systems, which may be involved in some way in these processes.

Intergeneric sexual interactions

With the very wide diversity of sexual systems in the fungi, each regulated by specific chemicals, and the equally wide diversity of fungal metabolites, it is not surprising that there are many examples of fungi affecting each other's sexual differentiation. One can speculate that these interactions can range from chance mimicking of an endogenous ligand triggering a sexual response, which may have little widespread significance, to evolved subversion of sexual regulation by antagonists during competition and by parasites during attack on their hosts. An example that illustrates this is that the effects of a1 hormone on induction of sexual structures in *Phytophthora* (discussed above) can be mimicked by volatile metabolites from antagonistic *Trichoderma* species (Brasier, 1975) termed homothallins I and II (Sakata & Rickards, 1980). Zearalenone A provides a further example of interspecies interaction. This is a well-characterized oestrogenic mycotoxin produced by *Gibberella zeae* and *Fusarium* species. Low concentrations of zearalenone stimulate sexual reproduction in a wide range of fungi (Nelson, 1971). Wolf and Mirocha (1973) suggest that it is an endogenous sexual regulator in *G.*

zeae, as at low concentrations it enhances perithecial production, but it inhibits at high concentrations, and its role as a specific sexual regulator is put into question by the observation that there is no correlation between zearalenone production by different strains and their formation of perithecia (Windels *et al.*, 1989). The involvement of trisporic acid biosynthesis in the mycoparasitism of *Parasitella parasitica* on *Absidia glauca*, discussed later, is a clear example of a sexual reaction being subverted for another purpose. The multiple interactions between hyphae of different Basidiomycete fungi, discussed earlier, may involve chance cross-talk between multiple complex systems of mating pheromones and receptors such as those that are being elucidated for *S. commune* and *C. cinereus* (discussed above). In addition, some of the sexual morphogens and growth factors that have effects on Ascomycetes, such as SF, mycosporines and linoleic acid, as well as zearalenone, have been reported to stimulate sexual development in a range of Agaricales (Dyer *et al.*, 1992).

HYPHAL INTERACTIONS AMONG MYCOPARASITES AND FUNGAL ANTAGONISTS

There is a wide range of antagonistic interspecies interactions among the fungi, ranging from parasitism, lysis and antibiosis through to competition. These interactions are being studied with the aim of developing biocontrol agents against plant pathogens (Chet *et al.*, 1997).

Mycoparasitism

Mycoparasites can be classified as necrotrophic, aggressively leading to cell death of a relatively wide range of hosts, and biotrophic, interacting with living host mycelium of a relatively small range of hosts. The recognition of host hyphae by parasite hyphae may occur at several points (Manocha & Sahai, 1993; Jeffries, 1997). For several species there is evidence for chemotropism, implying the presence of specific cell-surface receptors to sense specific chemicals diffusing from the host, followed by oriented growth, sometimes with atypical branching. When contact is made there is evidence in several cases for specific lectin–ligand or agglutinin–ligand adhesion. This may be followed by coiling growth to envelope the host hypha, or formation of appressoria, together with specific induction of lytic enzymes. There may or may not be penetration, sometimes with formation of haustoria by biotrophic mycoparasites.

The most investigated necrotrophic mycoparasites are *Trichoderma* species. These are actively parasitic on a range of plant pathogenic fungi, including species of *Pythium*, *Sclerotinia*, *Sclerotium*, *Fusarium* and *Verticillium*. Hyphae of *Trichoderma* species respond to hyphae of host fungi by growing towards them, coiling around them and penetrating and degrading the cell walls and contents. Hyphae of *Trichoderma hamatum* are observed to respond chemotropically to host hyphae of *Rhizoctonia solani* and *Pythium* species by branching atypically and growing towards them from some distance away (Chet *et al.*, 1981). The nature of the diffusible chemical signals involved is unclear. Dennis and Webster (1971) tested 80 isolates of *Trichoderma* against a range of test fungi including a Basidiomycete, an Ascomycete, a Zygomycete, an Oomycete and Mitosporic fungi. Most isolates coiled around most of the test fungi, in a variety of fashions. Vacuolation and sometimes bursting of host hyphae were observed with antibiotic-producing strains. The coiling response is not merely thigmotropism, as it was not observed following contact of *Trichoderma* hyphae with plastic threads. The involvement of specific lectins in surface recognition during mycoparasitism is suggested by Inbar and Chet (1994) who have characterized a lectin from *Sclerotium rolfsii*. When they coated nylon fibres with the lectin, they observed that hyphae of *Trichoderma harzianum* adhered to the fibres, often coiling around them. After contact is made between parasite and host hyphae, lytic enzymes, including chitinases, β-1,3-glucanases, proteases and lipases, are involved in the digestion of the host cell wall. Inbar and Chet (1995) have shown that specific chitinases are induced by *T. harzianum* during parasitism on *S. rolfsii* and also in the presence of the lectin-coated nylon fibres, and they suggest that recognition of the host is the first step in a cascade of antagonistic events during parasitism.

Similar observations have been made with other mycoparasitic fungi. Thus Lifshitz *et al.* (1984) observed hyphae of the Oomycete *Pythium nunn* growing towards host hyphae of other Oomycetes and massively coiling around them, leading to lysis. Hyphae of some host species were penetrated, while others were not. Specificity in production of lytic enzymes is also seen with *P. nunn*, with extracellular cellulases being produced in co-culture with host fungi, but not with non-hosts (Baker, 1987). Hyphae of *Coniothyrium minitans* parasitise hyphae of *Sclerotinia sclerotiorum* either directly or through the production of side branches with the formation of infection pegs. These penetrate the host hyphae, leading to granulation and vacuolation and intracellular growth of hyphae of *C. minitans* (Whipps & Gerlagh, 1992). There is evidence that hyphae of *C. minitans* utilize lytic enzymes, including glucanases and chitinases, to penetrate walls of both hyphae and the thick melanized rind of sclerotia of *S. sclerotiorum* (Jones *et al.*, 1974).

The biotrophic Zygomycete *Piptocephalis* species is obligately parasitic on hyphae of Mucorales. There is evidence for chemotropism to host hyphae, for example, of germ tubes of *Piptocephalis fimbriata* to hyphae of host *Mortierella vinacea* over a distance of 5 mm (Evans & Cooke, 1982). Preliminary characterization suggested that the attractants were of high molecular weight and were proteinaceous or associated with proteins. Chemotropism is also shown by germ tubes of the zygomycete biotrophic parasite *Dimargaris cristalligena* towards hyphae of their host *Cokeromyces*

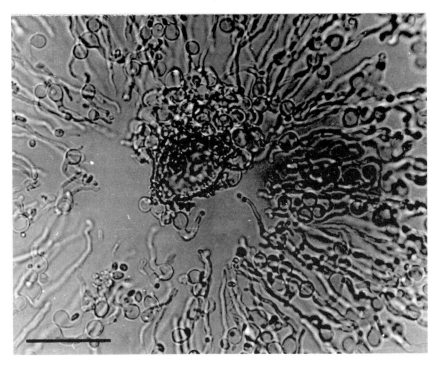

Fig. 3. Germ tubes of the zygomycete biotrophic parasite *Dimargaris cristalligena* growing towards a hyphal fragment of their host *Cokeromyces recurvatus*. Scale bar represents 25 mm. Micrograph by courtesy of Dr P. Jeffries.

recurvatus (Fig. 3; Pers. comm., P. Jeffries). Manocha and Chen (1991) describe the specific binding of hyphae and germ tubes of *Piptocephalis virginianum* to hyphae of host fungi, such as *Choanephera cucurbitarum*, but not to those of non-host species. They characterized two glycoproteins from host cells walls that were able to agglutinate ungerminated and germinating spores of host species.

The related mycoparasite, *Parasitella parasitica*, is a facultative parasite of many Mucorales. Its hyphae grow towards host hyphae, fuse with them, with the formation of characteristic structures, the 'sikyotic cells', which fuse with the host hyphae and release some nuclei into them. The final result is the formation of persistent 'sikyospores' (Kellner *et al.*, 1993). Wöstermeyer *et al.* (1995) show that this interaction is specific to mating type, with (+) *P. parasitica* fusing only with (−) *Absidia glauca*, and (−) *P. parasitica* fusing only with (−) *A. glauca*, and that it involves the collaborative biosynthesis of trisporic acid from prehormones from parasite and host. The parasitism is thus an abortive type of mating, and appears to involve two types of specificity, the mating-type interaction, presumably with chemotropism to the prohormones, and initial fusion, and the parasite–host recognition like that of *Piptocephalis* species.

Sporophagy

Fries and Swedjemark (1985) describe the phenomenon of sporophagy that they observed with vegetative hyphae of some hymenomycetes. For example, hyphae of *Coprinus comatus* turned and grew towards spores of *Leccinum aurantiacum*, in a homing response. They then branched to envelope the spores, and killed and consumed them. Of 136 species tested, 10 showed a strong activity and 15 showed weak activity. Of these, 9 and 14 species, respectively, were wood and litter decomposing fungi. Three with the strongest activity, *C. comatus*, *Lentinellus omphalodes* and *Pluteus cervinus*, were tested against spores of 22 hymenomycetes. White spores, for example, of *Russula* species, were left alone; dark-coloured spores, for example, of *Boletus* species, were attacked. Only living spores were attacked. This phenomenon was confined to spores, as no mycoparasitism or hyphal interference was observed with hyphae of four of the susceptible species.

Hyphal interference and other antagonisms

Many interactions have been observed between different species of fungi in rotting wood and in dung. Thus Ainsworth and Rayner (1991) observed a 'guerrilla strategy' of the wood-rotting hyphae of *Phanerochaete magnoliae*, when hyphae of a rival species, *Datronia mollis*, encountered them in co-culture. The *D. mollis* hyphae either underwent immediate lysis, or encoiled the hyphae of *P. magnoliae* and then lysed (Fig. 4). Ikediugwu and Webster (1970*a*,*b*) describe a form of antagonism between mycelia of coprophilous fungi that they termed hyphal interference. When hyphae of *Coprinus heptemerus* grew near to hyphae of a wide range of other fungi, including Basidiomycetes, Ascomycetes, Zygomycetes and Mitosporic fungi, they caused them to vacuolate and die. Direct contact was not required, as the phenomenon was exhibited, albeit to a lesser extent, when the hyphae were separated by a cellophane membrane. A wide range of other fungi were screened, using *Ascobolus crenulatus* as test organism. With one exception, the Mitosporic fungus *Stilbella erythrocephala*, only Basidiomycetes caused hyphal death in this fashion.

The examples of antagonistic hyphal interactions discussed above involve at least some degree of specificity and localization. There are, however, many examples of less specific antagonisms, ranging from competition for nutrients (Rayner & Webber, 1984) through to antifungal antibiotic production, for example, by *Trichoderma* species (Gnisalberti & Sivasithamparam, 1991). The antibiotics can have many effects. A metabolite produced by *Fusarium*

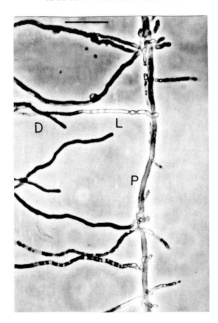

Fig. 4. Protoplasmic degeneration and lysis, L, in hyphae of *Datronia mollis*, D, following contact with a hypha of *Phanerochaete magnoliae*, P. Scale bar represents 50 mm. From Ainsworth and Rayner (1991).

oxysporum and several unrelated fungi that caused premature vacuolation when applied to hyphal tips was originally thought to be a fungal hormone controlling water distribution in hyphae, but proved to be bikaverin, a red pigment from *Fusarium* species (Cornforth *et al.*, 1971).

DEVELOPMENT OF MATURE HYPHAE WITHIN A CULTURE

When a spore germinates, typically it produces one or more germ tubes which elongate exponentially. This results from the autocatalytic effect of the increasing rate of uptake and metabolism of nutrients from the medium (Prosser, 1994). The extension rate eventually reaches a nearly constant value, probably when transport of material from the subapical region becomes limiting. Exponential growth of the colony is achieved by formation of subapical branches, each of which becomes an apically elongating hypha. Hyphae in a young colony, however, are juvenile in behaviour and undergo a poorly understood slow process of maturation (Gooday, 1995). Juvenile hyphae of many species are slower growing and narrower than are mature hyphae. Thus, in experiments with *Botrytis cinerea*, the mean extension rate of leading hyphae increased from 85 mm h^{-1} at 20 h to 330 mm h^{-1} at 44 h after inoculation (Table 1; Zhu & Gooday, 1992), and remained constant

Table 1. *Variations of apical growth rate and diameter with age of leading hyphae*

Fungus	Juvenile			Mature			Reference
	Age (h)	E (μm h^{-1})[a]	D (μm)[b]	Age (h)	E (μm h^{-1})	D (mm)	
B. cinerea	20	85	5.7	44	330	8.7	Zhu & Gooday (1992)
M. rouxii	12	315	5.0	26	340	8.4	Zhu & Gooday (1992)
N. crassa	25	410	6.8	40	1380	12.8	McLean & Prosser (1987)
C. albicans	12	19	2.6	72	46	3.4	Gow & Gooday (1982)
C. cinereus	–	16	3.0	–	269	5.9	Butler (1984)

[a] E, Mean extensions rate. [b] D, Mean diameter. From Gooday (1995).

thereafter. Over the same period, the mean diameter of the hyphae increased linearly from 5.7 to 8.7. There was a direct relationship between extension rate and square of diameter. In contrast, for *Mucor rouxii*, in the same study, extension rate increased dramatically from spore germination at 5.5 h to reach a value of over 300 mm h^{-1} at 10 h, after which it increased little over the next 36 h. Meanwhile, hyphal diameter increased more slowly, so that from 5.8 mm at 15 h it did not reach its maximum value of 8.4 mm until 26 h. There was no direct relationship between extension rate and square of diameter. Other examples of these phenomena of hyphal maturation include *N. crassa* (McLean & Prosser, 1987), *Candida albicans* (Gow & Gooday, 1982) and *Coprinus* species (Butler, 1984) (Table 1). A positive relationship between hyphal diameter and extension rate for a particular species has been reported for many fungal colonies: for both variables, values for leading hyphae are greater than those for primary branches, which in turn are greater than those for secondary branches (Prosser, 1994). Working with *N. crassa*, Steele and Trinci (1975) distinguished between 'undifferentiated' and 'differentiated' hyphae; the former being juvenile hyphae found in young colonies on solid media or in submerged culture, and the latter being wider and faster extending mature hyphae found at the edges of older colonies on solid media.

Juvenile hyphae of many fungi, as well as being slower growing and thinner than mature ones, have a limited repertoire of differentiation and hyphal interactions. Thus those of *M. mucedo* do not produce the sex hormone, trisporic acid, or respond to it to produce zygophores, as do mature hyphae (Gooday, 1968). Competence for sporulation of *Aspergillus nidulans* is acquired after 18–20 h growth of mycelium (Champe *et al.*, 1981; Pastushok & Axelrod, 1976). Acquisition of this competence for sporulation appears to be controlled genetically rather than environmentally, as it is unaffected by continuous replacement of the medium or by concentrations of limiting nutrients, and precocious mutants have been described that can sporulate earlier.

These observations of the phenomenon of maturity within a colony are equivalent to the phenomenon of quorum sensing within bacterial colonies. Thus, although the cytology and biochemistry of hyphae of filamentous Actinomycetes are totally different to those of fungal hyphae, there are some superficial properties in common, for example, similar growth kinetics, branching patterns, and sporulation on aerial hyphae. There is growing evidence for several systems of chemical communication between hyphae within colonies of Actinomycetes. Hyphae of many species of *Streptomyces* produce butyrolactone autoregulators, such as A-factor of *Streptomyces griseus*. When these molecules accumulate to nanomolar concentrations, they trigger the formation of antibiotics, pigments and aerial hyphae (Horinouchi & Beppu, 1994). The formation of aerial hyphae by *Streptomyces coelicolor* is dependent on the peptide SapB, present on hyphal surfaces and released into a zone surrounding the colonies. Production of SapB in turn is dependent on diffusible signals (Willey *et al.*, 1993). Communication between hyphae between different colonies can be seen when certain mutants deficient in SapB production complement each other and produce aerial mycelium when growing in close proximity (Chater, 1998).

The mechanisms of maturation of fungal vegetative hypha are unclear, but it may prove that, as for bacterial colonies, growth regulating chemicals have to reach critical concentrations to allow expression of particular genes. The observation in many cases that differentiation does not occur in submerged fungal cultures is consistent with this idea, as gradients of putative effectors could not then accumulate. One aspect of metabolic control of hyphal development is demonstrated by experiments with *cr-1* mutants of *N. crassa* (Pall & Robertson, 1986). These mutants are deficient in cyclic AMP. They form mycelia with a single size class of hyphae, 3–5 μm, in contrast to the hyphal hierarchy of their wild-type parent, which has leading hyphae of 14–20 μm, primary branches of 8–12 μm and secondary branches of 3–5 μm. When grown with 2–3 mM 8-bromocyclic AMP (an analogue of cyclic AMP), after 18–20 h some hyphal hierarchy was observed, with hyphal diameters of 3–13 μm. Thus it may be concluded that metabolism involving cAMP plays some part in the process of hyphal maturation.

CONCLUSIONS

The theme throughout this account has been chemical communication between hyphae. Although the diverse phenomena discussed involve a wide variety of different molecules, ranging from oxygen to peptides, the initial response of a hypha requires that it must have specific receptors coupled to appropriate signal transduction pathways. Much work has been done in the elucidation of signal transduction pathways in *S. cerevisiae*, and this must give the lead to understanding chemoreception in other fungi. In *S.*

cerevisiae, five different mitogen-activated kinases (MAP kinases) have been associated with biological responses to specific stimuli (Herskowitz, 1995; Madhani & Fink, 1998). One of these, Fus3p, is specifically involved in the mating pheromone responses. Its specificity is apparent as *fus3* null mutants have a reduced frequency of mating. Another MAP kinase, Kss1p, normally involved in regulation of filamentous growth in response to starvation, can partially substitute, to allow induction of mating-specific gene expression. A consequence, however, is that filamentous growth is also induced in these mutants. This indicates that Fus3p, as well as activating mating-specific genes, must be inhibiting inappropriate cross-talk from other MAP kinases. Such studies will be invaluable in development of our ideas about regulation of the fascinating range of hyphal interactions among the fungi; and in answering such questions as 'why do *S. commune* and *C. cinereus* have so many pheromones and pheromone receptors?'.

In both asexual and sexual interactions, after fusion has occurred, a further level of molecular events ensue within the cytoplasm, involving recognition of self/non-self and mating type. These events will lead to differential gene expression and perhaps a variety of intermolecular reactions. Here can be sought a lead to the advances recently made in understanding regulation of action of transcription factors during mating in Basidiomycetes and Ascomycetes (Casselton, 1997).

This is an exciting time to be a mycologist – there are an enormous range of fascinating biological phenomena to explore; and the tools are now being acquired to help understand their mechanisms.

REFERENCES

Ainsworth, A. M. & Rayner, A. D. M. (1986). Responses of living hyphae associated with self and non-self fusions in the basidiomycete *Phanerochaete velutina*. *Journal of General Microbiology*, **132**, 191–201.

Ainsworth, A. M. & Rayner, A. D. M. (1989). Hyphal and mycelial responses associated with genetic exchange within and between species of the basidiomycete genus *Stereum*. *Journal of General Microbiology*, **135**, 1643–59.

Ainsworth, A. M. & Rayner, A. D. M. (1990). Mycelial interactions and outcrossing in the *Coniophora puteana* complex. *Mycological Research*, **94**, 627–34.

Ainsworth, A. M. & Rayner, A. D. M. (1991). Ontogenetic stages from coenocyte to basidiome and their relationship to phenoloxidase activity and colonisation processes in *Phanerochaete magnoliae*. *Mycological Research*, **95**, 1414–22.

Aylmore, R. C. & Todd, N. K. (1984). Hyphal fusion in *Coriolus versicolor*. In *Ecology and Physiology of the Fungal Mycelium*, ed. D. H. Jennings & A. D. M. Rayner, pp. 103–25. Cambridge: Cambridge University Press.

Aylmore, R. C. & Todd, N. K. (1986*a*). Cytology of self-fusions in hyphae of *Phanerochaete velutina*. *Journal of General Microbiology*, **132**, 571–9.

Aylmore, R. C. & Todd, N. K. (1986*b*). Cytology of non-self hyphal fusions and somatic incompatibilty in *Phanerochaete velutina*. *Journal of General Microbiology*, **132**, 581–91.

Baker, R. (1987). Mycoparasitism: ecology and physiology. *Canadian Journal of Plant Pathology*, **9**, 370–9.

Barksdsale, A. W. (1967). The sexual hormones of the fungus *Achlya*. *Annals of the New York Academy of Sciences*, **144**, 313–19.

Beakes, G. W. (1994). Sporulation of lower fungi. In *The Growing Fungus*, ed. N. A. R. Gow & G. A. Gadd, pp. 339–66. London: Chapman & Hall.

Bégueret, J., Turcq, B. & Clavé, C. (1994). Vegetative incompatibility in filamentous fungi: *het* genes begin to talk. *Trends in Genetics*, **10**, 441–6.

Bistis, G. N. (1983). Evidence for diffusible, mating-type-specific trichogyne attractants in *Neurospora crassa*. *Experimental Mycology*, **7**, 292–9.

Brasier, C. M. (1975). Stimulation of sex organ formation in *Phytophthora* by antagonistic species of *Trichoderma*. 1. The effect *in vitro*. *New Phytologist*, **74**, 183–94.

Bu'Lock, J. D., Jones, B. E. & Winskill, N. (1976). The apocarotenoid system of sex hormones and prohormones in Mucorales. *Pure and Applied Chemistry*, **47**, 191–202.

Buller, A. H. R. (1933). *Researches on Fungi*, Vol. 5. London: Longmans Green.

Butler, G. M. (1984). Colony ontogeny in basidiomycetes. In *The Ecology and Physiology of the Fungal Mycelium*, ed. D. H. Jennings & A. D. M. Rayner, pp. 52–71. Cambridge: Cambridge University Press.

Caldwell, G. A., Naider, F. & Becker, J. M. (1995). Fungal lipopeptide mating pheromones: a model system for the study of protein prenylation. *Microbiological Reviews*, **59**, 406–22.

Carlile, M. J. (1995). The success of the hypha and mycelium. In *The Growing Fungus*, ed. N. A. R. Gow & G. M. Gadd, pp. 3–19. London: Chapman & Hall.

Casselton, L. A. (1997). Molecular recognition in fungal mating. *Endeavour*, **21**, 159–63.

Casselton, L. A. & Olesnicky, N. S. (1998). Molecular genetics of mating recognition in Basidiomycete fungi. *Microbiology and Molecular Biology Reviews*, **62**, 55–70.

Champe, S. P. & El-Zayat, A. A. E. (1989). Isolation of a sexual sporulation hormone from *Aspergillus nidulans*. *Journal of Bacteriology*, **171**, 3982–8.

Champe, S. P., Kurtz, M. B., Yager, L. N., Butnick, N. J. & Axelrod, D. E. (1981). Spore formation in *Aspergillus nidulans*: competence and other developmental processes. In *The Fungal Spore: Morphogenetic Controls*, ed. G. Turian & H. R. Hohl, pp. 255–76. London: Academic Press.

Champe, S. P., Nagle, D. L. & Yager, L. N. (1994). Sexual sporulation. In *Aspergillus: 50 Years On*, ed. S. D. Martinelli & J. R. Kinghorn, *Progress in Industrial Microbiology*, Vol. 29, pp. 201–27. Amsterdam: Elsevier.

Chater, K. F. (1998). Taking a molecular scalpel to the *Streptomyces* colony. *Microbiology*, **144**, 1465–78.

Chet, I., Harman, G. E. & Baker, R. (1981). *Trichoderma hamatum*: its hyphal interactions with *Rhizoctonia solani* and *Pythium* spp. *Microbial Ecology*, **7**, 29–38.

Chet, I., Inbar, J. & Hadar, Y. (1997). Fungal antagonists and mycoparasites. In *Environmental and Microbial Relationships*, ed. D. T. Wicklow & B. E. Söderström, Vol. IV, *The Mycota*, ed. K. Esser & P. A. Lemke, pp. 165–84. Berlin: Springer.

Cornforth, J. W., Ryback, G., Robinson, P. M. & Park, D. (1971). Isolation and characterization of a fungal vacuolation factor (bikavarin). *Journal of the Chemical Society C*, 2786–8.

Czempinski, K., Kruft, V., Wöstermeyer, J. & Burmester, A. (1996). 4-Dihydromethyltrisporate dehydrogenase from *Mucor mucedo*, an enzyme of the sexual hormone pathway: purification, and cloning of the corresponding gene. *Microbiology*, **142**, 2647–54.

Dennis, C. & Webster, J. (1971). Antagonistic properties of species-groups of *Trichoderma*. III. Hyphal interaction. *Transactions of the British Mycological Society*, **57**, 363–9.

Drinkard, L. C., Nelson, G. E. & Sutter, R. P. (1982). Growth arrest: a prerequisite for sexual development in *Phycomyces blakesleeanus*. *Experimental Mycology*, **6**, 52–9.

Dyer, P. S., Ingram, D. S. & Johnstone, K. (1992). The control of sexual morphogenesis in the Ascomycotina. *Biological Reviews*, **67**, 421–58.

Dyer, P. S., Ingram, D. S. & Johnstone, K. (1993). Evidence for the involvement of linoleic acid and other endogenous lipid factors in perithecial development of *Nectria haematococca* mating population VI. *Mycological Research*, **97**, 485–96.

Evans, G. H. & Cooke, R. C. (1982). Studies on Mucoralen mycoparasites. III. Diffusible factors from *Mortierella vinacea* Dixon-Stewart that direct germ tube growth of *Piptocephalis fimbriata* Richardson & Leadbeater. *New Phytologist*, **91**, 245–53.

Fries, N. (1981). Intra- and interspecific basidiospore homing reactions in *Leccinum*. *Transactions of the British Mycological Society*, **81**, 559–61.

Fries, N. (1983). Recognition reactions between basidiospores and hyphae in *Leccinum*. *Transactions of the British Mycological Society*, **77**, 9–14.

Fries, N. & Swedjemark, G. (1985). Sporophagy in hymenomycetes. *Experimental Mycology*, **9**, 74–9.

Gall, A. M. & Elliott, C. G. (1985). Control of sexual reproduction in *Pythium sylvaticum*. *Transactions of the British Mycological Society*, **84**, 629–36.

Gnisalberti, E. L. & Sivasithamparam, K. (1991). Antifungal antibiotics produced by *Trichoderma* spp. *Soil Biology and Biochemistry*, **23**, 1011–20.

Gooday, G. W. (1968). Hormonal control of sexual reproduction in *Mucor mucedo*. *New Phytologist*, **67**, 815–21.

Gooday, G. W. (1973). Differentiation in the Mucorales. *Symposium of the Society for General Microbiology*, **23**, 269–94.

Gooday, G. W. (1975). Chemotaxis and chemotropism in fungi and algae. In *Primitive Sensory and Communication Systems*, ed. M. J. Carlile, pp. 155–204. London: Academic Press.

Gooday, G. W. (1978). Functions of trisporic acid. *Philosophical Transactions of the Royal Society of London B*, **284**, 509–20.

Gooday, G. W. (1995). The dynamics of hyphal growth. *Mycological Research*, **99**, 385–94.

Gooday, G. W. (1998) Mating and sexual interactions in fungal mycelia. In *The Fungal Colony*, ed. N. A. R. Gow, G. D. Robson & G. M. Gadd, pp. 261–82. Cambridge: Cambridge University Press.

Gooday, G. W. & Adams, D. J. (1993). Sex hormones and fungi. *Advances in Microbial Physiology*, **34**, 69–145.

Gow, N. A. R. & Gooday, G. W. (1982). Growth kinetics and morphology of colonies of the filamentous form of *Candida albicans*. *Journal of General Microbiology*, **128**, 2187–94.

Gow, N. A. R. & Gooday, G. W. (1987). Effects of antheridiol on growth, branching and electrical currents in hyphae of *Achlya ambisexualis*. *Journal of General Microbiology*, **133**, 3531–5.

Gregory, P. H. (1984). The first benefactor's lecture. The fungal mycelium: an historical perspective. *Transactions of the British Mycological Society*, **82**, 1–11.

Herskowitz, I. (1995). MAP kinase pathways in yeast: for mating and more. *Cell*, **80**, 187–97.

Horinouchi, S. & Beppu, T. (1994). A-factor is a microbial hormone that controls cellular differentiation and secondary metabolism in *Streptomyces griseus*. *Molecular Microbiology*, **12**, 859–64.

Hutchinson, S., Sharma, P., Clarke, K. R. & MacDonald, I. (1980). Control of hyphal orientation in colonies of *Mucor hiemalis*. *Transactions of the British Mycological Society*, **75**, 177–91.

Ikediugwu, F. E. O. & Webster, J. (1970a). Antagonism between *Coprinus heptemerus* and other coprophilous fungi. *Transactions of the British Mycological Society*, **54**, 181–204.

Ikediugwu, F. E. O. & Webster, J. (1970b). Hyphal interference in a range of coprophilous fungi. *Transactions of the British Mycological Society*, **54**, 205–10.

Inbar, J. & Chet, I. (1994). A newly isolated lectin from the plant pathogenic fungus *Sclerotium rolfsii*: purification, characterization and its role in mycoparasitism. *Microbiology*, **140**, 651–7.

Inbar, J. & Chet, I. (1995). The role of recognition in the induction of specific chitinases during mycoparasitism by *Trichoderma harzianum*. *Microbiology*, **141**, 2823–9.

Jaffe, L. F. (1966). On autotropism in *Botrytis*: measurement technique and control by CO_2. *Plant Physiology*, **41**, 303–6.

Jeffries, P. (1997). Mycoparasitism. In *Environmental and Microbial Relationships*, ed. D. T. Wicklow & B. E. Söderström, Vol. IV, *The Mycota*, ed. K. Esser & P. A. Lemke, pp. 149–64. Berlin: Springer.

Jones, D., Gordon, A. H. & Bacon, J. S. D. (1974). Co-operative action by endo- and exo-β-1-3-glucanases from parasitic fungi in the degradation of cell-wall glucans of *Sclerotinia sclerotiorum* (Lib.) de Bary. *Biochemical Journal*, **140**, 47–55.

Kellner, M., Burmester, A., Wöstermeyer, A. & Wöstermeyer, J. (1993). Transfer of genetic information from the mycoparasite *Parasitella parasitica* to its host *Absidia glauca*. *Current Genetics*, **23**, 334–7.

Kemp, R. F. O. (1970). Inter-specific sterility in *Coprinus bisporus, C. congregatus* and other basidiomycetes. *Transactions of the British Mycological Society*, **54**, 488–9.

Ko, W. H. (1988). Hormonal heterothallism and homothallism in *Phytophthora*. *Annual Review of Phytopathology*, **26**, 57–73.

Kothe, E. (1997). Solving a puzzle piece by piece: sexual development in the basidiomycetous fungus *Schizophyllum commune*. *Botanica Acta*, **110**, 208–13.

Kronstad, J. W. & Staben, C. (1997). Mating type in filamentous fungi. *Annual Review of Genetics*, **31**, 245–76.

Lifshitz, R., Dupler, M., Elad, Y. & Baker, R. (1984). Hyphal interactions between a mycoparasite, *Pythium nunn* and several soil fungi. *Canadian Journal of Microbiology*, **30**, 1482–7.

Loubradou, G., Bégueret, J. & Turcq, B. (1996). An additional copy of the adenylate cyclase-encoding gene relieves developmental defects produced by a mutation in a vegetative incompatibility-controlling gene in *Podospora anserina*. *Gene*, **170**, 119–23.

McLean, K. M. & Prosser, J. I. (1987). Development of vegetative mycelium during colony growth of *Neurospora crassa*. *Transactions of the British Mycological Society*, **88**, 489–95.

McMorris, T. C. (1978). Antheridiol and the oogoniols, steroid hormones which control sexual reproduction in *Achlya*. *Philosophical Transactions of the Royal Society of London B*, **248**, 459–70.

McMorris, T. C., Toft, D. O., Moon, S. & Wang, W. (1993). Biological response of the female strain *Achlya ambisexualis* 734 to dehydro-oogoniol and analogues. *Phytochemistry*, **32**, 833–7.

Madhani, H. D. & Fink, G. R. (1998) The riddle of MAP kinase signaling specificity. *Trends in Genetics*, **4**, 151–5.

Manocha, M. S. & Chen, Y. (1991). Isolation and partial characterization of host cell surface agglutinin and its role in attachment of a biotrophic mycoparasite. *Canadian Journal of Microbiology*, **37**, 377–83.

Manocha, M. S. & Sahai, A. S. (1993). Mechanisms of recognition in necrotrophic and biotrophic mycoparasites. *Canadian Journal of Microbiology*, **39**, 269–75.

Mesland, D. A. M., Huisman, J. G. & Van den Ende, H. (1974). Volatile sexual hormones in *Mucor mucedo*. *Journal of General Microbiology*, **80**, 111–17.

Miyoshi, M. (1894) Über Chemotropismus der Pilze. *Botanische Zeitung*, **52**, 1–28.

Moore, D. (1995). Tissue formation. In *The Growing Fungus*, ed. N. A. R. Gow & G. M. Gadd, pp. 421–65. London: Chapman & Hall.

Müller, D. & Jaffe, L. (1965). A quantitative study of cellular rheotropism. *Biophysical Journal*, **5**, 317–35.

Mullins, J. T. (1994). Hormonal control of sexual dimorphism. In *The Mycota, Vol. I. Growth, Differentiation and Sexuality*, ed. J. G. H. Wessels & F. Meinhardt, pp. 413–21. Berlin: Springer-Verlag.

Musgrave, A. & Nieuwenhuis, D. (1975). Metabolism of radioactive antheridiol by *Achlya* species. *Archives of Microbiology*, **105**, 313–17.

Musgrave, A., Loes, E., Scheffer, R. & Oehlers, E. (1977). Chemotropism of *Achlya bisexualis* germ tubes to casein hydrolysate and amino acids. *Journal of General Microbiology*, **101**, 65–70.

Nelson, R. R. (1971). Hormonal involvement in sexual reproduction in the fungi with special reference to F-2, a fungal estrogen. In *Morphological and Biochemical Events in the Plant–parasite Interaction*, ed. S. Akai & S. Ouchi, pp. 181–205. Tokyo: The Phytopathological Society of Japan.

Oh, K. B., Nishiyama, T., Sakai, E., Matsuoka, H. & Kurata, H. (1997). Flow sensing in mycelial fungi. *Journal of Biotechnology*, **58**, 197–204.

O'Shea, S. F, Chaure, P. T., Halsall, J. R., Olesnicky, N. S., Leibbrandt, A., Connerton, I. F. & Casselton, L. A. (1998). A large pheromone and receptor gene complex determines multiple *B* mating type specificities in *Coprinus cinereus*. *Genetics*, **148**, 1081–90.

Palková, Z., Janderová, B., Gabriel, J., Zikánova, B., Pospísek, M. & Forstová, J. (1997). Ammonia mediates communication between yeast colonies. *Nature*, **390**, 532–6.

Pall, M. L. & Robertson, C. K. (1986). Cyclic AMP control of hierarchical growth pattern of hyphae in *Neurospora crassa*. *Experimental Mycology*, **10**, 161–5.

Pastushok, M. & Axelrod, D. E. (1976). Effect of glucose, ammonia and maintenance on the time of condiophore initiation by surface colonies of *Aspergillus nidulans*. *Journal of General Microbiology*, **94**, 221–4.

Pommerville, J. & Olsen, L. W. (1987). Evidence for a male-produced pheromone in *Allomyces macrogynus*. *Experimental Mycology*, **11**, 245–8.

Pommerville, J. C., Strickland, J. B., Romo, D. & Harding, K. E. (1988). Effects of analogs of the fungal sex pheromone sirenin on male gamete motility in *Allomyces macrogynus*. *Plant Physiology*, **88**, 139–42.

Prosser, J. I. (1994). Kinetics of filamentous growth and branching. In *The Growing Fungus*, ed. N. A. R. Gow & G. M. Gadd, pp. 301–18. London: Chapman & Hall.

Raper, J. R. (1952). Chemical regulation of sexual processes in the thallophytes. *Botanical Reviews*, **18**, 447–545.

Raper, J. R. (1966). *Genetics of Sexuality in Higher Fungi*. New York: Ronald Press.

Rayner, A. D. M. (1996). Interconnectedness and individualism in fungal mycelia. In *A Century of Mycology*, ed. B. C. Sutton, pp. 193–232. Cambridge: Cambridge University Press.

Rayner, A. D. M. & Webber, J. F. (1984). Interspecific mycelial interactions: an overview. In *A Ecology and Physiology of the Fungal Mycelium*, ed. D. H. Jennings & A. D. M. Rayner, pp. 383–417. Cambridge: Cambridge University Press.

Rayner, A. D. M., Powell, K. A., Thompson, W. & Jennings, D. H. (1985). Interspecific mycelial interactions: an overview. In *Developmental Biology of Higher Fungi*, ed. D. Moore, L. A. Casselton, D. A. Wood & D. H. Jennings, pp. 249–79. Cambridge: Cambridge University Press.

Robinson, P. M. (1973*a*). Autotropism in fungal spores and hyphae. *The Botanical Review*, **39**, 361–84.

Robinson, P. M. (1973*b*). Oxygen-positive chemotropic factor for fungi? *New Phytologist*, **72**, 1349–56.

Robinson, P. M. (1973*c*). Chemotropism in fungi. *Transactions of the British Mycological Society*, **61**, 303–13.

Sakata, K. & Rickards, R. W. (1980). Synthesis of homothallin II. In *Proceedings of 23rd Symposium of Natural Products*, pp. 165–72. Nagoya: Nagoya City University.

Saupe, S., Turcq, B. & Bégueret, J. (1995). A gene responsible for vegetative incompatibility in the fungus *Podospora anserina* encodes a protein with a GTP-binding motif and $G\beta$ homologous domain. *Gene*, **161**, 135–9.

Schuurs, T. A., Dalstra, H. J. P., Scheer, J. M. J. & Wessels, J. G. H. (1998). Positioning of nuclei in the secondary mycelium of *Schizophyllum commune* in relation to differential gene expression. *Fungal Genetics and Biology*, **23**, 150–61.

Siddiq, A. A., Ingram, D. S., Johnstone, K., Friend, J. & Ashby, A. M. (1989). The control of asexual and sexual development by morphogens in fungal pathogens. *Aspects of Applied Biology*, **23**, 417–26.

Silver, J. C., Brunt, S. A., Kyriakopolou, G., Borkar, M. & Nazarian-Armavil, V. (1993). Heat shock proteins in hyphal branching and secretion in steroid hormone induced fungal development. *Journal of Cellular Biochemistry Supplement*, **17**, 136.

Snetselaar, K. M., Bölker, M. & Kahmann, R. (1996). *Ustilago maydis* mating hyphae orient their growth towards pheromone sources. *Fungal Genetics and Biology*, **20**, 299–312.

Spellig, T., Bölker, M. Lottspeich, F., Frank, R. W. & Kahmann, R. (1994). Pheromones trigger filamentous growth in *Ustilago maydis*. *The EMBO Journal*, **13**, 1620–7.

Stadler, D. R. (1952). Chemotropism in *Rhizopus nigricans*: the staling reaction. *Journal of Cellular and Comparative Physiology*, **39**, 449–74.

Stadler, D. R. (1953). Chemotropism in *Rhizopus nigricans*. II. The action of plant juices. *Biological Bulletin*, **104**, 100–8.

Steele, G. C. & Trinci, A. P. J. (1975). Morphology and growth kinetics of differentiated and undifferentiated mycelia of *Neurospora crassa*. *Journal of General Microbiology*, **91**, 362–8.

Thomas, D. S. & Mullins, J. T. (1969). Cellulase induction and wall extension in the water mold *Achlya ambisexualis*. *Physiologia Plantarum*, **39**, 347–53.

Timberlake, W. E. & Orr, W. C. (1984). Steroid hormone regulation of sexual reproduction in *Achlya*. In *Biological and Development, Vol. 3b*, ed. R. F. Goldberg & K. Yamamoto, pp. 255–83. New York: Plenum Press.

Trinci, A. P. J., Saunders, P. T., Gosrani, R. & Campbell, K. A. S. (1979). Spiral growth of mycelial and reproductive hyphae. *Transactions of the British Mycological Society*, **72**, 283–92.

Vaillancourt, L. J. & Raper, C. A. (1996). Pheromones and pheromone receptors as mating-type determinants in Basidiomycetes. In *Genetic Engineering, Principles and Methods, Vol. 18*, ed. J. K. Setlow, pp. 219–47. New York: Plenum Press.

Vaillancourt, L. J., Raudaskoski, M., Specht, C. A. & Raper, C. A. (1997). Multiple genes encoding pheromones and a pheromone receptor define the Bβ1 mating-type specificity in *Schizophyllum commune*. *Genetics*, **146**, 541–51.

Van den Ende, H. (1984). Sexual interactions in the lower filamentous fungi. In *Encyclopedia of Plant Physiology, Vol. 17*, ed. H. F. Linskens & J. Heslop-Harrison, pp. 333–49. Berlin: Springer-Verlag.

Voorhees, D. A. & Peterson, J. L. (1986). Hypha-spore attractions in *Schizophyllum commune*. *Mycologia*, **78**, 762–5.

Ward, H. M. (1888). A lily disease. *Annals of Botany*, **2**, 319–82.

Watkinson, S. C. (1978). End-to-side fusions in hyphae of *Penicillium claviformae*. *Transactions of the British Mycological Society*, **70**, 451–3.

Werkman, T. A. (1976). Localization and partial characterization of a sex-specific enzyme in homothallic and heterothallic Mucorales. *Archives of Microbiology*, **109**, 209–13.

Wessels, J. G. H., Schuurs, T. A., Dalstra, H. J. P. & Scheer, J. M. J. (1998). Nuclear distribution and gene expression in the secondary mycelium of *Schizophyllum commune*. In *The Fungal Colony*, ed. N. A. R. Gow, G. D. Robson & G. M. Gadd, pp. 302–25. Cambridge: Cambridge University Press.

Whipps, J. M. & Gerlagh, M. (1992). Biology of *Coniothyrium minitans* and its potential for use in disease control. *Mycological Research*, **96**, 897–907.

Willey, J., Schwedock, J. & Losick, R. (1993). Multiple extracellular signals govern the production of a morphogenetic protein involved in aerial mycelium formation by *Streptomyces coelicolor*. *Genes and Development*, **7**, 895–903.

Windels, C. E., Mirocha, C. J., Abbas, H. K. & Weiping, X. (1989). Perithecium production in *Fusarium graminearum* populations and lack of correlation with zearalenone production. *Mycologia*, **81**, 272–7.

Wolf, J. C. & Mirocha, C. J. (1973). Regulation of sexual reproduction in *Gibberella zeae* (*Fusarium roseum* 'Graminearum') by F-2 (zearalenone). *Canadian Journal of Microbiology*, **19**, 725–34.

Wöstermeyer, J., Wöstermeyer, A., Burmester, A. & Czempinski, K. (1995). Relationships between sexual processes and parasitic interactions in the host–pathogen system *Absidia glauca–Parasitella parasitica*. *Canadian Journal of Botany*, **73** (Suppl. 1), S243–50.

Zhu, W-Y. & Gooday, G. W. (1992). Effects of nikkomycin and echinocandin on differentiated and undifferentiated mycelia of *Botrytis cinerea* and *Mucor rouxii*. *Mycological Research*, **96**, 371–7.

SIGNALLING IN DINOFLAGELLATES

JOANN M. BURKHOLDER[1] AND JEFFREY J. SPRINGER[2]

Department of Botany[1] and Department of Marine, Earth and Atmosphere Sciences, College of Agriculture and Life Sciences[2], North Carolina State University, Raleigh, NC 27695, USA

INTRODUCTION

The ability to cope with changing environmental conditions is a fundamental aspect of organic evolution. Dinoflagellate cells are continually exposed to environmental stimuli such as temperature gradients and varying light regimes, to which they must respond. To maintain ecological fitness, dino- flagellates must be able to receive and process these signals so as to optimize their access to a favourable physical/chemical environment, avoid predators, and communicate with other organisms.

Signalling is defined here as the means by which dinoflagellates utilize a suite of ions, molecules, and associated pathways to convey and respond to information gathered from their internal/external environment. Signalling pathways in these single-celled or colonial organisms have been documented or proposed for a number of endogenous biological processes (Mittag, 1998; Tsim *et al.*, 1997). However, the majority of mechanisms by which dino- flagellates selectively respond to signalling molecules are not well understood and represent a pioneer area in protozoology.

Dinoflagellate response to an incoming signal requires, first, a receptor or binding site for that particular signalling ion or molecule. Receptors among dinoflagellates, as in other organisms, may vary both qualitatively and quantitatively. Maintenance of only a limited number of receptors would enable dinoflagellates to restrict the range of signals affecting them, allowing for a greater signal-to-noise ratio.

Alternatively, these predominantly unicellular organisms also may main- tain a consortium of different receptors, allowing simultaneous sensitivity to many extracellular signals (Morgan, 1989). By acting together, signals evoke responses that are antagonistic or synergistic, that is, more than simply the sum of the effects that each signal would evoke on its own. In this way, a relatively small number of signals could be used in different combinations to influence subtle and complex controls over cellular responses including, but not limited to, alterations in morphology, cell behaviour, and toxin produc- tion (Fig. 1). This review summarizes the current status of knowledge on signalling in dinoflagellates, including comparisons with other groups of organisms for which signalling pathways are better understood, and recom- mendations for future research directions.

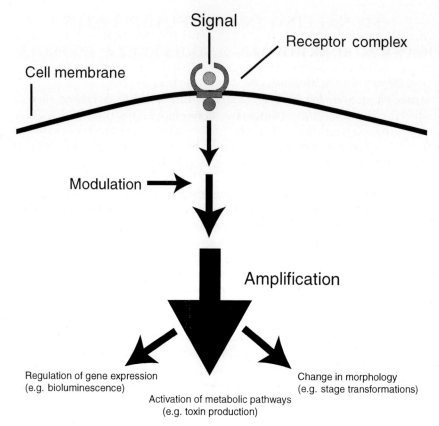

Fig. 1. A hypothetical signalling cascade in dinoflagellates. A receptor protein located on the cell surface transduces an extracellular signal into an intracellular signal, initiating a signalling cascade. The signal is then relayed to the cell interior where it is amplified and distributed.

CELL SIGNALLING PATHWAYS AND RECEPTORS

Known signalling pathways by which external stimuli are converted into internal cellular responses are few in number, yet they regulate many physiological and biochemical processes in eukaryotes (Potter, 1990). Dinoflagellates are believed to have a variety of cell signalling pathways and receptors, as have been documented in other protozoans (Kincaid, 1991; Smyth *et al.,* 1988; Van Houten, 1990).

Two fundamentally different types of receptors are known to respond to two basic classes of signals (Morgan, 1989). The first of these signal classes consists of molecules that are too large or too hydrophilic to cross the target cell's plasmalemma. Receptors for these types of signals must be located in a position which allows binding at an extracellular site. Receptors must then be able to transmit information about the binding agent to an effector system located on the opposite side of the plasma membrane. A second, smaller class of signals consists of molecules that are sufficiently small and hydrophobic to diffuse across the plasmalemma. Receptors for these signal molecules lie inside the target cell, and generally consist of enzymes or gene regulatory proteins.

Mechanisms for transduction of signals across the plasmalemma involve two general groups of receptors which differ in the nature of the intracellular signal that they generate when the extracellular signal molecule binds to them. The first group contains G protein-linked receptors that are enzymatically non-active, but structurally linked with downstream signalling complexes. Such receptors are composed of sensory, transmembrane, and specialized intracellular domains. Signal transduction involves activation of a membrane-bound protein (G protein subunit), which is released to diffuse in the plane of the plasmalemma, initiating a cascade of other effects. The second group of cell surface receptor proteins includes receptors with catalytic or ion channel activity. Binding of these enzyme-linked or ion channel-linked receptors, unlike signal transduction involving G protein receptors, results in complex structural rearrangements of the receptor molecule that lead to alterations of enzyme activity or changes of membrane permeability for certain ions. In processes involving enzyme-linked receptors, for example, an enzyme activity is switched on at the cytoplasmic end of the receptor, which generates a variety of further signals including molecules that are released into the cytosol.

Dinoflagellates are considered to utilize both general types of cell surface receptor proteins. However, with exception of certain signalling pathways involved in circadian rhythms of photosynthetic dinoflagellates, and in mutualistic endosymbiosis of dinoflagellates and corals (pp. 351–2), the current knowledge base about signalling in these organisms is extremely limited. The response of apparent G-protein receptors has been examined in several species; less information is available for enzyme-linked receptors and carbohydrate mediators such as lectins. The role of signal amplification enzymes and secondary messenger molecules such as c-AMP and inositol phosphates in intracellular recognition events remains to be examined.

G-protein receptors

When a propagating signal (for example, an extracellular signalling molecule) binds to a transmembrane receptor, the receptor protein encountered undergoes a conformational change that alters the intracellular face of the receptor and enables it to interact with a G protein located on the interior side of the plasmalemma. A new intracellular signal is generated in response. The subsequent events in the intracellular transduction process relay messages that are carried by signalling molecules until an enzyme is activated, the expression of a gene is switched on, or a change in morphology is effected. G-protein-linked receptors form the largest family of cell-surface receptors, with hundreds of such receptors already identified in mammalian cells (Spiegel, 1994). These signal molecules are as varied in structure as in function – they can be proteins, small peptides, or derivatives of amino acids or fatty acids, and for each there is a different receptor or set of receptors. Their evolutionary origins are ancient, since G-protein-linked receptors occur in distantly related organisms and are highly conserved in eukaryote cells, from unicellular algae to vertebrates (Jones, 1994).

In dinoflagellates, the formation of cysts can be induced by indoleamines such as melatonin (N-acetyl-5-methonytryptamine; Wong & Wong, 1994) and related compounds which most likely interact with G protein-coupled receptors (Table 1). Involvement of G proteins or G-protein coupled receptors has been demonstrated by the use of mastoparan, which is a peptide toxin from wasp venom that mimics an agonist-bound receptor to activate corresponding G proteins (Higashijima et al., 1988). Indoleamines also can mediate photoperiod control of asexual cell encystment in dino-flagellates (Balzer & Hardeland, 1991; Tsim et al., 1996; Fig. 2). The indoleamine receptor in dinoflagellates is believed to be distinct from the mammalian melatonin receptor that has been cloned to date (Tsim et al., 1996). Upon removal of indoleamines, dinoflagellates have been observed to excyst and become motile within 24 hours, indicating that the encystment response can be reversible.

Table 1. *The minimal concentration of indoleamines or mastoparan required for complete encystment of dinoflagellates*

Treatment	Concentration (mM)	
	G. tamarensis	*C. cohnii*
Melatonin	1	5
5-Methoxytryptamine	0.05	0.5
2-Iodomelatonin	0.5	1
N-Acetylserotonin	2	>10
6-Hydroxymelatonin	>2	>10
Serotonin	1	10
Mastoparan	0.05	0.1
Proteinase K-treated mastoparan	No effect	ND
Trypsinized mastoparan	No effect	ND

Cultures of *G. tamarensis* and *C. cohnii* were seeded at 1×10^3 cells/ml and 1×10^4 cells/ml, respectively, and treated with various concentrations (from 0.5 μM to 10 mM) of the indicated drugs for 3 days. *G. tamarensis* was maintained at 17 °C with a light and dark cycle of 12 hours each at a light intensity of 4500 lux. *C. cohnii* was kept at 25 °C in complete darkness. The highest concentration of proteinase K-treated or trypsinized mastoparan tested was 0.5 mM. ND indicates not determined. Results shown are the minimal concentrations of drugs required to induce complete encystment from a representative experiment: similar results were obtained from two additional experiments.

From Tsim et al. (1996).

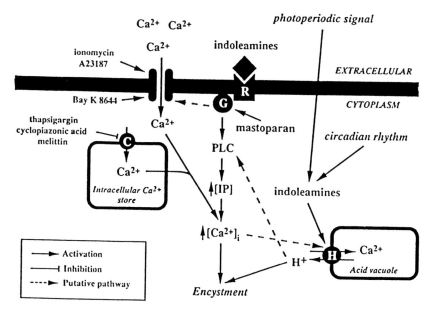

Fig. 2. Proposed signal transduction pathway of indoleamine-induced encystment in dinoflagellates. R, putative indoleamine receptor; G, putative G protein; PLC, putative phospholipase C; C, putative Ca^{2+}-ATPase; H, V type H^+-ATPase. (From Tsim *et al.* 1997.)

G-proteins such as indoleamines are known to influence encystment of the photosynthetic dinoflagellate, *Gonyaulax polyedra*. This species encysts under short-day conditions (L:D 10:1 hours; 15 °C, 800 lux; Balzer & Hardeland, 1991). Alteration of the dark period by 2 hours (for example, L:D/L:D 8:2:2:12, or 2:2:8:12) can inhibit encystment. Thus, encystment likely is controlled by photoperiod rather than by light deficiencies, and the photoperiod response apparently is mediated by indoleamines. Melatonin levels in *G. polyedra* have been found to be higher than in the mammalian pineal gland (Poeggeler *et al.*, 1991). Melatonin content is maximal at night, when encystment is inhibited. Treatment with melatonin (10^{-4} M) during the normally non-induced dark period, however, can mimic a short-day response and promote *G. polyedra* encystment. The photoperiodic signalling pathway involves melatonin synthesis, deactylation of melatonin to 5-methoxytryptamine, protein transfer from acid vacuoles, and acidification of the cytoplasm (Hardeland *et al.*, 1995). This G-protein effect is also temperature controlled; melatonin additions do not induce encystment at temperatures $\geqslant 20$ °C.

Enzyme-linked receptors

Enzyme-linked cell surface receptors respond to extracellular signal proteins that regulate cell nutrition, growth and proliferation (animal systems, reviewed by Carpenter, 1987; Yarden & Ullrich, 1988). Receptors can mediate direct, rapid effects on the cytoskeleton in controlling cell movement and shape alterations (Harris, 1989; van Haastert & Devreotes, 1993), of potential importance in stage transformations of dinoflagellates with complex life cycles such as *Pfiesteria piscicida* (Burkholder & Glasgow, 1997; work in progress). The extracellular signals to which these receptors respond typically are proteins attached to surfaces, rather than diffusable growth factors (Lauffenburger & Linderman, 1993). When an external signal molecule binds to the extracellular domain of an enzyme-linked receptor, the enzyme activity of the receptor's intracellular domain becomes activated (Cadena & Gill, 1992).

The major signalling pathway from receptor tyrosine kinase to the nucleus involves activation of a receptor tyrosine kinase which, in turn, may activate a small intracellular signalling protein that is tethered to the cytoplasmic face of the plasma membrane. This activated protein promotes activation of a phosphorylation cascade, in which the final protein kinase of the cascade phosphorylates certain gene regulation proteins so that their regulation of gene transcription is altered and a change in the pattern of gene expression is effected.

A cAMP-dependent protein tyrosine kinase has been identified in certain marine dinoflagellates (Dawson *et al.*, 1997). The enzyme in *Prorocentrum lima* appears to consist of two different isoforms in the R_2C_2 configuration. The enzyme was believed to be regulated by protein phosphatase-1, which possesses a classical protein kinase consensus phosphorylation site.

Diverse collections of signalling proteins may be recruited by enzyme-linked receptors; for example, a phospholipase may function similarly to phospholipase C in activating the inositol phospholipid signalling pathway (Morgan, 1989). The presence of a phospholipase C signalling cascade in dinoflagellates and its possible linkage to indoleamine-induced encystment has been suggested (Tsim *et al.*, 1997).

Many cell-surface and plasma membrane proteins are induced or have increased expression due to environmental stresses such as nutrient limitation, which inhibit growth (Price & Morel, 1990; Palenik & Koke, 1995). Certain of these cell-surface markers change with cell age; some may be expressed in late log phase as cells first become nutrient-stressed, whereas others may be expressed in stationary phase (for review see Palenik & Wood, 1997). Extracellular enzyme activity has been reported in marine flagellates (Nagata & Kirchman, 1992), suggesting the presence of a feedback mechanism by which these organisms can track extracellular nutrient concentrations. The presence of cell-surface enzyme activity in dinoflagellates has been suggested as a marker indicative of phosphate stress (Palenik & Morel, 1991; Dyhrman & Palenik, 1997). Like other microorganisms (Jannson *et al.*, 1988), both photosynthetic and heterotrophic dinoflagellates also have plasma membrane-bound phosphatases that, when excreted into the external medium in response to P limitation, enable the cells to hydrolyse phosphate for uptake from organic compounds such as phosphomono-esterases (Rivkin & Swift, 1979; Wynne, 1981; Boni *et al.*, 1989; Dyhrman & Palenik, 1997).

Aside from internal and external P supplies, certain other environmental stimuli can trigger rapid phosphate uptake. For example, sudden introduction of suspended sediments, simulating sediment loading to phytoplankton communities during storm events, has been observed to stimulate rapid excretion (within seconds) of thick mucilaginous coverings over unarmored (gymnodinoid) dinoflagellates in temporary cyst formation (Burkholder, 1992). During this process, these dinoflagellates showed high phosphate uptake as indicated by track light microscope-autoradiography with ^{33}P-phosphate. The mucus covering would have prevented direct contact with the sediment particles, thereby maintaining the integrity of the cells' delicate outer membranes so that the cells could emerge intact from the cysts after the water column had cleared. The increased P uptake would have increased phosphorus supplies and associated energy reserves for use in surviving the adverse conditions imposed by episodic sediment loading. Signalling pathways and receptors for this rapid phosphorus uptake and mucus production (probably via the Golgi apparatus) in dinoflagellate response to sudden water-column sediment loading are not known.

Carbohydrate-linked mediators at the cell surface

Lectins are carbohydrate-binding proteins that agglutinate cells or precipitate glycoconjugates (glycoproteins, glycolipids, or polysaccharides that are deployed on the cell surface (Goldstein *et al.*, 1980). Because of their binding specificity, lectins can serve as recognition molecules within a cell, between cells, or between organisms (Chrispeels & Raikhel, 1991). These compounds are believed to provide a consistent mechanism for recognition and attachment at the cell surface (Sharon & Lis, 1989).

Accordingly, agglutinins and/or lectin receptors have been shown to function in cell recognition processes such as symbioses with marine invertebrates or algae (Reisser, 1992; p. 351, this volume).

Glycoconjugates have been found on the cell surface(s) of some dinoflagellate species, and may be involved in intercellular recognition events (Sawayama *et al.*, 1993). In other protists (for example, Chlorophyceae), glycoconjugates have been shown to be involved in cell-to-cell recognition, adhesion, morphogenesis, and wall assembly (Imam *et al.*, 1984; Goodenough *et al.*, 1986; Schlipfenbacher *et al.*, 1986; Samson *et al.*, 1987). The presence of lectin-like compounds and lectin receptors has been confirmed in eight of nine examined dinoflagellate species, which exhibited lectin activity in the form of haemagglutination (Hori *et al.*, 1996). Moreover, 11 of 12 examined dinoflagellate species showed receptor activity for macroalgal or terrestrial plant lectins (Hori *et al.*, 1996; Table 2).

Intercellular recognition events involved in prey acquisition are common among dinoflagellates (for example, Gaines & Elbrächter, 1987). Heterotrophy is well developed even in many photosynthetic species (Carlsson & Granéli, 1997). Obvious prey recognition commonly occurs in dinoflagellates, as in other microorganisms, prior to feeding, and involves specific receptors at the cell surface (Ryter & de Chastellier, 1983). Strong chemosensory stimulation by certain prey species has been documented (that is, by the ambush–predator dinoflagellate, *Katodinium* [*Gymnodinium*] *fungiforme* toward wounded ciliates – Spero & Moree, 1981 and Spero, 1985; by the parasitic dinoflagellate, *Amoebophyra ceratii*, toward the photosynthetic dinoflagellate, *Gymnodinium sanguineum* – Coats & Bockstahler, 1994; and by zoospores of the toxic estuarine dinoflagellate, *Pfiesteria piscicida*, toward fresh finfish epidermis – Burkholder & Glasgow, 1997). In recent observations, lobose amoebae of *P. piscicida* (length 120–450 μm) have been shown to attract various cryptomonad and ciliate prey to their cell surfaces. The live prey continue to move but remain adjacent to these amoebae until their pseudopodia slowly trap and engulf them (Glasgow *et al.*, 1998; J. M. Burkholder & J. J. Springer, unpublished observations).

Research has been undertaken on a limited number of dinoflagellate species to determine signalling pathways and receptors involved in these strong intercellular, prey attraction behaviours. For example, more than 30 ephemeral substances found in fresh fish secreta, excreta and mucus have been tested for chemosensory stimulation of *P. piscicida*, with certain of these compounds having been shown to be strong attractants (PhD thesis in progress, authors' laboratory). A *Crypthecodinium cohnii*-like dinoflagellate shows especially strong preference for a certain prey species within the unicellular red algal genus, *Porphyridium* (Rhodophyceae; Ucko *et al.*, 1989). Under nitrate and sulphate limitation, the *Porphyridium* sp. cell wall constituents contain higher content of methylhexose and lower levels of glucose and xylose. These cells are not as readily consumed as prey that were grown under nitrate/sulphate-replete conditions (Ucko *et al.*, 1994).

The physical process of feeding in dinoflagellates such as *P. piscicida* zoospores, the *C. cohnii*-like dinoflagellate, and various other species often involves myzocytosis, in which the prey cell contents are ingested by a 'feeding tube' or peduncle, leaving behind the prey's outer covering (Schnepf & Deichgräber, 1984). The feeding tube is retracted after each prey cell (minus the outer covering) is consumed (for example, Ucko *et al.*, 1997; videotapes of *P. piscicida* from authors' laboratory). In the *C. cohnii*-like species (and likely in other dinoflagellates with identical feeding behaviour), the feeding tube retraction is known to be affected by cytochalasin D and probably depends on actin (Ucko *et al.*, 1997). Actual uptake of the prey contents also is altered by addition of cytochalasin D, and the feeding tube is

Table 2. *Response of live microalgal cells to macroalgal and terrestrial plant lectins*

| | | Addition of lectin | | | | | | |
| | | Macroalgal lectin | | | Terrestrial plant lectin | | | |
Microalgal species n	No addition	Hypnin A (control)	Solnin B	Boonins	CFA	WGA	DBA	UEA-1
Dinophyceae								
Alexandrium cohorticula	A	C/F	C	C	C	C	C	C
A. tamarense	A	C	C/F	C/F	C/E	C	C	C
A. catenella	A	C	C	C	B	C	A	A
Amphidinium carterae	C	C	C	C	C	C	C	C
Coolia monotis	A	C	C	C	C	B	B/C	B/C
Heterocapsa sp.	A	A	C/F	C/F	A	C/F	A	A
Gymnodinium mikimotoi	A	C/E/F	C/E	C/E	C/E	C/E	C/F	C/F
G. catenatum	A	C	C	C	A	C	C	C
Gymnodinium sp.	A	C/F	C/E/F	C/E/F	C/E/F	C/E/F	C/E/F	C/F
Prorocentrum lima	A	C/F	C/F	C/F	A	C/D	C/D	C/D
P. balticum	A	C	C	C	A	C/F	C/F	C/F
P. micans	A	C	C	C	A	C/F	C/F	C/F

A, active motility; B, impaired motility; C, disappearance of motility; D, agglutination; E, abnormal morphology; F, rupture or lysis.
Modified from Hori et al. (1996).

enlarged; thus, actomyosin may be involved in the myzocytotic process, as well. The retracted feeding tube forms a microtubular basket that is associated with characteristic elongate vesicles that may be either electron-transparent or electron-dense. The electron-transparent vesicles apparently serve as a membrane source during feeding tube protrusion and internal channel formation. The membrane-bound, electron-dense vesicles may secrete the enzymes that locally dissolve the prey cell wall and/or chemically impair the prey so that escape is not possible. In prey acquisition as well as many other intercellular recognition phenomena that are commonly indicated from dinoflagellate behaviour, much about the signalling processes in these organisms remains to be investigated.

Cyclins and cell division

Eukaryote cell cycle control involves the orderly progression of DNA replication and mitosis. At cell cycle checkpoints, progression of the process is halted in response to failure of certain events. Two classes of proteins, cyclins and cyclin-dependent kinases (CDKS), are the central components of cell cycle control in eukaryotes (Draetta, 1990; Sherr, 1993). They form complexes that control both initiation of DNA synthesis and entry into mitosis (Murray & Kirschner, 1989). The cyclins are expressed differentially during the cell cycle, but are activated by critical phosphorylations and dephosphorylations only when they are associated with G1 or mitosis-specific substrates.

Although amoeboid stages of dinoflagellates have typical eukaryote nuclear structure, flagellated stages have a 'mesokaryote' nucleus with permanently condensed chromosomes during interphase, an extranuclear spindle, and a nuclear envelope that remains intact throughout the cell cycle (Taylor, 1987; Rizzo, 1991). Histones and nucleosomes are absent (Rizzo, 1991). Many of the mitosis-specific substrates that are phosphorylated by cdc2 kinases in other eukaryotes are absent in flagellated stages of dino-flagellates, raising questions about the nature of molecular mechanisms that regulate dinoflagellate cell division. The presence of cyclins and cyclin-dependent kinases (including cdc2 kinase) has been demonstrated in dino-flagellates (for example, *Crypthecodinium* [*Cryptothecodinium*] *cohnii* and *Gambierdiscus toxicus* (Rodriquez *et al.*, 1994; Van Dolah *et al.*, 1995; Leveson *et al.*, 1997). The cdc2-like kinase in *G. toxicus* was found to be expressed constitutively throughout the cell cycle. The cdc2-like kinase is activated specifically during mitosis, suggesting that cell division in this dinoflagellate is controlled by the universal mechanisms known for other eukaryotes (Van Dolah *et al.*, 1995).

The dinoflagellate cell cycle, especially in photosynthetic species, is under the control of a circadian clock (Homma & Hastings, 1989). Dinoflagellates increasingly have been the focus of research to understand the evolution of

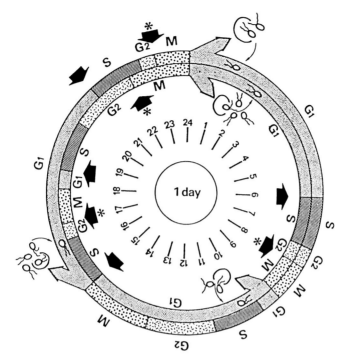

Fig. 3. A representative diagram of successive cell cycles in *Crypthecodinium cohnii* over 24 h. In this particular example, one vegetative cell performed two successive complete cell cycles (16 h) and released four daughter cells. One of these new swimming cells released two daughter cells 10 h later (external circle of the diagram). During this time, other swimming cells gave an inverse alternation (internal circle). Different diagrams could be possible with other alternations. Transition points G1S ('Start' point) are represented by arrows and G2M by arrows plus star. (From Bhaud *et al.*, 1994.)

eukaryote cell cycle control. *Crypthecodinium cohnii*, a heterotrophic species, has been used extensively to investigate cytologic events of the dinoflagellate cell cycle (Kubai & Ris, 1969; Perret *et al.*, 1993). The histone H1 kinase is activated in mitotic cells (Bhaud *et al.*, 1994); and, in general, regulation of cell cycles in dinoflagellates and higher eukaryotic organisms appears comparable, despite the relatively unique features of mesokaryotic dinoflagellates (Fig. 3).

CIRCADIAN RHYTHMS

Signalling mechanisms associated with circadian rhythms have been examined in a few representative dinoflagellate species. Both planktonic and benthic dinoflagellates have been shown to exhibit pronounced circadian rhythms in photosynthesis, vertical migration, bioluminescence, and other behaviour (Sweeney & Hastings, 1957; Esaias & Curl, 1972; Sweeney, 1984).

Circadian rhythms in photosynthesis and bioluminescence have been detected in dinoflagellate species such as *Gonyaulax polyedra* (Hastings *et al.*, 1961), *Ceratium furca* (Prezelin *et al.*, 1977), and *Pyrocystis fusiformis* (Sweeney, 1981; Fig. 4). At the ultrastructural level, circadian rhythms also have been shown to influence chloroplast movement in *Pyrocystis lunula* (Töpperwien & Hardeland, 1980); intracellular location of bioluminescent 'microsources' in *Pyrocystis noctiluca* (Hardeland & Nord, 1984); and spontaneous bioluminescence in *P. elegans, P. acuta,* and *P. lunula* (Hardeland, 1982; Colepicolo *et al.*, 1993). Species within the genus *Gonyaulax* have

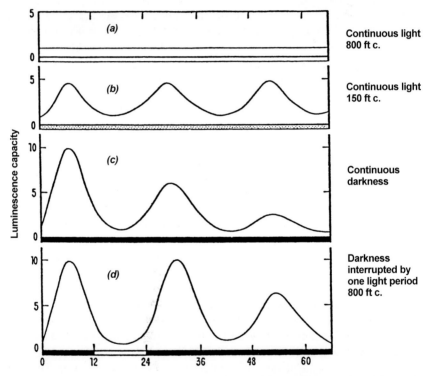

Fig. 4. Diagrammatic representation of the changes in luminescence capacity of *Gonyaulax polyedra* taken from continuous bright light (800 foot–candles) and maintained at the same temperature (20 °C) in the light and dark regimes indicated by bars along the abscissa. In curve (*a*), the cells remain in bright light; in curve (*b*), the light intensity was reduced to 150 foot–candles at 0 hours; in curve (*c*), the cells were transferred to darkness at 0 hours; and curve (*d*), the darkness was interrupted with a single light period (800 foot–candles) between 12 and 24 hours. The difference in the magnitude of the luminescence capacity of the first cycles of curves (*b*) and (*c*) exemplifies the inhibitory effects of light of intermediate intensities comparable to those employed in determining the action spectrum of photoinhibition. The maximum light inhibition obtainable is represented in curve (*a*). The differences in the magnitude of luminescence capacity at comparable times during the second cycles of curves (*c*) and (*d*) illustrate the photoenhancement brought about by the preceding light period in curve (*d*). (From Sweeney *et al.*, 1959.)

shown circadian rhythms in photosynthetic capacity (Hastings *et al.*, 1961), cell division (Sweeney & Hastings, 1958), and relative membrane potential (Adamich *et al.*, 1976). Multiple circadian rhythmicity has enabled examination of crosstalk and repetitiveness among the signalling pathways for different physiological processes (for example, between photosynthesis and bioluminescence; see below).

Photosynthesis

Photosynthetic dinoflagellates generally are considered to utilize C_3 photosynthesis (Calvin cycle; Knoetzel & Rensing, 1990). Photosystems I and II, and the reaction centres of chlorophyll *a* follow circadian rhythms in the representative species, *G. polyedra* (Knoetzel & Rensing, 1990). The photosystems appear to utilize typical signalling pathways as in other C_3 photosynthetic plants. They have maintained circadian rhythm when photosynthetic dinoflagellates (*G. polyedra* and other tested species) have been transferred from a 12:12 L:D cycle to continuous light after 4 days, although the photosynthetic maximum in the 12:12 L:D period was reduced over time (for example, Sweeney, 1969).

The circadian rhythm for photosystems I and II share the same oscillator as in bioluminescence (below), although photosynthesis and bioluminescence are 180° out of phase. The photosynthetic period for *G. polyedra* has been reported as 22.5 to 23.0 hours; electron flow decreases when bioluminescence is maximal; and electron flow through the photosystems parallels photosynthesis (as oxygen evolution). Knoetzel and Rensing (1990) hypothesized that the organization of the complex pigment photosynthetic apparatus may control the rhythmicity. This hypothesis was based on observations that the thylakoid membranes vary in organization from stacks of two (subjective day) to three (subjective night). Moreover, circadian changes occur in chlorophyll *a* fluorescence, following light availability. As a result, the oscillator controls the distribution of photosynthetic units within the thylakoids, and some units were found to be uncoupled from the electron transport pathway at night.

Reports have varied about the potential for rhythmicity in respiration of photosynthetic dinoflagellates, as indicated by *G. polyedra*. Knoetzel and Rensing (1990) described oxygen evolution during subjective day (circadian time 30) as twice that during subjective night (circadian time 42; but see Hastings, 1961, who used more limited techniques and reported no apparent rhythm for oxygen consumption in this species).

Bioluminescence

The best understood circadian rhythm in dinoflagellates is bioluminescence, a reaction in which the substrate luciferin is oxidized by luciferase, usually producing a 0.1-second flash of bluish-green light (maximal wavelength *ca.* 474–478 nm; Swift & Taylor, 1967; Sweeney, 1987). Light may be emitted as a flash that is induced by mechanical, chemical or electrical means; a flash also can occur simultaneously; or a glow (small amount of light for an extended period) can be produced without apparent stimulation (von der Heyde *et al.*, 1992). The quantity of light emitted is influenced by the species and the cell size, with larger cells producing more light per flash. Nutritional status can also influence the quantity of light emitted (Roenneburg, 1996).

In dinoflagellates, the substrate/enzyme reaction occurs in scintillon organelles. The reaction has three basic components: luciferin (substrate, LH_2), luciferin binding protein (LBP), and luciferase (Morse *et al.*, 1989). The binding protein, a dimer of two identical subunits (72 kDa), protects the substrate from auto-oxidation. LBP binds luciferin at an alkaline pH; at acidic pH, the confirmation of the LBP is (reversibly) altered, and this change initiates the onset of the luciferase enzyme activity. For example, binding at pH 6 is weaker than at pH 8, and the reaction at the lower pH yields 100-fold more enzyme activity. LBP synthesis is maximal in early night, just before cessation of oxygen production by photosynthesis. During the dark phase, LBP comprises 1% of the total cell protein, decreasing to 0.1% of the total during the day. Scintillons generally contain a 20-fold surplus of LBP and LH_2 relative to the quantity of luciferase available. At night, nearly all of the substrate is bound, thus preventing its inactivation by auto-oxidation. Small changes in cellular pH release luciferin, enabling its reaction with the enzyme and with oxygen (Morse *et al.*, 1989).

'Flash' bioluminescent events have been found to be related to the number of scintillons, the level of LBP, and luciferase activity (von der Heyde *et al.*, 1992). In contrast, 'glow' bioluminescent events have been related to scintillon breakdown and division. Two different periods for flash versus glow events have been clearly discerned, indicative of the presence of two 'clocks'.

Both the proteins and the substrate involved in dinoflagellate biolumines-cence are under translational control, wherein a *trans* factor binds a *cis* acting element at the 5′ or 3′ untranslated region of the LBP's mRNA. The binding area consists of a 22-nucleotide region that is uridine- and guanosine-rich (CCTR, believed to be a dimer). Translation begins at the onset of night, when LBP is synthesized and CCTR binding activity decreases. Following an increase in binding activity during the night, CCTR binding declines late in the dark period with cessation of LBP synthesis. CCTR represses LBP mRNA during the day and therefore, acts as a translational regulator. It has been hypothesized that this regulator may be necessary because of possible steric hinderance at the 5′ or 3′ untranslated region of the LBP mRNA (Mittag *et al.*, 1996).

The circadian rhythm for bioluminescence in dinoflagellates has been observed to phase-shift, depending on the timing of white light. When dinoflagellates capable of bioluminescence have been given light in the early night, bioluminescence is delayed (Hastings, 1996). When given light late in the dark period, the timing of subsequent bioluminescence is advanced. A light pulse given to cells that had been acclimated to constant darkness produces a delay in the early subjective night, and an advance during late subjective night (Hastings, 1996). The circadian rhythm is also sensitive to the availabilty of red and blue light (Roenneberg & Hastings, 1988). *Gonyaulax* was found to have a blue light-sensitive input system that was activated three hours into subjective night (producing delays), and a red/ blue-sensitive system that produced delays during subjective day.

Phase shifts in bioluminescence are influenced by temperature as well as light timing/quality. In *G. polyedra*, bioluminescence is inversely proportional to temperature over the range of 13–27 °C (Sweeney, 1987). When the temperature of cultured *G. polyedra* acclimated at 20 °C is lowered to *ca.* 10 °C, bioluminescence is emitted without mechanical or chemical stimulation. However, cultures grown at such lowered temperatures do not spontaneously emit light until the temperature is lowered to 4 °C. *Pyrocystis fusiformis* responds similarly in spontaneously emitting light at low temperatures, and at temperatures $\geqslant 27$ °C (Sweeney, 1987).

Swimming behaviour

Certain dinoflagellates show circadian rhythms in vertical migration through the water column (for example, pelagic species – Eppley *et al.*, 1968), shallow coastal environs (for example, *Prorocentrum* spp. which phase-shift diurnally from a benthic to a planktonic habit – Faust & Steidinger, 1995); and tide pools (for example, *Peridinium gregarium* – Lombard & Capon, 1971). The aggregation of pelagic dinoflagellate populations, for example, *G. polyedra*, occurs daily at the same location in the water column following an endogenous rhythm. Under experimental conditions, period length of the endogenous aggregating rhythm in *G. polyedra* was >24 hours with blue light, and >24 hours with red or yellow light (Roenneburg *et al.*, 1989). In a tide pool dinoflagellate, endogenous aggregation rhythms have been related to the timing of tidal cycles, and the behaviour appears to be advantageous in avoiding washout during high tide (Lombard & Capon, 1971). Other factors that apparently influence the establishment of circadian rhythms for aggregation in dinoflagellate populations are geotaxis and bioconvection (Levandowsky & Kaneta, 1987).

Aside from these factors, the circadian rhythms of photosynthesis, bioluminescence, and vertical migration are influenced by nutrient supplies. Nitrate is considered a nonphotic signal for dinoflagellates in marine waters because it affects circadian rhythm amplitude, phase, and period (Roenneburg *et al.*, 1989). The activity of nitrate reductase in *G. polyedra* has been found to have a circadian rhythm with oscillations related to changes in daily synthesis and breakdown (Ramalho *et al.*, 1995). Nitrate reductase is usually active during the light period, when it is localized in the chloroplast (Fritz *et al.*, 1996). In addition, nitrate deficiency has promoted shortened photosynthetic circadian rhythms in *G. polyedra*, and has influenced vertical migration activity (Eppley *et al.*, 1968; Fritz *et al.*, 1996). When nitrate is limiting in surface waters, *G. polyedra* migrates during the night to deeper, more nitrate-replete depths (Packard, 1973; Cullen & MacIntyre, 1998). Although many photosynthetic organisms actively take up nitrate during light periods, *G. polyedra* can switch to reliance on respiratory energy for active uptake of nitrate in the dark (Packard, 1973). An analogous phenomenon appears to occur for certain dinoflagellates in stratified freshwaters, which exploit deeper waters for N during dark periods in response to N limitation (for example, *Gymnodinium bogoriense*; Lieberman *et al.*, 1994). Such organisms also show active uptake of phosphate at depth during the night.

Multiple oscillators

The photosynthetic dinoflagellate, *G. polyedra*, is thus known to exhibit more than one circadian rhythm, as illustrated by the rhythms described for bioluminescence versus vertical migration or aggregation (Roenneberg, 1996). Two pacemakers are believed to be involved which are distinct in their rhythmic periods. They show phase responses to light and dark pulses, as well as temperature compensation (Morse *et al.*, 1989). The overall circadian system, controlled by multiple oscillators, clearly is equipped with multiple feedback loops that interact with environmental variables (Roenneberg, 1996). For example, inhibition of photosynthesis affects the phase and the period of the oscillator. In swimming behaviour and aggregation, an indicated preference for certain light intensities changes during the course of a diurnal cycle. The multiple oscillator system enables a high degree of flexibility in adapting to temporal environmental changes.

MUTUALISTIC ENDOSYMBIOSIS

Dinoflagellates participate in an array of endosymbioses wherein they function as hosts (freshwater to marine species, harboring bacteria or kleptochloroplasts from algal prey – Doucette *et al.*, 1997; Fields & Rhodes, 1991; Larsen, 1992), or as symbionts (marine species – mutualistic, commensal, or parasitic in certain foraminiferans, radiolarians, sponges, cnidarians, turbellarian flatworms, and molluscs). These interactions range from temporary/casual to mutualistic/obligatory (McEnery & Lee, 1981; Larsen, 1992). Signalling in cell-to-cell recognition and subsequent interactions among these varied symbioses have not been examined or are poorly understood, with the exception of the ecologically significant symbioses of dinoflagellates with cnidarian corals. This highly species-specific and stable

interaction, of ancient origin, provides the foundation of coral reef ecosystems throughout the world's subtropical and tropical oceans.

Stable hereditary endosymbioses, exemplified by the most advanced coral/
dinoflagellate symbioses, are characterized by a high degree of specificity
between the host and symbiont. Success of the interaction depends on a
network of exchanged chemical signals that ensure cell-to-cell recognition
and reaction of two organisms united as one unit (Reisser, 1992). The host
and symbiont exchange various types of signals at different organizational
levels that include ultrastructure, physiology, and behaviour. Communication between the interacting species is based on complex phenomena involving permanent recognition, specificity, and regulation within the integrated
host/symbiont unit. The symbiotic interaction is initiated when certain
strains of the dinoflagellate, *Symbiodinium* sp. send signals that are recognized as acceptable by host corals. The symbionts are enclosed in perialgal
vacuoles, whereas other captured algae and bacteria are sequestered into
digestive vacuoles. Certain strains of *Symbiodinium* sp. show characteristic
circadian rhythms in motility, of interest since some coral hosts (for example,
Tridacna squamosa) acquire motile dinoflagellate symbionts at much higher
rates (Fitt *et al.*, 1981; Lerch & Cook, 1984).

Many of the mechanisms for cell-to-cell recognition and types of signals
exchanged in the coral/dinoflagellate endosymbiosis remain to be elucidated.
As in other algae, signalling molecules embedded in the cell surface are
believed to play a decisive role in communication between the potential
partners. In endosymbioses involving the alga *Chlorella*, for example, the cell
wall surface of different *Chlorella* species and strains differ in surface charge
and binding capacity with different lectins and antibodies (Reisser *et al.*,
1982, 1988). When the wall surface of a cell that previously triggered
formation of a perialgal vacuole is changed, either by cell wall degrading
enzymes or by coupling of lectins or antibodies, the signalling pattern is
changed. As a consequence, the host animal encloses the former symbiont in
a digestive vacuole and consumes it. Binding experiments with various lectins
suggest that a specific pattern of surface components is common to algae
which induce a stable host/symbiont interface (Reisser *et al.*, 1988). These
specfic molecules on the surface of algal cells may interact with counterpart
cell surface recognition modulators on host cells during contact, regulating
selective uptake and, hence, specificity (Trench, 1987). Dinoflagellates
isolated from different marine hosts have distinct cell wall surface patterns,
as demonstrated by antibody techniques (Kinzie & Chee, 1982). Techniques
with antibodies and lectins similarly have indicated the presence of a
carbohydrate recognition pattern on cell wall surfaces of *Chlorella* sp.
symbionts in specific endosymbioses with *Hydra viridis* (Pool, 1979).

Extracellular transport of macromolecules such as glycoproteins may also
be serve as signalling molecules for cell-to-cell recognition and expressed
specificity in the coral/dinoflagellate endosymbiosis (Trench, 1988, 1992;

Table 3. *Protein, neutral sugar, sugar amine, and uronic acid composition of total exudates from five species of symbiotic dinoflagellates*

Species	Protein	Neutral sugar	Glucosamine	Galactosamine	Uronic acid
Symbiodinium microadriaticum	48.0	173	4.1	8.0	25.0
S. kawagutii	73.0	29	3.8	2.6	3.6
S. goreauii	126.0	231	3.4	2.4	26.0
S. pilosum	120.8	230	2.6	5.1	7.8
Symbiodinium sp. (no. 11)	660.0	134	7.1	4.6	43.0

Protein, neutral sugars, and uronic acids are in units of mg/g^{-1} protein. Protein estimated from HPLC analysis assuming 100 g mole^{-1} amino acids.

From Markell & Trench (1993).

Table 3). Symbiotic dinoflagellate cells that are isolated in culture have been found to release large molecular-weight glycoproteins (Markell *et al.*, 1992). Each of four isolates of *Symbiodinium* produced exudates that were distinct from those of the other isolates (Markell *et al.*, 1992). The potential function of these exudates as molecular signals between these microalgal symbionts and their invertebrate hosts is intriguing, given the known role of other secreted macromolecules such as the fibronectins which facilitate cell adhesion and regulate morphogenesis (Ruoslahti, 1988).

SUMMARY AND RECOMMENDATIONS

Dinoflagellates dominate the plankton of the subtropics in the world oceans (Taylor, 1987; South & Whittick, 1987). They contribute to both the microflora and microfauna of both marine and freshwater food webs (Taylor, 1987; Pollingher, 1987). 'Red tide' and other toxic species significantly impact both fish and human health (Burkholder & Glasgow, 1997), while other, beneficial members of this group are important participants in forming the Earth's coral reefs. Despite their ecological and economic significance, the signalling pathways that control cellular and communication processes in dinoflagellates remain, in general, very poorly understood.

The information contained in this review was synthesized primarily from scattered literature, often with brief mention, preliminary findings, or results from beginning stages of research on signal reception and transduction in these protists. In photosynthetic dinoflagellates, the output of circadian signalling pathways has been proposed to mediate cellular processes including encystment, cell division, and bioluminescence. Within these broad topics, however, relatively little is yet understood about cellular and communication processes in dinoflagellates, emphasizing a small number of species. The signalling involved in many fundamental cellular processes such as stage transformations, nutrient acquisition, chromosome structure, and predator/ prey recognition remain mostly unknown. Even signalling pathways which govern the selective responses that control cell-to-cell recognition for one of the most widely known, fascinating, and ecologically significant endosymbioses – the coral/dinoflagellate association – have only begun to be examined.

Determination of mechanisms for cell-surface recognition was beyond reach in this field until the advent of DNA/RNA probes, confocal microscopy, and related molecular applications that have enabled focus at the appropriate microscale to begin to resolve biochemical and molecular interactions (Reisser, 1992). Epifluorescence microscopy coupled with lectin specificity has provided a means to probe rapidly and precisely the spatial relationships of complex molecular distributions on cell surfaces (Surek & von Sengbusch, 1981; von Sengbusch & Muller, 1983). While these techniques have been applied to dinoflagellates mostly to facilitate species identification, they should be rigorously applied to gain fundamental information about controls on cellular and communication processes in this important group of protists. The insights gained from such research additionally will strengthen understanding about classical pathways of signal transduction in marine and freshwater organisms, as exemplified by these evolutionary ancient organisms (Loeblich, 1976; Dawson et al., 1997).

REFERENCES

Adamich, M., Laris, P. C. & Sweeney, B. M. (1976). *In vivo* evidence for a circadian rhythm in membranes of *Gonyaulax. Nature* (London), **261**, 583–5.

Balzer, I. & Hardeland, R. (1991). Photoperiodism and effects of indoleamines in a unicellular alga, *Gonyaulax polyedra. Science*, **253**, 795–7.

Bhaud, Y., Barbier, M. & Soyer-Gobillard, M. O. (1994). A detailed study of the complex cell cycle of the dinoflagellate protist *Crypthecodinium cohnii* as studied *in vivo* and by cytofluorimetry. *Journal of Cell Science*, **100**, 675–82.

Boni, L., Carpene, E., Wynne, D. & Reti, M. (1989). Alkaline phosphatase activity in *Protogonyaulax tamarensis. Journal of Plankton Research*, **11**, 879–85.

Burkholder, J. M. (1992). Phytoplankton and episodic suspended sediment loading: Phosphate partitioning and mechanisms for survival. *Limnology and Oceanography*, **37**, 974–88.

Burkholder, J. M. & Glasgow, H. B., Jr. (1997). *Pfiesteria piscicida* and other toxic *Pfiesteria*-like dinoflagellates: behavior, impacts, and environmental controls. *Limnology and Oceanography*, **42**, 1052–75.

Cadena, D. L. & Gill, G. N. (1992). Receptor tyrosine kinases. *FASEB Journal*, **6**, 2332–7.

Carlsson, P. & Granéli, E. (1997). The ecological significance of phagotrophy in photosynthetic flagellates. In *Physiological Ecology of Harmful Algal Blooms*, ed. D. M. Anderson, A. D. Cembella & G. M. Hallegraeff. NATO ASI Series **41**, pp. 508–24.

Carpenter, G. (1987). Receptors for epidermal growth factor and other polypeptide mitogens. *Annual Review of Biochemistry*, **56**, 881–914.

Chrispeels, M. J. & Raikhel, N. V. (1991). Lectins, lectin genes, and their role in plant defence. *Plant Cell*, **3**, 1–9.

Coats, D. W. & Bockstahler, K. R. (1994). Occurrence of the parasitic dinoflagellate *Amoebophyra ceratii* in Chesapeake Bay populations of *Gymnodinium sanguineum*. *Journal of Eukaryotic Microbiology*, **41**, 586–93.

Colepicolo, P., Roenneberg, T., Morse, D., Taylor, W. R. & Hastings, J. W. (1993). Circadian regulation of bioluminescence in the dinoflagellate *Pyrocystis lunula*. *Journal of Phycology*, **29**(2), 173–9.

Cullen, J. J. & MacIntyre, J. G. (1998). Behavior, physiology and the niche of depth-regulating phytoplankton. In *Physiological Ecology of Harmful Algal Blooms*, ed. D. M. Anderson, A. D. Cembella & G. M. Hallegraeff, pp. 559–79. NATO ASI Series, Vol. G 41. Berlin: Springer-Verlag.

Dawson, J. F., Ostergaard, H. L., Klix, H., Boland, M. P., & Holmes, C. F. P. (1997). Evidence for reversible tyrosine protein phosphorylation in the okadaic acid-producing marine dinoflagellate *Prorocentrum lima*. *Journal of Eukaryotic Microbiology*, **44**(2), 89–95.

Doucette, G. J., Kodama, M., Franca, S. & Gallacher, S. (1997). Bacterial interactions with harmful algal bloom species: bloom ecology, toxigenesis, and cytology. In *Physiological Ecology of Harmful Algal Blooms*, ed. D. M. Anderson, A. D. Cembella & G. M. Hallegraeff, pp. 619–47. NATO ASI Series, Vol. G 41. Berlin: Springer-Verlag.

Draetta, G. (1990). Cell cycle control in eukaryotes: molecular mechanisms of cdc2 activation. *Trends in Biochemical Science*, **15**, 378–83.

Dyhrman, S. T. & Palenik, B. P. (1997). The identification and purification of a cell-surface alkaline phosphatase from the dinoflagellate *Prorocentrum minimum* (Dinophyceae). *Journal of Phycology*, **33**, 602–12.

Eppley, R. W., Holm-Hansen, O., & Stickland, J. D. H. (1968). Some observations on the vertical migration of dinoflagellates. *Journal of Phycology*, **4**, 333–40.

Esias, W. E. & Curl, H. C., Jr. (1972). Effect of dinoflagellate bioluminescence on copepod ingestion rates. *Limnology and Oceanography*, **17**, 901–6.

Faust, M. A. & Steidinger, K. A. (1995). Ecology of benthic dinoflagellates. In *Harmful Algal Blooms*, ed., P. Lassus, G. Arzul, E. Erard, P. Gentien & C. Marcaillou, pp. 855–7. Paris: Technique et Documentation – Lavoisier, Intercept Ltd.

Fields, S. D. & Rhodes, R. G. (1991). Ingestion and retention of *Chroomonas* spp. (Cryptophyceae) by *Gymnodinium acidotum* (Dinophyceae). *Journal of Phycology*, **27**, 525–9.

Fitt, W. K., Chang, S. S. & Trench, R. K. (1981). Motility patterns of different strains of the symbiotic dinoflagellate *Symbiodinium* (= *Gymnodinium*) *microadriaticum* (Freudenthal) in culture. *Bulletin of Marine Science*, **31**, 436–43.

Fritz, L., Stringher, C. & Colepicolo, P. (1996). Imaging oscillation in *Gonyaulax*: a chloroplast rhythm of nitrate reductase visualized by immunocytochemistry. *Biological Journal of Microbiology Research*, **29**(1), 111–17.

Gaines, G. & Elbrächter, M. (1987). Heterotrophic nutrition. In *The Biology of Dinoflagellates*, ed. F. J. R. Taylor. Botanical Monographs, Vol. 21, pp. 224–68. Oxford: Blackwell.

Glasgow, H. B. J., Lewitus, A. J. & Burkholder, J. M. (1998). Feeding behaviour of the ichthyotoxic estuarine dinoflagellate, *Pfiesteria piscicida*, on amino acids, algal prey, and fish vs. mammalian erythrocytes. In *Harmful Algae*, ed, B. Reguera, J. Blanco, M. L. Fernández & T. Wyatt, pp. 394–7. Xunta de Galicia and Inter-governmental Oceanographic Commission of UNESCO.

Goldstein, I. J., Hughes, R. C., Monsigny, M., Osawa, T. & Sharon, N. (1980). What should be called a lectin? *Nature* (London), **285**, 66.

Goodenough, U. W., Gebhart, B., Mechan, R. P. & Heuuser, J. E. (1986). Crystal of *Chlamydomonas reinhardii* cell wall: polymerization, depolymerization, and purification of glycoprotein monomers. *Journal of Cell Biology*, **103**, 405–18.

Hardeland, R. (1982). Circadian rhythms of bioluminescence in two species of *Pyrocystis* (Dinophyta). Measurements in cell populations and in single cells. *Journal of Interdisciplinary Cycle Research*, **13**, 49–54.

Hardeland, R. & Nord, P. (1984). Visualization of free-running circadian rhythm in the dinoflagellate *Pyrocystis noctiluca*. *Marine Behavioural Physiology*, **11**, 199–207.

Hardeland, R., Balzer, I., Poeggeler, B., Fuhrberg, H., Uria, G., Behrmann, R., Wolf, T. J., Meyer, T. J. & Reiter, R. J. (1995). On the primary functions of melatonin in evolution: mediation of photoperiodic signals in a unicell, photooxidation, and scavenging of free radicals. *Journal of Pineal Research*, **18**, 104–11.

Harris, E. H. (1989). *The Chlamydomonas Sourcebook: A Comprehensive Guide to Biology and Laboratory Use*. San Diego: Academic Press.

Hastings, J. W. (1996). Chemistries and colors of bioluminescent reactions: a review. *Gene*, **173**(1), 5–11.

Hastings, J. W., Astrachan, L. & Sweeney, B. M. (1961). A persistent daily rhythm in photosynthesis. *Journal of General Physiology* **45**, 69–76.

Hauser, D. C. R., Levandowsky, M., Hunter, S. H., Chunosoff, L. & Hollwitz, J. S. (1975). Chemosensory responses by the heterotrophic marine dinoflagellate *Crypthecodinium cohnii*. *Microbial Ecology*, **1**, 246–54.

Higashijima, T., Uzu, S., Nakajima, T. & Ross, E. M. (1988). Mastoparan, a peptide toxin from wasp venom, mimics receptors by activating GTP-binding regulatory proteins (G proteins). *Journal of Biological Chemistry*, **263**: 6491–4.

Homma, K. & Hastings, J. W. (1989). Cell growth kinetics, division asymmetry and volume control at division in the marine dinoflagellate *Gonyaulax polyedra*: a model of circadian clock control of the cell cycle. *Journal of Cell Science*, **92**, 303–18.

Hori, K., Ogata, T., Kamiya, H. & Mimuro, M. (1996). Lectin-like compounds and lectin receptors in marine microalgae: hemagglutination and reactivity with purified lectins. *Journal of Phycology*, **32**, 783–90.

Imam, S. H., Bard, R. F. & Tosteson, T. R. (1984). Specificity in marine microbial surface interactions. *Applied and Environmental Microbiology* **48**, 833–9.

Jansson, M., Olsson, H. & Pettersson, K. (1988). Phosphatases: origin, characteristics and function in lakes. *Hydrobiologia*, **170**, 157–75.

Jones, T. L. Z. (1994). G proteins in nonmammalian species. In *G Proteins*, ed. A. M. Spiegel, pp. 6–17. Austin, TX, USA: R. G. Landes Co.

Kincaid, R. L. (1991). Signalling mechanisms in microorganisms: common themes in the evolution of signal transduction pathways. In *Advances in Second Messenger and Phosphoprotein Research*, ed. P. Greengard & G. A. Robinson, pp. 165–84. New York: Raven Press.

Kinzie, R. A. III & Chee, G. S. (1982). Strain-specific differences in surface antigens of symbiotic algae. *Applied and Environmental Microbiology*, **44**(5), 1238–40.

Knoetzel, J. & Rensing, L. (1990). Characterization of the photosynthetic from the marine dinoflagellate *Gonyaulax polyedra*. I. Pigment and polypeptide composition of the pigment–protein complex. *Journal of Plant Physiology*, **136**, 271–9.

Kubai, D. F. & Ris, H. (1969). Division in the dinoflagellate *Gyrodinium cohnii* (Schiller) – a new type of nuclear reproduction. *Journal of Cell Biology*, **40**, 508–28.

Larsen, J. (1992). Endocytobiotic consortia with dinoflagellate hosts. In *Algae and Symbioses*, ed. W. Reisser, pp. 427–42. Bristol, UK: Biopress Limited.

Lauffenburger, D. A. & Linderman, J. J. (1993). Signal transduction. In *Receptors: Models for Binding, Trafficking, and Signalling*, pp. 181–235. New York: Oxford University Press.

Lerch, K. A. & Cook, C. B. (1984). Some effects of photoperiod on the motility rhythm of cultured zooxanthellae. *Bulletin of Marine Science*, **34**(3), 477–83.

Levandowsky, M. & Kaneta, P. J. (1987). Behavior in dinoflagellates. In *The Biology of Dinoflagellates*, ed. F. J. R. Taylor, pp. 360–97. Botanical Monographs, Volume 21. Palo Alto (CA): Blackwell Scientific Publications Inc.

Leveson, A., Wong, F. & Yong, J. T. Y. (1997). Cyclins in a dinoflagellate cell cycle. *Molecular and Marine Biotechnology*, **6**(3), 172–9.

Lieberman, O. S., Shilo, M. & van Rijn, J. (1994). The physiological ecology of a freshwater dinoflagellate bloom population: vertical migration, nitrogen limitation, and nutrient uptake kinetics. *Journal of Phycology*, **30**, 964–71.

Loeblich, A. R. III (1976). Dinoflagellate evolution: speculation and evidence. *Journal of Protozoology*, **23**, 13–28.

Lombard, E. H. & Capon, B. (1971). Observations on the tidepool ecology and behavior of *Peridinium gregarium*. *Journal of Phycology*, **7**, 188–94.

McEnery, M. E. & Lee, J. J. (1981). Cytological and fine structural studies of 3 species of symbiont-bearing larger foraminifera from the Red Sea. *Micropaleontology*, **27**(1), 71–83.

Markell, D. A., Trench, R. K. & Iglesias-Prieto, R. (1992). Macromolecules associated with the cell walls of symbiotic dinoflagellates. *Symbiosis*, **12**, 19–31.

Mittag, M. (1996). Exploring the signaling pathway of circadian bioluminescence. *Physiologia Plantarum*, **96**, 727–32.

Mittag, M. (1998). Molecular mechanisms of clock-controlled proteins in phytoflagellates. *Protist*, **149**, 101–7.

Morgan, N. G. (1989). Receptors and their regulation. In *Cell Signalling*, ed. N. G. Morgan, pp. 9–31. New York: Guilford Press.

Morse, D., Markovic, P. & Roenneberg, T. (1996). Several clocks may simplify the circadian system of *Gonyaulax*. *Brazilian Journal of Medical and Biological Research*, **29**(1), 101–3.

Morse, D., Pappenheimer, A. M. & Hastings, J. W. (1989). Role of a luciferin-binding protein in the circadian bioluminescent reaction of *Gonyaulax polyedra*. *Journal of Biological Chemistry*, **264**(20), 11 822–6.

Murray, A. W. & Kirschner, M. W. (1989). Dominoes and clocks: the union of two views of the cell cycle. *Nature* (London), **246**, 614–21.

Nagata, T. & Kirchmann, D. L. (1992). Release of macromolecular organic complexes by heterotrophic marine flagellates. *Marine Ecology Progress Series*, **83**, 233–40.

Packard, T. T. (1973). The light dependence of nitrate reductase in marine phytoplankton. *Limnology and Oceanography*, **18**(3), 466–9.

Palenik, B. & Morel, F. M. M. (1991). Amine oxidases of marine phytoplankton. *Applied and Environmental Microbiology*, **57**(8), 2440–3.

Palenik, B. & Koke, J. A. (1995). Characterization of a nitrogen-regulated protein identified by cell surface biotinylation of a marine phytoplankton. *Applied and Environmental Microbiology*, **61**(9), 3311–15.

Palenik, B. & Wood, A. M. (1997). Molecular markers of phytoplankton physiological status and their application at the level of individual cells. In *Molecular Approaches to the Study of the Oceans*, ed. K. Cooksey. London: Chapman & Hall.

Perret, E., Davoust, J., Albert, M., Besseau, L. & Soyer-Gobllard, M.O. (1993). Microtubule organization during the cell cycle of the primitive eukaryote dinoflagellate *Crypthecodinium cohnii*. *Journal of Cell Science* **104**, 639–51.

Poeggeler, B., Balzer, I., Hardeland, R. & Lerchl, A. (1991). Pineal hormone melatonin oscillates also in the dinoflagellate *Gonyaulax polyedra*. *Naturwissenschaften*, **78**, 268–9.

Pollingher, U. (1987). Ecology of dinoflagellates: B. Freshwater ecosystems. In *The Biology of Dinoflagellates*, ed. F. J. R. Taylor, pp. 502–9. Botanical Monographs Volume 21. Palo Alto (CA): Blackwell Scientific Publications, Inc.

Pool, R. R., Jr. (1979) The role of algal antigenic determinants in the recognition of potential algal symbionts by cells of *Chlorohydra*. *Journal of Cell Science*, **35**, 367–79.

Potter, B. V. L. (1990). Synthesis and biology of second messenger analogues. In *Transmembrane Signalling: Intracellular Messengers and Implications for Drug Development*, ed. R. Nahorski, pp. 207–39. Wiley-Interscience.

Prezelin, B. B., Meeson, B. W. & Sweeney, B. M. (1977). Characterization of photosynthetic rhythms in marine dinoflagellates. 1. Pigmentation, photosynthetic capacity and respiration. *Plant Physiology*, **60**(3), 384–7.

Price, N. M. & Morel, F. M. M. (1990). Cadmium and cobalt substitution for zinc in a marine diatom. *Nature*, **344** (6267), 658–60.

Ramalho, C. B., Hastings, J. W. & Colepicolo, P. (1995). Circadian oscillation of nitrate reductase activity in *Gonyaulax polyedra* is due to changes in cellular protein levels. *Plant Physiology*, **107**(1), 225–31.

Reisser, W. (1992). Basic mechanisms of signal exchange, recognition, specificity, and regulation in endosymbiotic systems. In *Algae and Symbioses*, ed. E. Reisser, pp. 657–74. Bristol, UK: Biopress Ltd.

Reisser, W., Klein, T. & Becker, B. (1988). Studies on phycoviruses. 1. On the ecology of viruses attacking chlorellae exsymbiotic from a European strain of *Paramecium bursaria*. *Archiv fur Hydrobiologie*, **111**(4), 575–83.

Reisser, W., Radunz, A. & Wiessner, W. (1982). Participation of algal surface structures in the cell recognition process during infection of aposymbiotic *Paramecium bursaria* with symbiotic chlorellae. *Cytobios*, **33**(129), 39–50.

Rivkin, R. B. & Swift, E. (1979). Diel and vertical patterns of alkaline phosphatase activity in the oceanic dinoflagellate *Pyrocystis noctiluca*. *Limnology and Oceanography*, **24**, 107–16.

Rizzo, P. J. (1991). The enigma of the dinoflagellate chromosome. *Journal of Protozoology*, **38**(3), 246–52.

Rodriquez, M., Cho, J. W., Sauer, H. W. & Rizzo, P. J. (1994). Evidence for the presence of a cdc2-like protein kinase in the dinoflagellate *Crypthecodinium cohnii*. *Journal of Eukaryotic Microbiology*, **40**, 91–6.

Roenneberg, T. (1996). The complex circadian system of *Gonyaulax polyedra*. *Physiologia Plantarum*, **96**, 733–7.

Roenneberg, T. & Hastings, J. W. (1988). Two photoreceptors control the circadian clock of a unicellular alga. *Naturwissenschaften*, **74**, 206–7.

Roenneberg, T., Colfax, G. & Hastings, J. W. (1989). A circadian rhythm of population behavior in *Gonyaulax polyedra*. *Journal of Biological Rhythms*, **4**(2), 201–16.

Ruoslahti, E. (1988). Fibronectin and its receptors. *Annual Review of Biochemistry*, **57**, 375–413.

Ryter, A. & De Chastellier, C. (1983). Phagocytic–pathogenic microbe interactions. *International Review of Cytology*, **85**, 287–319.

Samson, M. R., Klis, F. M., Homan, H. L., van Egmond, P., Musgrave, A. & van den Ende, H. (1987). Composition and properties of the sexual agglutinins of the flagellated green alga *Chlamydomonas eugametos*. *Planta*, **170**, 314–22.

Sawayama, S., Sako, Y. & Ishida, Y. (1993). Inhibitory effects of concanavalin A and tunicamycin on sexual attachment of *Alexandrium catenella* (Dinophyceae). *Journal of Phycology*, **29**, 189–90.

Schlipfenbacher, R., Wenzel, S., Lottspeich, F. & Sumper, M. (1986). An extremely hydroxyproline-rich glycoprotein is expressed in inverting *Volvox* embryos. *FEBS Letters*, **209**, 57–62.

Schnepf, E. & Deichgräber, G. (1984). 'Myzocytosis', a kind of endocytosis with implications to compartmentalization in endosymbiosis: observations in *Paulsenella* (Dinophyta). *Naturwissenschaften*, **71**, 218–19.

Sharon, N. & Lis, H. (1989). *Lectins*. London: Chapman & Hall, pp. 127.

Sherr, C. J. (1993). Mammalian G_1 cyclins. *Cell*, **73**, 1059–65.

Smyth, R. D., Saranak, J. & Foster, K. W. (1988). Algal visual systems and their photoreceptor pigments. *Progress in Phycological Research*, **6**, 254–86.

South, G. R. & Whittick, A. (1987) *Introduction to Phycology*. Boston: Blackwell Scientific Publications.

Spero, H. J. (1985). Chemosensory capabilities in the phagotrophic dinoflagellate, *Gymnodinium fungiforme*. *Journal of Phycology*, **21**, 181–4.

Spero, H. J. & Moree, M. (1981). Phagotrophic feeding and its importance in the life cycle of the holozoic dinoflagellate, *Gymnodinium fungiforme*. *Journal of Phycology*, **17**, 43–51.

Spiegel, A. M. (1994). G protein-coupled receptors. In *G Proteins*, ed. A. M. Spiegel, pp. 6–17. Austin, TX, USA: R.G. Landes Co.

Surek, B. & von Sengbusch, P. (1981). The localization of galactosyl residues and lectin receptors in the mucilage and the cell walls of *Cosmocladium saxonicum* (Desmidiaceae) by means of fluorescent probes. *Protoplasma*, **108**, 149–61.

Sweeney, B. M. (1969). Transducing mechanisms between circadian clock and overt rhythms in *Gonyaulax*. *Canadian Journal of Botany*, **47**, 299–308.

Sweeney, B. M. (1981). The circadian rhythms in bioluminescence, photosynthesis, and organellar movement in the large dinoflagellate, *Pyrocystis fusiformis*. In *International Cell Biology 1980–1981*, ed. H. G. Schweiger, pp. 807–14. Berlin & New York: Springer-Verlag.

Sweeney, B. M. (1984). Circadian Rhythmicity in Dinoflagellates. In *Dinoflagellates*, ed. D. L. Spector, pp. 343–64. Academic Press.

Sweeney, B. M. (1987). Bioluminescence and circadian rhythms. In *The Biology of Dinoflagellates*, ed. F. J. R. Taylor, pp. 269–81. Botanical Monographs, Volume 21. Palo Alto (CA): Blackwell Scientific Publications, Inc.

Sweeney, B. M. & Hastings, J. W. (1957). Characteristics of the diurnal rhythm of luminescence in *Gonyaulax polyedra*. *Journal of Cellular and Comparative Physiology*. **49**, 115–28.

Sweeney, B. M. & Hastings, J. W. (1958). Rhythmic cell division in populations of *Gonyaulax polyedra*. *Journal of Protozoology*, **5**, 217–24.

Swift, E. & Taylor, W. R. (1967). Bioluminescence and chloroplast movement in the dinoflagellate *Pyrocystis lunula*. *Journal of Phycology*, **3**, 77–81.

Taylor, F. J. R. (1987). Ecology of dinoflagellates: A. General and marine ecosystems. In *The Biology of Dinoflagellates*, ed. F. J. R. Taylor, pp. 399–502. Botanical Monographs, Volume 21. Palo Alto (CA): Blackwell Scientific Publications, Inc.

Töpperwien, F. & Hardeland, R. (1980). Free-running circadian rhythm of plastid movement in individual cells of *Pyrocystis lunula* (Dinophyta). *Journal of Interdisciplinary Cycle Research*, **11**, 325–9.

Trench, R. K. (1987). Dinoflagellates in non-parasitic symbioses. In *The Biology of Dinoflagellates*, ed. F. J. R. Taylor, pp. 530–70. Botanical Monographs, Volume 21. Palo Alto (CA): Blackwell Scientific Publications, Inc.

Trench, R. K. (1988). Specificity in dinomastigote–marine invertebrate symbioses: an evaluation of hypotheses of mechanisms involved in producing specificity. In *Cell to Cell Signals in Plant, Animal, and Microbial Symbiosis*, ed. S. Scannerini *et al*. NATO ASI Series, volume H17, pp. 325–46. Berlin: Springer-Verlag.

Trench, R. K. (1992). Microalgal–invertebrate symbiosis, current trends. In *Encyclo-*

pedia of Microbiology, ed. J. Lederberg. Volume 3, pp. 129–42. New York: Academic Press.

Tsim, S. T., Yung, L. Y, Wong, J. T. Y. & Wong, Y. H. (1996). Possible involvement of G proteins in indoleamine-induced encystment in dinoflagellates. *Molecular and Marine Biotechnology*, **5**(2), 162–7.

Tsim, S. T., Wong, J. T. Y. & Wong, Y. H. (1997). Calcium ion dependency and the role of inositol phosphates in melatonin induced encystment of dinoflagellates. *Journal of Cell Science*, **110**, 1387–93.

Ucko, M., Cohen, E., Gordin, H. & Arad (Malis), S. (1989) Relationship between the unicellular red microalga *Porphyridium* sp. and its predator, the dinoflagellate *Gymnodinium* sp. *Applied and Environmental Microbiology*, **55**, 2990–4.

Ucko, M., Elbrächter, M. & Schnepf, E. (1997). A *Crypthecodinium cohnii*-like dinoflagellate feeding myzocytotically on the unicellular red alga *Porphyridium* sp. *European Journal of Phycology*, **32**, 133–40.

Ucko, M., Geresh, S., Simon-Berkovitch, B. & Arad (Malis), S. (1994). Predation by a dinoflagellate on a red microalga with a cell wall modified by sulfate and nitrate starvation. *Marine Ecology Progress Series*, **104**, 292–8.

Van Dolah, F. M., Leighfield, T. A., Sandel, H. D. & Hsu, C. K. (1995). Cell division in the dinoflagellate *Gambierdiscus toxicus* is phased to the diurnal cycle and accompanied by activation of the cell cycle regulatory protein, CDC2 kinase. *Journal of Phycology*, **31**, 395–400.

Van Haastert, P. J. M. & Devreotes, P. N. (1993). Biochemistry and genetics of sensory transduction in *Dictyostelium*. In *Signal Transduction: Prokaryotic and Simple Eukaryotic Ssytems*, ed. J. Kurjan & B. L. Taylor, pp. 329–52. San Diego: Academic Press.

Van Houten, J. (1990). Chemosensory transduction in *Paramecium*. In *Biology of the Chemotactic Response*, ed. J. P. Armitage & J. M. Lackie, pp. 297–321. UK: Cambridge University Press.

Von der Heyde, F., Wilkens, A. & Rensing, L. (1992). The effect of temperature on the circadian rhythm of flashing and glow in *Gonyaulax polyedra*: are the two rhythms controlled by two oscillators? *Journal of Biological Rhythms*, **7**(2), 115–23.

Von Sengbusch, P. & Muller, U. (1983). Distribution of glycoconjugates at algal cell surfaces as monitored by FITC-conjugated lectins. *Protoplasma* **114**, 103–12.

Wong, J. T. Y. & Wong, Y. H. (1994). Indoleamine induced encystment in dinoflagellates. *Journal of the Marine Biological Association of the United Kingdom*, **74**, 467–9.

Wynne, D. (1981). The role of phosphatases in the metabolism of *Peridinium cinctum* from Lake Kinneret. *Hydrobiologia*, **83**, 93–9.

Yarden, Y. & Ullrich, A. (1988). Growth factor receptor tyrosine kinases. *Annual Review of Biochemistry*, **57**, 443–78.

INDEX

References to tables/figures are shown in italics.